Green Low-Carbon Technology for Metalliferous Minerals

Green Low-Carbon Technology for Metalliferous Minerals

Editor

Lijie Guo

MDPI • Basel • Beijing • Wuhan • Barcelona • Belgrade • Manchester • Tokyo • Cluj • Tianjin

Editor
Lijie Guo
Beijing General Research
Institute of Mining and
Metallurgy (BGRIMM)
Beijing, China

Editorial Office
MDPI
St. Alban-Anlage 66
4052 Basel, Switzerland

This is a reprint of articles from the Special Issue published online in the open access journal *Metals* (ISSN 2075-4701) (available at: https://www.mdpi.com/journal/metals/special_issues/green_metalliferous).

For citation purposes, cite each article independently as indicated on the article page online and as indicated below:

LastName, A.A.; LastName, B.B.; LastName, C.C. Article Title. *Journal Name* **Year**, *Volume Number*, Page Range.

ISBN 978-3-0365-5797-7 (Hbk)
ISBN 978-3-0365-5798-4 (PDF)

Contents

About the Editor

Lijie Guo

Lijie Guo, PhD, is an international green mining specialist with over 18 years of experience in mining engineering who is currently working as the deputy director and professor at the Beijing General Research Institute of Mining and Metallurgy, China. He is a committee member of the international mine fill council. He has dedicated himself to developing innovative research for the sustainable and green development of global mineral resources and focused on improving practices related to the management of mining residue materials, particularly mine tailings. He has more than 80 publications in international peer-reviewed journals and conferences and 2 monographs.

metals MDPI

Editorial

Green Low-Carbon Technology for Metalliferous Minerals

Lijie Guo [1,2]

[1] Beijing General Research Institute of Mining and Metallurgy, Beijing 100160, China; guolijie@bgrimm.com
[2] National Centre for International Research on Green Metal Mining, Beijing 102628, China

1. Introduction and Scope

Metalliferous minerals play a central role in the global economy. They will continue to provide the raw materials we need for industrial processes. Significant challenges will likely emerge if the climate-driven green and low-carbon development transition of metalliferous minerals exploitation is not managed responsibly and sustainably. Green low-carbon technology is vital to promote the development of metalliferous mineral resources shifting from extensive destructive mining to clean and energy-saving mining in future decades. Global mining scientists and engineers have conducted a lot of research in related fields such as green mining, ecological mining, energy-saving mining, and mining solid waste recycling, and have achieved a great deal of innovative progress and achievements.

This Special Issue intends to collect the latest developments in the green low-carbon mining field, written by well-known researchers who have contributed to the innovation of new technologies, process optimization methods, or energy-saving techniques in metalliferous minerals development.

Various research topics are available to address this issue related to the role of green low-carbon technology for metalliferous minerals, including: green low-carbon technologies and systems; green low-carbon mining optimization methods; frontiers in mining with backfill; mine waste and heat management; the geo-mechanical behavior of mine backfill; energy-saving techniques in mining; alternative byproduct materials for green mining; green low-carbon development criteria of mining; and case studies of green low-carbon mining.

Citation: Guo, L. Green Low-Carbon Technology for Metalliferous Minerals. *Metals* **2022**, *12*, 1719. https://doi.org/10.3390/met12101719

Received: 13 September 2022
Accepted: 16 September 2022
Published: 14 October 2022

2. Contributions

In the present Special Issue, 'Green Low-Carbon Technology for Metalliferous Minerals', of *Metals*, 17 manuscripts were published.

The first paper, by Gao et al. [1], provides an experimental study for the disposal of arsenic-containing tailings with low carbon and high efficiency. In their research, sodium sulphate (Na_2SO_4), sodium hydroxide (NaOH), calcium nitrate $Ca(NO_3)_2$, and calcium hydroxide $Ca(OH)_2$ were independently added to metallurgical slag-based binder (MSB) solidification/stabilisation (S/S)-treated tailings (MSTs) to enhance the MST arsenic S/S performance. The microstructure analyses showed that the high specific surface area and amorphous properties of calcium–sodium aluminosilicate hydrate facilitated the adsorption or solid-solution formation of As(V) and As(III). As(V) formed an inner-sphere complex in ettringite, whereas As(III) formed an outer-sphere complex, and the relatively larger size and charge of As(V) compared with SO_4^{2-} restrict substitution inside channels without affecting the ettringite structure under high loading of As(V). The added $Ca(OH)_2$ promoted the hydration reaction of MSBs and facilitated the formation of a Ca–As(V) precipitate with low solubility. This work is beneficial for the application of cement-free MSB in the S/S process. The study carried out by Cui et al. [2] explores efficient methods for enhancing leaching efficiency, which is critical for bioleaching technology to deal with sulfide concentrate. In their study, a novel artificial microbial community was established to augment the bioleaching efficiency and recovery of copper (Cu) and zinc (Zn). This method of Cu and Zn recovery using a new-type artificial microbial community is expected to be an environmentally-friendly and efficient bioleaching technology solution, which has the potential of large-field engineering application in the future.

Regarding energy-saving techniques in mining, Tan et al. [3] provide a case study of structural parameter optimization for large-spacing sublevel caving in Chengchao Iron Mine. The findings of this study can further the goal of green and efficient mining and provide a theoretical reference for the popularization and application of pillarless sublevel caving with large structural parameters at home and abroad. The case study of the dispatch optimization model for haulage equipment between stopes based on mine short-term resource planning carried out by Li et al. [4] constructed an optimization model for short-term resource planning to maximize the total revenue during the planning period. Results show that, based on the improved optimization algorithm, using short-term production planning schemes to guide mine production operations can increase the haulage equipment utilization rate, thereby increasing mine production revenue.

Tailings ponds are one of the three major production facilities in metal mines. Through statistical analysis, Lin et al. [5] reveal the causes and regional distribution patterns of 342 tailings dam failures globally from 1915 to 2021. Causes of tailings pond failures differed among regions. Most tailings pond failures in Asia and Europe were related to hydroclimate, while earthquakes mainly triggered those in South America. Their study provides theoretical data for the pre-design and the safe and stable operation of global tailings ponds, which will help prevent global tailings pond failures. Furthermore, using complex network theory, Zhen et al. [6] studied the propagation laws of reclamation risk in tailings ponds.

Wang & Gan [7] studied the influence degree of particle settlement factors in pipe transportation of backfill slurry. In their study, a particle tracking module of the software was used to simulate the transport process, and the influences on the sedimentation rate were analyzed considering the slurry concentration, particle size, and flow velocity. Li et al. [8] established a mathematical model for the particle-size distribution of mine tailings using twelve tailing materials sourced from metal mines around China. Wang et al. [9] studied the characteristics of spiral pipes that increased resistance and reduced pressure, and the amendment equation of the stowing gradient. In their study, an equation for calculating the effective stowing gradient is obtained, which provides a convenient method for engineering applications and industrial design. Xie et al. [10] established an identification model of tailing settlement velocity to reasonably and accurately acquire the settlement interface and velocity of tailings. Their results show that the model has high recognition accuracy, has a rapid and efficient recognition process, and the relative error can be controlled within 3%. It can be used as a new technology for measuring the settling velocity of tailings. Zhu et al. [11] studied the response of floc networks in cemented paste backfill to a pumping agent, which provides insights into the effects of floc and liquid networks on the performance of paste, and it is of engineering significance in terms of realizing safe and efficient CPB operations.

Regarding the geomechanical behavior of mine backfill, Zhang et al. [12] studied the strength development and engineering performance of coal-based solid waste paste filling material. Peng et al. [13] examined the factors influencing the strength distribution of in situ cemented tailings backfill. In their study, it was found that solid materials settle and become redistributed during the flow of backfill slurries in stopes. With respect to tailings, coarse particles have a larger flowing resistance and are deposited near the feeding point, leading to the median sizes of particles increasing first and then decreasing thereafter. Le et al. [14] established a numerical model for the compressive behavior of backfill. The effects of the particle size of the granular backfill and the height and buried depth of mined-out stopes on surface subsidence have been clarified. The research results are of great significance for guiding backfill mining and evaluating surface subsidence and movement.

In the paper by Liu et al. [15], systematic research was conducted on the dynamic mechanical properties of limestone under confining pressure after high-temperature treatment.

A corresponding constitutive model was established, which provides technical support for the green and low-carbon mining of these resources. In the paper by Wang et al. [16], a three-dimensional numerical model using PLAXIS 3D was established to monitor and analyze the dynamic response for an open-pit mine with inside inclined shafts under train loading. In the paper by Shi et al. [17], the blasting dust concentration and particle size distribution during tunnel construction by drilling and blasting were studied to guide the dust reduction technology in tunnel blasting to quickly remove the dust generated during the blasting process and improve the engineering construction efficiency.

3. Conclusions and Outlook

As the guest editor, I sincerely thank all the experts and scholars for submitting excellent papers to this Special Issue, and the reviewers for providing many constructive comments for the Special Issue.

I also want to thank the editor Zach Ma for his great help and contribution to preparing this Special Issue.

Overall, we hope the contribution of this collection builds up the understanding and interest of all researchers and practitioners who are focused on the recent developments and advancements in studying green low-carbon technology for metalliferous minerals. It is hoped that the results of this issue can provide a reference for future research and contribute to promoting the development of green low-carbon technology for metalliferous minerals.

Funding: This research was funded by the National Key Research and Development Program of China (Grant No. 2021YFC2900600) and the National Natural Science Foundation of China (Grant No. 52274122).

Conflicts of Interest: The author declares no conflict of interest.

References

1. Gao, W.; Li, Z.; Zhang, S.; Zhang, Y.; Fu, P.; Yang, H.; Ni, W. Enhancing Arsenic Solidification/Stabilisation Efficiency of Metallurgical Slag-Based Green Mining Fill and Its Structure Analysis. *Metals* **2021**, *11*, 1389. [CrossRef]
2. Cui, X.; Yuan, X.; Li, H.; Che, X.; Zhong, J.; Wang, L.; Liu, Y.; Hu, X.; Zhang, Q.; Jin, R.; et al. Copper and Zinc Recovery from Sulfide Concentrate by Novel Artificial Microbial Community. *Metals* **2021**, *12*, 45. [CrossRef]
3. Tan, Y.; Guo, M.; Hao, Y.; Zhang, C.; Song, W. Structural Parameter Optimization for Large Spacing Sublevel Caving in Chengchao Iron Mine. *Metals* **2021**, *11*, 1619. [CrossRef]
4. Li, N.; Feng, S.; Ye, H.; Wang, Q.; Jia, M.; Wang, L.; Zhao, S.; Chen, D. Dispatch Optimization Model for Haulage Equipment between Stopes Based on Mine Short-Term Resource Planning. *Metals* **2021**, *11*, 1848. [CrossRef]
5. Lin, S.-Q.; Wang, G.-J.; Liu, W.-L.; Zhao, B.; Shen, Y.-M.; Wang, M.-L.; Li, X.-S. Regional Distribution and Causes of Global Mine Tailings Dam Failures. *Metals* **2022**, *12*, 905. [CrossRef]
6. Zhen, Z.; Zhang, Y.; Hu, M. Propagation Laws of Reclamation Risk in Tailings Ponds Using Complex Network Theory. *Metals* **2021**, *11*, 1789. [CrossRef]
7. Wang, C.; Gan, D. Study and Analysis on the Influence Degree of Particle Settlement Factors in Pipe Transportation of Backfill Slurry. *Metals* **2021**, *11*, 1780. [CrossRef]
8. Li, Z.; Guo, L.; Zhao, Y.; Peng, X.; Kyegyenbai, K. A Particle Size Distribution Model for Tailings in Mine Backfill. *Metals* **2022**, *12*, 594. [CrossRef]
9. Wang, W.; Mu, H.; Mei, G.; Guo, L.; Lu, X.; Wang, A.; Sun, R. The Characteristics of Spiral Pipe Increasing Resistance and Reducing Pressure and the Amendment Equation of Stowing Gradient. *Metals* **2022**, *12*, 1105. [CrossRef]
10. Xie, J.; Qiao, D.; Han, R.; Wang, J. Tailings Settlement Velocity Identification Based on Unsupervised Learning. *Metals* **2021**, *11*, 1903. [CrossRef]
11. Zhu, J.; Wu, S.; Cheng, H.; Geng, X.; Liu, J. Response of Floc Networks in Cemented Paste Backfill to a Pumping Agent. *Metals* **2021**, *11*, 1906. [CrossRef]
12. Zhang, J.; Yang, K.; He, X.; Wei, Z.; Zhao, X.; Fang, J. Experimental Study on Strength Development and Engineering Performance of Coal-Based Solid Waste Paste Filling Material. *Metals* **2022**, *12*, 1155. [CrossRef]
13. Peng, X.; Guo, L.; Liu, G.; Yang, X.; Chen, X. Experimental Study on Factors Influencing the Strength Distribution of In Situ Cemented Tailings Backfill. *Metals* **2021**, *11*, 2059. [CrossRef]
14. Le, Z.-H.; Yu, Q.-L.; Pu, J.-Y.; Cao, Y.-S.; Liu, K. A Numerical Model for the Compressive Behavior of Granular Backfill Based on Experimental Data and Application in Surface Subsidence. *Metals* **2022**, *12*, 202. [CrossRef]
15. Liu, L.; Li, R.; Qin, H.; Sun, W. Experimental SHPB Study of Limestone Damage under Confining Pressures after Exposure to Elevated Temperatures. *Metals* **2021**, *11*, 1663. [CrossRef]

16. Wang, Y.; Ni, S.-T.; Yang, F.-W.; Wang, Z.-X.; Zhang, H.; Ma, K.; Li, X.-J. Monitoring and Analysis of Dynamic Response for Open-Pit Mine with Inside Inclined Shafts under Train Loading. *Metals* **2021**, *11*, 1681. [CrossRef]
17. Shi, J.; Zhang, W.; Guo, S.; An, H. Numerical Modelling of Blasting Dust Concentration and Particle Size Distribution during Tunnel Construction by Drilling and Blasting. *Metals* **2022**, *12*, 547. [CrossRef]

metals

MDPI

Article

Experimental Study on Strength Development and Engineering Performance of Coal-Based Solid Waste Paste Filling Material

Jiqiang Zhang [1,2,3], Ke Yang [1,2,3], Xiang He [2,3,*], Zhen Wei [1,2,3], Xinyuan Zhao [1,2,3] and Juejing Fang [1,2,3]

1 State Key Laboratory of Mining Response and Disaster Prevention and Control in Deep Coal Mines, Anhui University of Science and Technology, Huainan 232001, China; austzhangjq@163.com (J.Z.); yksp2003@163.com (K.Y.); sdweizhen0302@163.com (Z.W.); zhaoxinyuan0402@hotmail.com (X.Z.); juejing_fang@163.com (J.F.)
2 Institute of Energy, Hefei Comprehensive National Science Center, Hefei 230031, China
3 National & Local Joint Engineering Research Center of Precision Coal Mining, Anhui University of Science and Technology, Huainan 232001, China
* Correspondence: xianghe_cumtb@126.com

Abstract: To explore the strength development characteristics and engineering performance of different coal-based solid waste filling materials cemented into filling body, coal gangue was used as coarse material, fly ash, desulfurization gypsum, gasification slag, and furnace bottom slag as fine material, and cement as a gelling agent. The uniaxial compressive strength (UCS) and bleeding rate of coal-based solid waste cemented backfill (CBSWCB) were tested by an orthogonal experiment, and the influencing factors of mechanical properties and strength development were analyzed. The multiple generalized linear model of strength and bleeding rate was established, and the optimal filling material ratio was determined. The engineering performance index of CBSWCB with the optimal ratio was tested. The results show the following points: (1) the concentration and content of desulfurization gypsum had a great influence on the early compressive strength of CBSWCB, while fly ash, gasification slag, and furnace bottom slag had little influence on the early compressive strength. (2) High concentration, high content of fly ash and furnace bottom slag, low content of desulfurization gypsum, and gasification slag can significantly improve the early strength. High concentration and high content of fly ash, low content of gasification slag, furnace bottom slag, and desulfurization gypsum are beneficial to the later strength increase. (3) Under the optimal ratio scheme, the bleeding rate of CBSWCB was 1.6%, the slump was 16.6 cm, the cohesion was general, the segregation resistance was good, the initial setting time was 5.42 h, the final setting time was 7 h, and the early strength after curing for 8 h reached 0.24 MPa.

Keywords: filling mining; coal-based solid waste; orthogonal experiment; strength development; regression analysis; engineering performance

Citation: Zhang, J.; Yang, K.; He, X.; Wei, Z.; Zhao, X.; Fang, J. Experimental Study on Strength Development and Engineering Performance of Coal-Based Solid Waste Paste Filling Material. *Metals* **2022**, *12*, 1155. https://doi.org/10.3390/met12071155

Academic Editor: Antoni Roca

Received: 25 May 2022
Accepted: 4 July 2022
Published: 6 July 2022

1. Introduction

With the rapid development of China's economy, the demand for coal resources has increased sharply, and ecological and environmental problems, such as surface subsidence, groundwater pollution, and gangue discharge caused by large-scale mining, have become increasingly prominent [1] (Figure 1). In recent years, filling coal mining technology has become one of the effective technical ways to solve the above-mentioned problems. Among them, paste filling makes solid waste into a paste in proportion and transports it to the goaf. It has developed rapidly in black, non-ferrous, gold, coal, and other systems [2], which effectively solves the coal-based solid waste generated during the deep processing of coal (Figure 2). It avoids environmental pollution, resource waste, surface subsidence, and other problems caused by mining, and achieves good social and economic benefits [3]. However, the comprehensive application of coal-based solid waste in China started late, and the comprehensive utilization path is relatively simple, and large-scale industrial utilization has not yet been achieved. At this stage, the method of land occupation and stacking is mostly adopted, occupying a large number of land resources, and the heavy metals contained in it will pollute the environment, and the added value of deep-processing products is generally low, which will increase the production cost of enterprises [4,5]. Taking desulfurized gypsum as an example, it is mainly used as a cement retarder, paper gypsum board, gypsum block, and other new wall materials. By 2019, the output will reach 71.5 Mt, and the utilization rate has not yet reached 80%, but the utilization rate of Germany, Japan, and other countries has already reached 100%, which is still a large gap compared with developed countries. Therefore, the reduction, harmlessness, and resource utilization of coal-based solid waste are imperative.

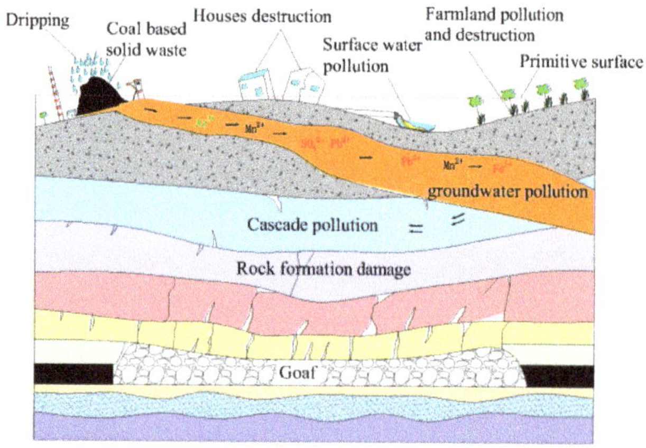

Figure 1. Ecological and environmental problems caused by large-scale mining of coal resources.

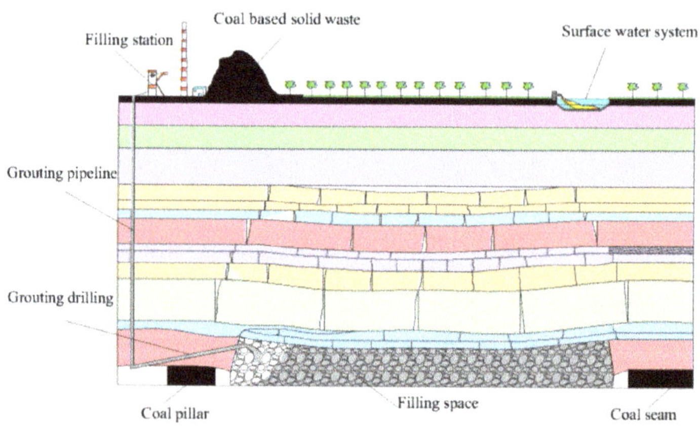

Figure 2. Application of coal-based solid waste in paste filling.

CBSWCB is formed from coal-based solid waste filling materials after a reasonable proportion design to support the rock mass around the stope. Therefore, exploring the strength characteristics, composition and mechanical properties of CBSWCB is the research foundation of backfill mining, and the ultimate goal of backfill mining is to choose low-cost, high-strength, low-pollution, and fluid-filling materials. In recent years, coal-green-filling mining has been widely used. Many scholars at home and abroad have carried out multi-dimensional research on filling materials, mainly focusing on the selection of filling materials and optimization of the ratio [6–9], the influence of mechanical properties [10–12], backfill mechanics constitutive models [13,14] and failure mechanisms [15,16].

At present, certain experimental studies have been carried out on coal-based solid waste filling materials. The filling materials are concentrated in coal gangue, gypsum, fly ash, slag, tailings, etc. [17–20], and the single factor analysis is mainly conducted through orthogonal experiments, regression analysis, visual modeling, neural network learning, response surface methodology, and other methods to analyze and optimize the factors affecting the strength development of the proportioning materials. Ref. [21] used fly ash, coal gangue, ordinary silicate cement, and water as the main filling materials, and found that the quality ratio of filling materials was 10%:20%:50%:20% without additives, which achieved a good filling effect. Ref. [22] tried to replace coal gangue with waste concrete to prepare filling paste, studied its initial flow properties, rheological properties, flow properties after standing, bleeding rate, and compressive strength, and determined the reasonable range of coarse aggregate replacing gangue.

Ref. [23] explored the evolution process and reaction mechanism of geopolymers synthesized by circulating fluidized bed fly ash, and investigated the influence of process parameters such as activator modulus, solid-liquid ratio, curing temperature, and time on the mechanical properties of geopolymers, and optimized the preparation conditions through a neural network. Ref. [24] explored the sensitive influence factors of the slump, compressive strength, and elastic modulus of the gangue-fly ash-tailings cementation through an orthogonal experiment, and they carried out a mix proportion design and determined the optimal mix proportion on this basis. Ref. [25] studied the influence of admixture content and tailings content, and slurry concentration on the fluidity and strength of semi-hydrated phosphogypsum fillings through orthogonal experiments. It was found that slurry concentration had an obvious impact on the strength and fluidity in the early stage, and admixture content had a great impact on the later stage. The water content of cemented paste filling (CPB) technology can ensure that the paste filling material does not ooze out, separate, or solidify when the water content reaches the lowest level during pipeline transportation [26]. The simultaneous unconfined compressive strength (UCS) test is a simple and the most commonly used method to assess the strength of CBSWB, but the information provided by the UCS test is limited. Based on the UCS test, understanding its hydraulic properties and exploring the hydraulic properties of CBSWB can provide a reasonable theoretical basis for the safe and economical structural design of CBSWB [27]. Xinguo Zhang et al. [28] used coal gangue to prepare cementitious paste filling material (CPB), added fly ash to partially replace ordinary Portland cement (OPC), and conducted a UCS test, slump test, bleeding test, and segregation through an orthogonal test, A series of hydraulic-mechanical tests were carried out. On this basis, the selected filling material was applied to the mine, and its filling performance was monitored. Dan Ma et al. [29] studied the evolution of the hydraulic and mechanical properties of coal-based solid waste CPB under different particle size distributions and cementing materials, and analyzed the deterioration mechanism of the mechanical properties. Although the above research has carried out extensive and detailed research, the material selection is generally single, and the research on the interaction between gasification slag and furnace bottom slag as paste filling materials is not in depth enough. At the same time, there are few studies on filling coal mining in the Ningdong coal power base, and the influencing factors of CBSWCB on the performance and strength development of paste filling engineering need to be further studied.

The Coal-Thermal Power-Chemical Industry Base in East Ningxia is a large-scale coal base of 100 million tons and a coal power base of tens of millions of kilowatts in China. In 2018, the coal output was about 91.55 million tons, the coal chemical production capacity was 22.25 million tons, and the thermal power capacity was 14.95 million kilowatts. In recent years, it has produced more than 90 million tons of coal-based solid wastes, such as coal gangue, fly ash, desulfurization gypsum, gasification slag, and furnace bottom slag [30], which is increasing year by year. Renjiazhuang Coal Mine is located on the edge of the Mu Us Desert 20 km northeast of Lingwu City in Ningxia, 30 km west of the Yellow River and facing Yinchuan City, and adjacent to the Hongshiwan Coal Mine in the north. Meihuajing Coal Mine is located in the Yuanyang Lake mining area of the Coal-Thermal Power-Chemical Industry Base in East Ningxia. Yuanyang Lake Power Plant is adjacent to Meihuajing Coal Mine, and the straight line distance between Yuanyang Lake Power Plant and Renjiazhuang Coal Mine is 25.4 km. The straight line distance between Meihuajing Coal Mine and Yuanyang Lake Power Plant is 2.5 km. Convenient transportation, geographical conditions, and sufficient sources of coal-based solid waste materials are conducive to the implementation of coal mine filling and mining. Taking the filling mine of Renjiazhuang Coal Mine in the Coal-Thermal Power-Chemical Industry Base in East Ningxia as the engineering background, this paper selected coal gangue as aggregate, fly ash, gasification slag, furnace bottom slag, and desulfurization gypsum as fine material and cement as cementitious material, established the multiple regression model of CBSWCB through an orthogonal experiment, explores the influence of coal-based solid waste filling material on the mechanical properties of CBSWCB, and analyzes its strength development characteristics. On this basis, the proportion optimization was carried out, and the engineering performance index of the optimization results was tested, to provide a reliable reference for guiding the paste filling mining of coal mine.

2. Test Materials and Schemes

2.1. Material Composition

The chemical composition of the dried coal-based solid waste filling material was determined by X-ray fluorescence (XRF), and the mineralogical composition was determined by X-ray diffraction (XRD). As shown in Figures 3 and 4, the main component of coal gangue is SiO_2, accompanied by some silica-based compounds. A large amount of SiO_2 ensures the high hardness and deformation resistance of gangue. Fly ash is the dry discharged ash of the Yuanyanghu Power Plant. The main components are SiO_2, Fe_2O_3, and CaO. The content of CaO and SiO_2 reaches 92.27%, which reduces the polymerization degree of glass in fly ash and increases its activity. The gasification slag is mainly composed of SiO_2, the bottom slag is mainly composed of SiO_2, $Al_2O_3 \bullet SiO_2$, and Fe_2O_3, and the desulfurization gypsum is mainly composed of $CaSO_4$ and $CaSO_4 \bullet 2H_2O$.

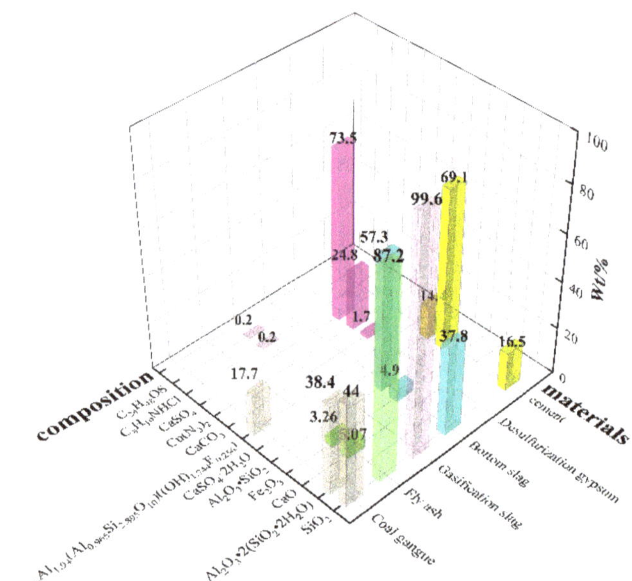

Figure 3. Analysis of main components of coal-based solid waste filling materials.

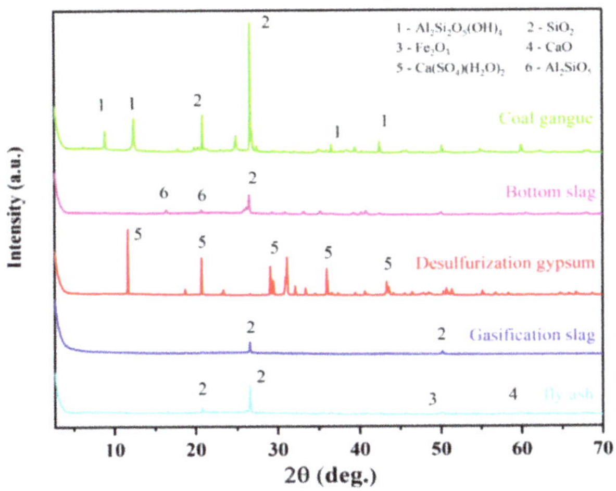

Figure 4. X-ray diffraction (XRD) of coal-based solid waste filling material.

Through the scanning electron microscope (SEM) on six kinds of experimental material microstructure tests, as shown in Figure 5, the gangue microscopic form is mainly a block, flake and other irregular forms. This irregular block form resting on the plane does not easily flow, the gangue large particle structure formed for the maintenance of the initial specimen provides a certain strength, and in the late maintenance stage, the generation of crystalline and gel material attached to the gangue and other large particles occurs. The pores are formed by the particles. The large particles of the morphological structure mean that it is easier for it to play its structural role, increasing the compressive strength of the filled specimen. The fly ash microscopic form is more regular, showing spherical particles and these spherical particles are conducive to their own transport and flow; it is easier for them to penetrate the pores between the large particles, increasing the opportunity to react with the cement hydration products, in addition to desulfurization. The microscopic form of gypsum is irregular in size and prismatic block form, and its particle size is moderate and uniform, which makes the particle composition of the mixture more reasonable; the microscopic form of gasification slag is not uniform, and the surface is rough and irregular, and the rough surface increases its specific surface area, which is more conducive to the adhesion of hydration gel and other substances on its surface, thus making the bonding between the particles of the mixture more tight. The microscopic morphology of furnace slag is irregularly lumpy, with obvious sintering traces on the surface, and the microscopic surface is smooth with fine pores, which are conducive to the entry of free water, and the outward migration of free water in the middle and later stages of maintenance is conducive to the continuation of hydration reaction inside the specimen, thus increasing the strength of the specimen. The microscopic morphology of No. 425 ordinary silicate cement is irregularly lumpy, and the particle size is on the high side. The microscopic form of No. 425 ordinary silicate cement is irregularly lumpy, the particle size is small, and the surface is rough, so the tiny particles will enter the pores between the large particles faster and mix with other materials fully in the mixing process so that the hydration product $Ca(OH)_2$ can react with other components more easily and generate calcium alumina and other crystalline substances, which will increase the strength of the specimen.

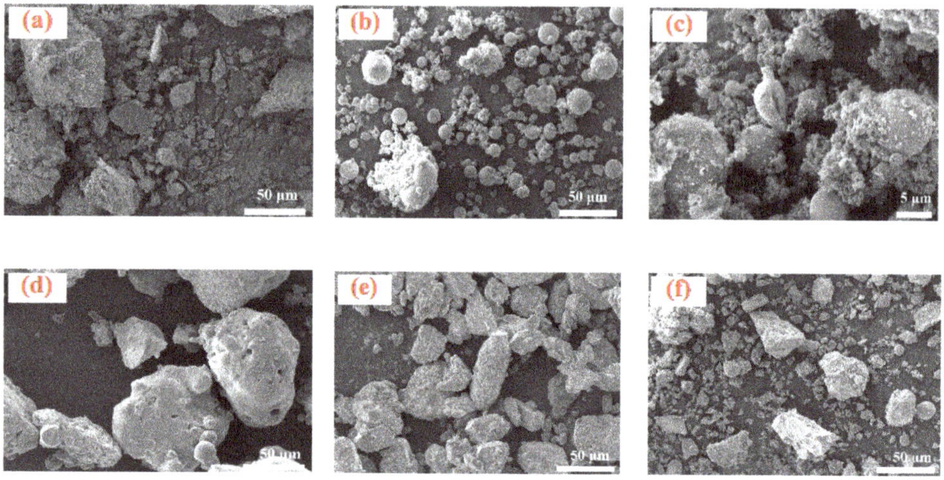

Figure 5. SEM microstructure of coal-based solid waste filling material, with (**a**) coal gangue, (**b**) fly ash, (**c**) gasification slag, (**d**) furnace bottom slag, (**e**) desulfurization gypsum, and (**f**) cement in the figure.

2.2. Test Scheme

An L16(4^5) orthogonal test is adopted. Without special instructions, the ratios involved in this paper are all mass ratios. Considering mainly exploring the filling performance of coal-based solid waste, we took the quality of coal gangue and cement as invariants, the cement mixture was 5% of the total mass of coal-based solid waste, and we set five research factors, namely A (concentration) and B (ash gangue ratio), C (gasification slag: coal gangue), D (bottom slag: coal gangue); E (desulfurized gypsum: coal gangue). Studies have shown that 75% to 80% of paste filler slurry concentration is beneficial to the stability of the roof and surrounding rock in the goaf [3]. Fly ash is coal-based solid waste. Due to its porous structure and spherical particle size, it has good permeability in a loose state and is a good cementitious material. Desulfurization gypsum has a small particle size, stable composition, and less harmful impurities. Fly ash and desulfurized gypsum account for a large proportion, which can meet the requirements of large-scale green economic filling. Gasification slag and bottom slag have pozzolanic properties, and blast slag has been gradually used at home and abroad to replace some cement. Considering the specific filling requirements and output, a gradient design is made for these two kinds of coal-based solid waste. The mixing amount of the two is close to but less than the above three kinds of coal-based solid waste. After comprehensive consideration of how to save the consumption of coal-based solid waste filling materials and test cost, four horizontal gradients were specifically set, as shown in Table 1.

Table 1. L16(4^5) orthogonal experimental factors and levels.

Level	Factor				
	A (Concentration/%)	B (Ash Gangue Ratio)	C (Gasification Slag: Coal Gangue)	D (Bottom Slag: Coal Gangue)	E (Desulfurized Gypsum: Coal Gangue)
L1	74	0.3	0.2	0.1	0.2
L2	76	0.4	0.25	0.15	0.3
L3	78	0.5	0.3	0.2	0.4
L4	80	0.6	0.35	0.25	0.5

2.3. Sources of Experiment Materials and Their Preparation

The experiment uses coal gangue as the coarse material, fly ash, gasification slag, furnace bottom slag, desulfurized gypsum as the fine material, and 425 ordinary Portland cement as a cementitious material. Among them, coal gangue comes from Renjiazhuang and Meihuajing coal mines, and fly ash, desulfurization gypsum, gasification slag and furnace bottom slag come from Yuanyang Lake Power Plant. The geographical location and material sources of the mining area are shown in Figure 6. At the same time, the particle size of coal-based solid waste filling materials has an important influence on the strength of coal-based solid waste cemented backfill (CBSWCB). In order to obtain the best strength effect of CBSWCB, the gradation of five kinds of coal-based solid waste took place. The gangue rock blocks were crushed and sieved into three particle size intervals, including (0.0–1.25) mm, (1.25–2.5) mm (2.5–5.0) mm of rock particles, and fly ash, gasification slag, furnace bottom slag. The large blocks in the desulfurized gypsum were crushed and sieved into particles with a particle size of less than 2.5 mm. The particle size composition of gangue is shown in Table 2.

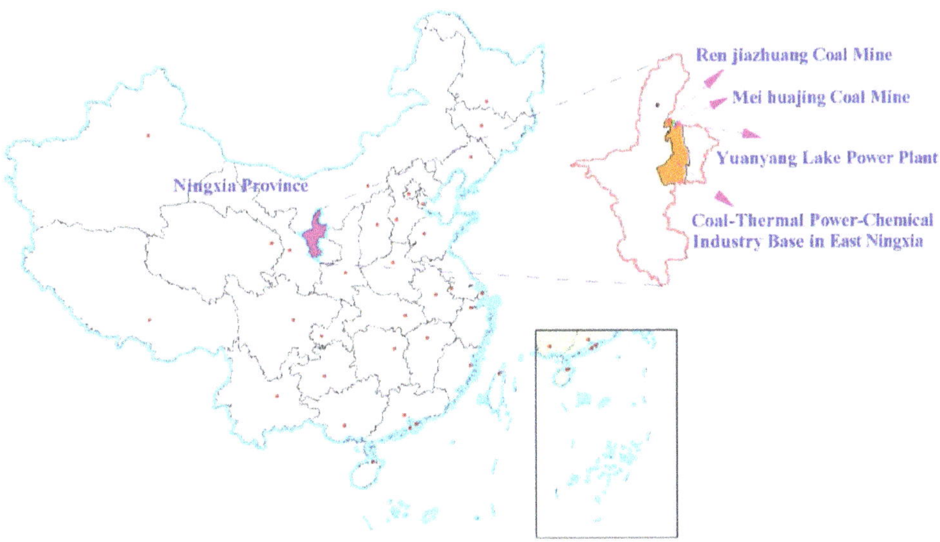

Figure 6. The geographical location of the mining area and sources of coal-based solid waste materials.

Table 2. Particle size composition of gangue.

Particle Size (mm)	Mass Percentage (%)
0–1.25	29.8
1.25–2.5	42.5
2.5–5.0	27.7

Particle size distribution, or particle gradation, is an important parameter for the properties of powder materials, such as fly ash, gasification slag, furnace bottom slag, and desulfurized gypsum. According to the SEM image, the particle size distribution of the four kinds of coal-based solid waste materials is counted. As shown in the Figure 7, the particle size of fly ash is mainly distributed in 5–10 μm, the particle size of gasification slag is mainly distributed in 0–3 μm, and the particle size of furnace bottom slag is mainly distributed in 0–3 μm. The particle size is mainly distributed in 10–30 μm, and the particle size of desulfurized gypsum is mainly distributed in 5–25 μm.

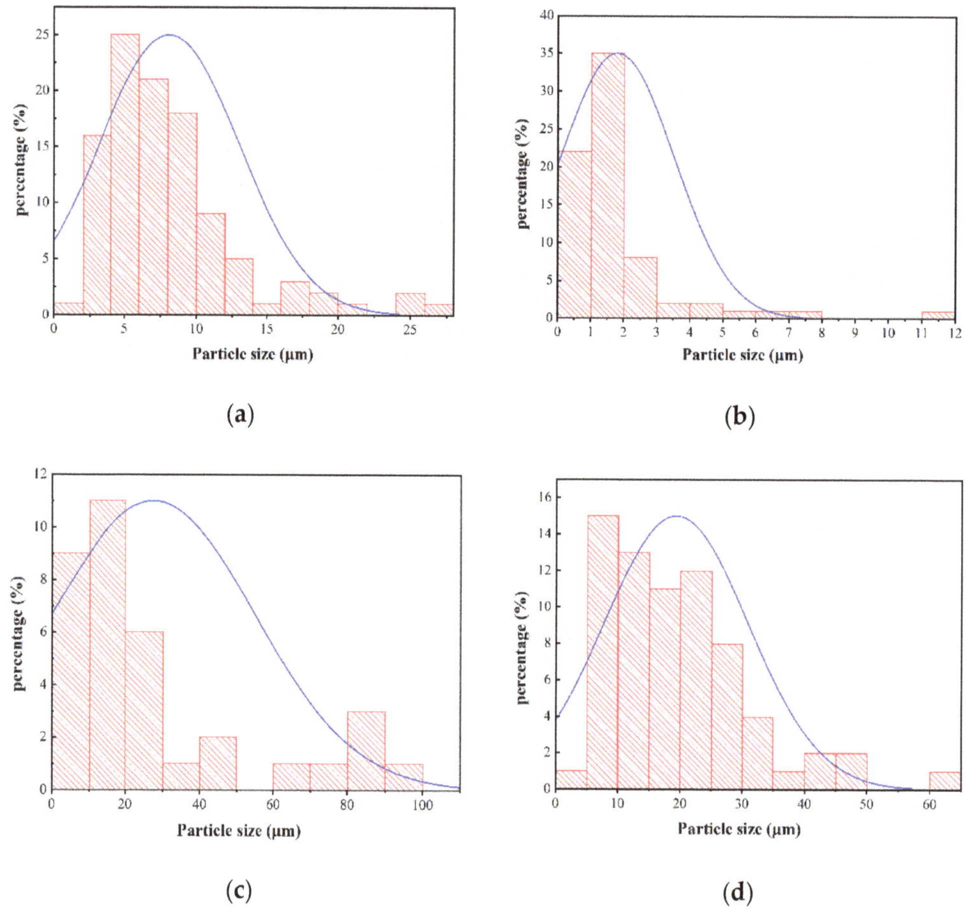

Figure 7. Particle size distribution diagram of coal-based solid waste materials, including (**a**) fly ash, (**b**) gasification slag, (**c**) bottom slag, (**d**) desulfurized gypsum.

According to the GB\T50080-2016 national standard [31], the quality of coal gangue is fixed, and according to the design plan of various coal-based solid waste materials, the test water and cement were weighed in turn, we used a cement paste mixer to fully stir the mixture, and casted the mixed cementitious material into a standard cube mold of 70.7 × 70.7 × 70.7 mm. We used the vibrating table to reduce the air and gap in the mold, then used the scraper to level the mold surface, and placed the leveled mold indoors, as shown in Figure 8. At the same time, part of the mixed cementitious material was added into the plexiglass container with a height of about 120 mm to ensure that the height of the measured filling surface was consistent and then it was sealed with fresh-keeping film. After standing indoors for 24 h, the bleeding rate was measured and recorded. The 48 sets of experimental data obtained were recorded. Three standard cube molds of 70.7 × 70.7 × 70.7 mm were made in each group of experiments, and a total of 144 standard cube molds were made in the experiment. After indoor curing for 12 h, we took the sample out of the mold and placed it in a constant temperature and humidity curing box at 20 °C and humidity of 90 ± 5% until it was taken out during the test, as shown in Figure 9. It completed the uniaxial compressive strength test on an electro-hydraulic servo rock pressure tester (loading rate is 0.05 KN/s) according to the national standard [32] of GB\T23561.12-2010.

Figure 8. Partial specimen of CBSWCB.

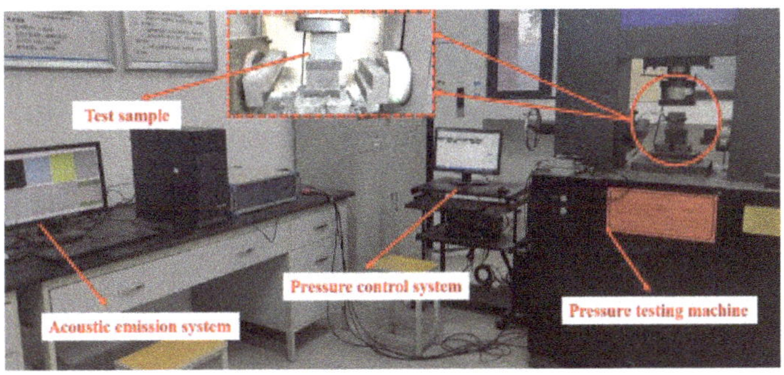

Figure 9. Test system.

While considering the coal-based solid waste filling material dosage and test cost, the optimization model was predicted through the development and regression analysis of coal-based solid waste cemented filling strength, and the required coal gangue, fly ash, gasification slag, furnace bottom slag, and desulfurization gypsum were re-screened according to the optimized proportioning scheme of filling materials obtained, and the secondary experiments were conducted according to GB\T50080-2016. Slump, expansion, initial and final setting time, and strength tests were conducted for the optimized proportioning test to verify the engineering performance and strength development characteristics, which will be described in detail in the later part of this paper.

3. Results

3.1. The Development of Backfill Strength

3.1.1. Sensitivity Analysis of Filling Body Strength

The bleeding rate and uniaxial compressive strength (3 d, 7 d, 28 d) of the coal-based solid waste material backfill were tested by the L16(4^5) orthogonal experiment (3 d, 7 d, 28 d) and the three datasets of bleeding rate are averaged, and the experimental results are shown in Table 3. The arithmetic average of the uniaxial peak compressive strength under the same value of a single factor was taken to obtain the influence of a single factor on the strength development of CBSWCB under the synergistic influence of multiple factors, as shown in Figures 10 and 11.

Table 3. Results of L16 (45) orthogonal experiment of CBSWCB.

Test No.	Factor					UCS (MPa)			Bleeding Rate (%)
	A	B	C	D	E	3 d	7 d	28 d	
S1	74	0.3	0.2	0.1	0.2	0.184	0.272	0.254	1.98
S2	74	0.4	0.25	0.15	0.3	0.161	0.27	0.31	2.39
S3	74	0.5	0.3	0.2	0.4	0.188	0.361	0.35	2.42
S4	74	0.6	0.35	0.25	0.5	0.175	0.245	0.261	4.76
S5	76	0.3	0.25	0.2	0.5	0.147	0.369	0.448	2.78
S6	76	0.4	0.2	0.25	0.4	0.173	0.396	0.604	3.16
S7	76	0.5	0.35	0.1	0.3	0.193	0.345	0.411	1.22
S8	76	0.6	0.3	0.15	0.2	0.343	0.621	0.796	1.15
S9	78	0.3	0.3	0.25	0.3	0.368	0.764	0.626	1.23
S10	78	0.4	0.2	0.2	0.2	0.324	0.583	0.704	2.01
S11	78	0.5	0.35	0.15	0.5	0.276	0.571	0.433	1.49
S12	78	0.6	0.25	0.1	0.4	0.365	0.971	1.723	0.5
S13	80	0.3	0.35	0.15	0.4	0.348	0.601	0.616	4.89
S14	80	0.4	0.3	0.1	0.5	0.317	0.643	0.697	0.55
S15	80	0.5	0.25	0.25	0.2	0.454	0.809	0.839	1.08
S16	80	0.6	0.2	0.2	0.3	0.406	0.582	0.574	0.53

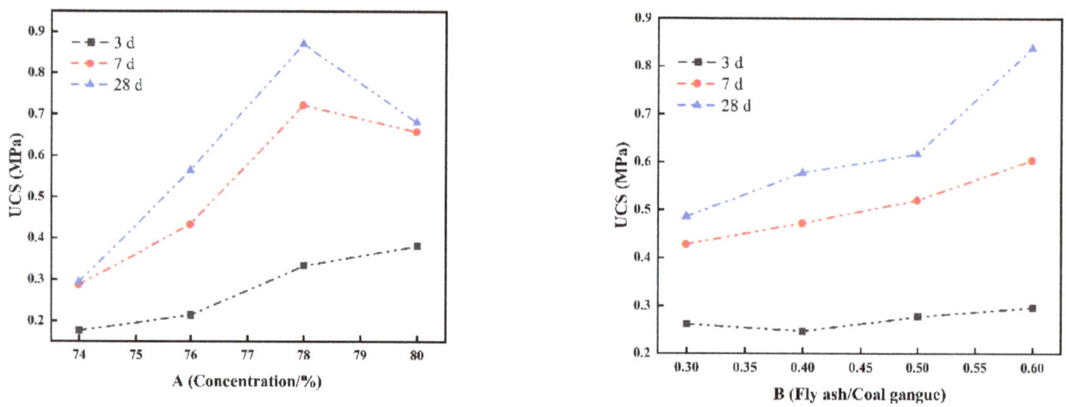

Figure 10. Influence of concentration and ash/gangue ratio on strength development.

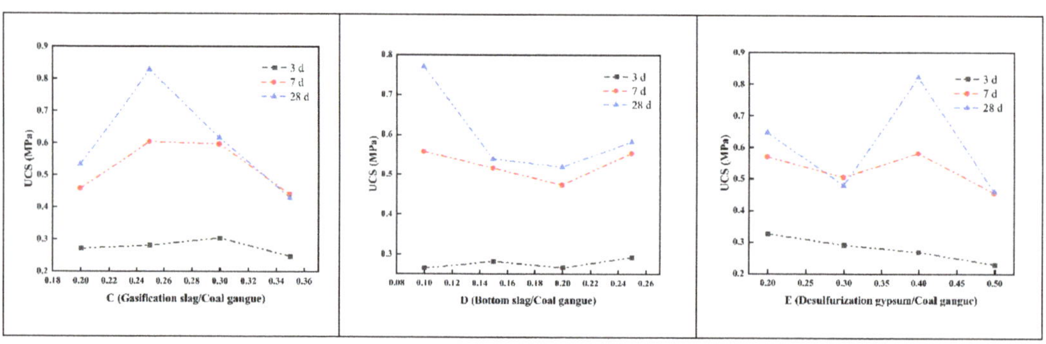

Figure 11. Influence of gasification slag, bottom slag and desulfurized gypsum on strength development.

In Figure 10, the compressive strength of CBSWCB shows a trend of first increasing and then decreasing with the increase in concentration. When the concentration is lower than 78%, the compressive strength increases rapidly and decreases when it is higher than 78%. At the same time, the compressive strength increases rapidly before 7 days, increases slowly in the later stage, and finally tends to ease. It can be observed from the 3 d strength that fly ash has little effect on the early strength. With the increase in fly ash content in different curing ages, the compressive strength gradually increases, and there is large room for the rise in late strength.

In Figure 11, with the increase in the content of gasified slag, the compressive strength of CBSWCB increases first and then decreases. When the factor C is 0.25, it reaches the maximum, then decreases rapidly, and finally tends to be stable. At the same time, with the increase in curing age, the compressive strength gradually decreases and the change rate gradually increases. It can be observed that the content of gasified slag has an important negative impact on the later strength. With the increase in the content of the furnace bottom slag, the compressive strength decreases first and then increases, and the change in 3 d strength is small. It can be observed that the content of the furnace bottom slag has little effect on the early strength. The content of desulfurized gypsum is negatively correlated with the 3 d compressive strength. With the increase in curing age, the compressive strength decreases first, then increases, and then decreases with the increase in the content of desulfurized gypsum, but on the whole, the content of desulfurized gypsum is negatively correlated with the compressive strength.

3.1.2. Analysis of Filling Body Strength Range

The range analysis of compressive strength after different curing ages (3 d, 7 d, and 28 d) at different levels of each factor of CBSWCB is carried out. As shown in Table 4, the order of influencing factors on 3 d compressive strength is concentration > desulfurization gypsum content > fly ash content > gasification slag content > furnace bottom slag content. The order of influencing factors on 7 d compressive strength is concentration > gasification slag content > fly ash content > desulfurization gypsum content > furnace bottom slag content. The order of influencing factors on 28 d compressive strength is concentration > gasification slag content > separated gypsum content > fly ash content > furnace bottom slag content.

AiBiCiDiEi is used to characterize the optimal test scheme combination, where A, B, C, D, and E represent the five factors, respectively, and i (for example, 1, 2, 3, 4) represents the different level gradients of the corresponding factors. The optimal test scheme combination of 3 d compressive strength in this test is $A_4B_4C_3D_4E_1$. The best test scheme combination of 7 d compressive strength is $A_3B_4C_2D_1E_3$. The optimal test scheme combination of 28 d compressive strength is $A_3B_4C_2D_1E_3$. According to the analysis in Table 4, on the whole, the concentration and the content of desulfurized gypsum have a great impact on the early compressive strength, while fly ash, gasification slag, and furnace bottom slag have little impact on the early compressive strength, and the influence degree of the three is close. High concentration and high content of fly ash and furnace bottom slag, low content of desulfurization gypsum, and gasification slag can significantly improve the early strength. High concentration and high content of fly ash, low content of gasification slag, furnace bottom slag, and desulfurization gypsum is conducive to the increase in later strength. At the same time, the compressive strength of 3 d to 7 d increases rapidly. With the increase in curing age, the influence on the compressive strength gradually decreases and finally tends to be stable. Comprehensive consideration of the optimal test scheme combination of this test is $A_3B_4C_2D_1E_3$, that is, the concentration is 78%, the ash: gangue ratio is 0.6, the gasification slag: coal gangue ratio is 0.25, the furnace bottom slag: coal gangue ratio is 0.1, and the desulfurization gypsum: coal gangue ratio is 0.4.

Table 4. Range analysis of compressive strength of coal based solid waste backfill at different curing ages.

Factor	Level	3 d UCS (MPa)	7 d UCS (MPa)	28 d UCS (MPa)
A	74	0.177	0.287	0.29375
	76	0.214	0.43275	0.56475
	78	0.33325	0.72225	0.8715
	80	0.38125	0.65875	0.6815
B	0.3	0.26175	0.427958333	0.486
	0.4	0.24735	0.473	0.57875
	0.5	0.27775	0.5215	0.61725
	0.6	0.29655	0.60475	0.8385
C	0.2	0.27175	0.45825	0.534
	0.25	0.28175	0.60475	0.83
	0.3	0.304	0.59725	0.61725
	0.35	0.248	0.4405	0.43025
D	0.1	0.26475	0.55775	0.77125
	0.15	0.282	0.51575	0.53875
	0.2	0.26625	0.47375	0.519
	0.25	0.2925	0.5535	0.5825
E	0.2	0.32625	0.57125	0.64825
	0.3	0.29085	0.50645	0.48025
	0.4	0.2685	0.58225	0.82325
	0.5	0.22875	0.457	0.45975
Significance		A > E > B > C > D	A > C > B > E > D	A > C > E > B > D
Optimal level		$A_4B_4C_3D_4E_1$	$A_3B_4C_2D_1E_3$	$A_3B_4C_2D_1E_3$

3.2. *Regression Analysis of Filling Body*

Intuitive analysis can not analyze the influence the degree of various factors on the test results in the test process. To make up for this defect, the regression analysis method is used to perform regression analyses of the bleeding rate and uniaxial compressive strength of 3 d, 7 d, and 28 d.

3.2.1. Regression Analysis of the Bleeding Rate

To facilitate the regression analysis and simplify the regression model, multiple linear regression analysis was carried out on the test data of the bleeding rate measured in Table 3 through IBM SPSS statistics software. It was found that the determination coefficient R^2 of the regression model was 0.49 and the significance coefficient was 0.177, indicating that there was collinearity independence between the independent variables and the fitting of the regression model was poor. By using MATLAB 2020 to perform multiple generalized linear analyses, we can obtain the following CBSWCB generalized linear model as shown in Formula (1):

$$\begin{gathered} Y_B = 207.3x_1 + 118x_2 + 62.2x_3 - 330.4x_4 + 351.7x_5 - 1.3x_1{}^2 - 264.5x_2{}^2 - 393.5x_3{}^2 \\ -389.3x_4{}^2 - 429.3x_5{}^2 + 10.4x_1x_2x_3x_4x_5 + 41.1x_1x_2x_3x_4 - 9.1x_1x_2x_3x_5 - 3.6x_1x_2x_4x_5 - 806 \\ R^2 = 0.97 \end{gathered} \quad (1)$$

where Y_B is the bleeding rate, x_1 is A (concentration), x_2 is B (ash gangue ratio), x_3 is C (gasification slag/coal gangue), x_4 is D (furnace bottom slag/coal gangue), x_5 is e (desulfurized gypsum/coal gangue). The bleeding rate data in Table 3 are interpolated with scattered nodes through MATLAB software, and the five factors are taken as independent variables and the bleeding rate as dependent variables. As shown in Figure 12, a visual model is constructed to intuitively analyze the effect of various factors on the bleeding rate.

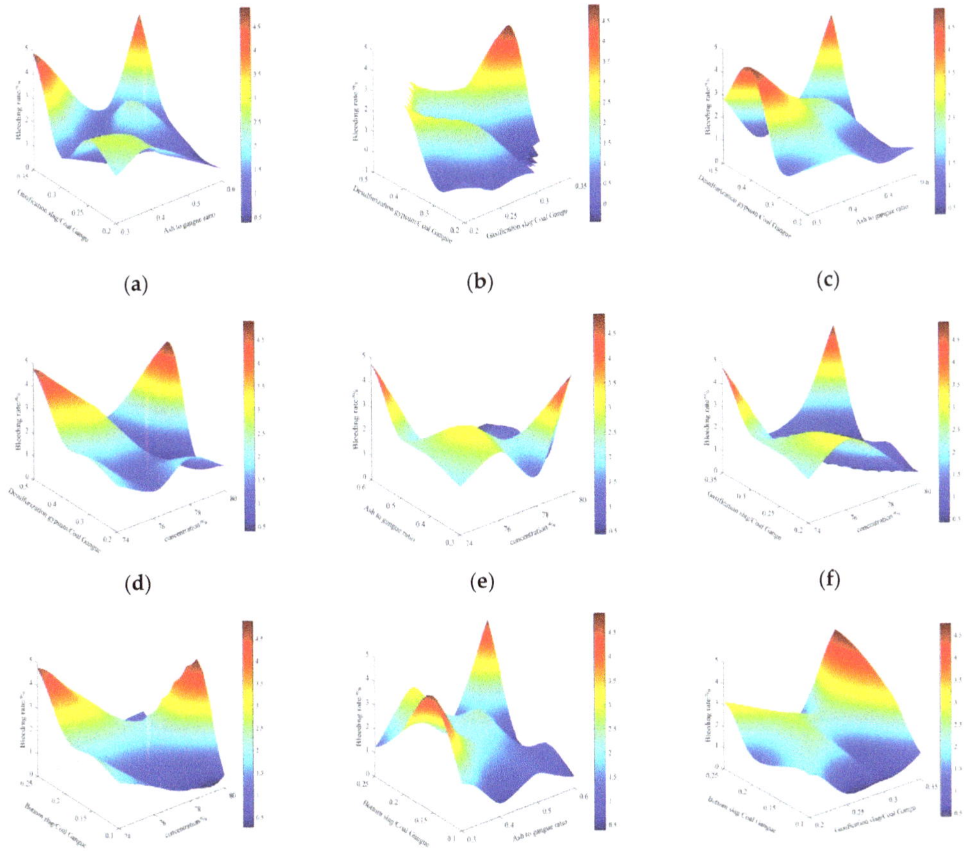

(a) (b) (c) (d) (e) (f) (g) (h) (i)

Figure 12. Effect of coal-based solid waste filling material on bleeding rate. (**a**) Factors of B and C; (**b**) factors of C and E; (**c**) factors of B and E; (**d**) factors of A and E; (**e**) factors of A and B; (**f**) factors of A and C; (**g**) factors of A and D; (**h**) factors of B and D; (**i**) factors of C and D.

Factor C in Figure 12a significantly affects the bleeding rate. When B = 0.3 and C = 0.35, the bleeding rate reaches the maximum, indicating that appropriately increasing the content of gasification slag can promote the increase in bleeding rate. In Figure 12b, with the increase in factors C and E, the bleeding rate shows an increasing trend as a whole, reaching the maximum when E is 0.4–0.45 and C = 0.35, both of which positively affected the bleeding rate. In Figure 12c, when the ash gangue ratio is less than 0.5, with the decrease in factor B and the increase in E, the bleeding rate gradually increases, and there is a certain negative correlation between them. In Figure 12d, when the concentration is constant, the bleeding rate shows an upward trend with the increase in factor E, indicating that the desulfurization gypsum significantly affects the bleeding rate. In Figure 12e, with the increase in factors A and B, the bleeding rate gradually decreases, indicating that A and B simultaneously negatively affect the bleeding rate. In Figure 12f, when the concentration is less than 76%, the negative changes in factors A and C will promote the filling body bleeding. When the concentration is greater than 76%, the two changes in the same direction will increase the bleeding body rate. In Figure 12g, it can be observed that factors A and D have a significant negative correlation and affect the bleeding rate, and their negative changes can promote an increase in the bleeding rate. In Figure 12h, when factors B and D change in the same direction, the bleeding rate increases significantly, reaching the maximum when B = 0.6 and D = 0.25. In Figure 12i, with the increase in factors C and D, the bleeding rate gradually increases, reaching the maximum when C = 0.35 and D are 0.2–0.25, indicating that factors C and D positively affect the bleeding rate.

Combined with the comprehensive comparative analysis of the established regression model, it is found that the amount of furnace bottom slag has a greater impact on the bleeding rate. Appropriately reducing the amount of furnace bottom slag can promote an increase in bleeding rate, which is closely related to the loose and porous structure of furnace bottom slag, more carbon particles, and high loss of ignition, resulting in strong water absorption. The content of desulfurized gypsum has little impact on the bleeding rate, and the concentration and content of fly ash negatively affect the bleeding rate, while gasification slag, furnace bottom slag, and desulfurized gypsum positively affect the bleeding rate, but the effects of the three are not very different.

3.2.2. Regression Analysis of Compressive Strength

The uniaxial compressive strength obtained from the L16(4^5) orthogonal test is analyzed by multiple regression. Through linear regression, it is found that the determination coefficient R^2 of the 3d compressive strength linear regression model is 0.9007, 7d is 0.6356 and 28 d is 0.4448. With the increase in curing age, the regression model coefficient gradually decreases and the significance of the model decreases. Therefore, MATLAB is used for generalized linear fitting. Due to many design factors, the generalized linear model is simplified, the significance level is set as 0.05, and the multiple linear regression model of CBSWCB uniaxial compressive strength is proposed, as shown in Formula (2). Formulas (3)–(5) are multiple generalized linear models of 3 d, 7 d, and 28 d UCS, respectively.

$$\begin{aligned} Y_i = {} & b_1x_1 + b_2x_2 + b_3x_3 + b_4x_4 + b_5x_5 + b_6{x_1}^2 + b_7{x_2}^2 + b_8{x_3}^2 + b_9{x_4}^2 + b_{10}{x_5}^2 \\ & + b_{11}x_1x_2x_3x_4x_5 + b_{12}x_1x_2x_3x_4 + b_{13}x_1x_2x_3x_5 + b_{14}x_1x_2x_4x_5 + b_{15} \end{aligned} \tag{2}$$

In the above equation, Y_i is the uniaxial compressive strength, where i = 3, 7, 28, respectively, representing the uniaxial compressive strength of coal-based solid waste filling at 3 d, 7 d, and 28 d. x_1 is A (concentration), x_2 is B (ash gangue ratio), x_3 is C (gasification slag/coal gangue), x_4 is D (furnace bottom slag/coal gangue), x_5 is e (desulfurized gypsum/coal gangue). b_k is the multiple regression coefficients ($k = 1, 2, \cdots, 15$).

$$\begin{aligned} Y_3 = {} & 2.65x_1 - 1.84x_2 + 11.36x_3 - 2.95x_4 + 2.2x_5 - 0.02{x_1}^2 - 0.49{x_2}^2 - 24.19{x_3}^2 - 7.64{x_4}^2 \\ & -4.92{x_5}^2 - 104.58 \\ & R^2 = 0.99 \end{aligned} \tag{3}$$

$$\begin{aligned} Y_7 = {} & 10.87x_1 - 3.68x_2 + 12.3x_3 - 14.66x_4 + 18.45x_5 - 0.07{x_1}^2 - 3.36{x_2}^2 - 43.82{x_3}^2 \\ & -9.98{x_4}^2 - 24.45{x_5}^2 - 0.8x_1x_2x_3x_4x_5 + 2.27x_1x_2x_3x_4 + 0.014x_1x_2x_3x_5 - 0.07x_1x_2x_4x_5 \\ & -422.85 \\ & R^2 = 0.97 \end{aligned} \tag{4}$$

$$\begin{aligned} Y_{28} = {} & 33.2x_1 + 10.8x_2 - 6.3x_3 - 56.461x_4 + 68.1x_5 - 0.2{x_1}^2 - 27.5{x_2}^2 - 53.4{x_3}^2 \\ & -12{x_4}^2 + 2.8x_1x_2x_3x_4x_5 + 6.8x_1x_2x_3x_4 - x_1x_2x_3x_5 - 1.3x_1x_2x_4x_5 - 1290.5 \\ & R^2 = 0.94 \end{aligned} \tag{5}$$

3.3. *Engineering Performance Test of Filling Body*

The engineering performance of CBSWCB includes transportation performance and mechanical performance. Transportation performance is the ability to transport coal-based solid waste materials to the goaf of the working face, to avoid problems such as slurry segregation and blocking the pipeline. Mechanical performance is the ability to control the deformation of overlying strata after filling, that is, compressive strength and deformation resistance [33].

Slump can well reflect the cohesion and friction resistance of cementitious materials. Bleeding rate and water retention can reflect the stability of CBSWCB and the segregation capacity of slurry. A large bleeding rate and poor water retention will loosen the surface of the filling body and reduce its durability. Cohesion can reflect the uniformity of the filling body and the stratification and segregation of pumped slurry. The initial and final setting time is an important engineering performance index. If the initial setting time is too short, it will cause problems, such as the blockage of slurry transmission pipeline, the tension of filling operation, and difficulty of filling body self-reliance. If the final setting time is too long, it will seriously affect the normal production and operation of the coal mine.

In this test, indicators such as slump, bleeding rate, initial setting time, final setting time, cohesion, water retention, and early strength are selected to evaluate the engineering performance of new coal-based solid waste filling materials. Combined with the field experience of filling mining at home and abroad, the slump is between 18 and 23 cm, the bleeding rate between 1.5% and 5%, which are considered to have "good" and "excellent" transportability, respectively. The concentration of paste filling slurry is 76–80%, which can ensure that no bleeding in the stope occurs and effectively avoids the problem of cementation of ore and rock [3]. Early strength refers to the strength required by the filling material to maintain its structural stability. The high productivity of general coal mines requires that the early strength is the comprehensive strength within 8 h after filling, which is mainly estimated by empirical Formula (6) [34].

$$h^2 = a\sigma^3 \tag{6}$$

where "h" is the mining height, "a" is the empirical coefficient, the filling body of the coal mine is generally taken as 600, and "σ" is the early strength. The mining height of the filling face in Renjiazhuang Coal Mine is 3.8 m. To ensure the filling effect and normal operation of the working face, the early strength of CBSWCB needs to be higher than 0.29 Mpa.

4. Discussion

4.1. Analysis of Factors Influencing the Development of Backfill Strength

During different curing ages, the compressive strength of CBSWCB first increased and then decreased with the increase in concentration, and gradually increased with the increase in fly ash content (Figure 10), which may be the spherical surface of fly ash. Smaller spherical particles adhere unevenly, and there are pores between the small spherical particles and the matrix. The compressive strength gradually increases with the adsorption and compaction of the pores in the later stage, which reduces the ability of the slurry to produce plastic deformation, resulting in glue. The cementing effect is better in the coagulating material, and the compressive strength is increased. The compressive strength of CBSWCB first increases and then decreases with the increase in gasification slag content. With the increase in furnace bottom slag content, it first decreases and then increases. As the content of desulfurized gypsum increases, it first decreases. The ups and downs of rising and falling are shown in Figure 11.

After the CBSWCB strength sensitivity analysis in Section 3.1.2, it was found that gasification slag and furnace bottom slag have different effects on the early strength and late strength of CBSWCB, and the current research on gasification slag and bottom slag as paste filling materials is not enough. In order to further study the influence of the interaction of different material factors on CBSWCB, the control variable method was used to analyze the influence mechanism of the interaction between gasification slag and bottom slag on the strength of CBSWCB. The factors remain unchanged, and the griddata function in MATLAB software was used to plot to explore the effect of gasification slag and furnace bottom slag content at different curing ages on the compressive strength of coal-based solid waste backfill, as shown in Figure 13. The high-strength areas of 3 d and 7 d are concentrated in factor C of 0.22–0.27, D of 0.1–0.13, and 0.22–0.25 in the area, with the increase in the curing age, the high-intensity area shifts to the area with a factor C of 0.22–0.27 and D of 0.1–0.12, and finally reaches the maximum when C is 0.25 and D is 0.1. It can be observed that the influence of gasification slag content on the strength of CBSWCB is lower than that of the furnace bottom slag. Low content of gasification slag and high content of furnace bottom slag is beneficial to increase the compressive strength. This may be due to the high residual carbon content of the gasification slag. The high content of the gasification slag will hinder the gelation reaction between it and cement or lime. At the same time, under the co-stimulation of the hydration reaction of other materials, for the furnace bottom slag, the Al_2O_3, Fe_2O_3, and active SiO_2 react with calcium silicate and calcium aluminate gel, resulting in the formation of a dense structure with a certain strength of CBSWCB. As the curing age increases, the pores left by the surface of the furnace bottom slag burnt are packed tightly, and the influence of the content of gasification slag and furnace bottom slag on the compressive strength gradually decreases.

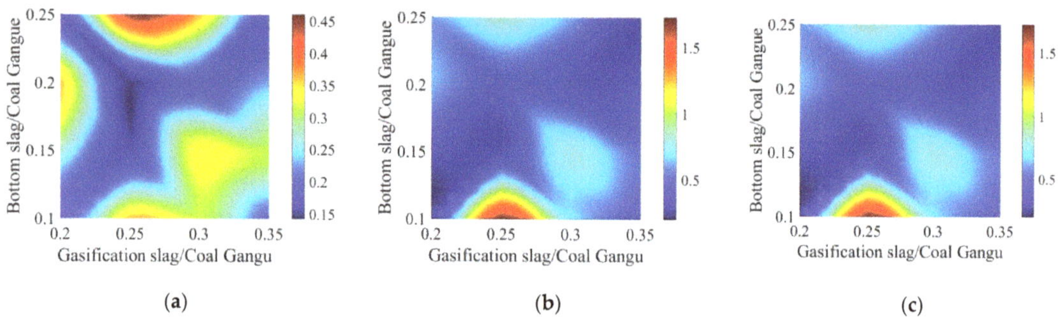

Figure 13. Influence of C and D factors on compressive strength of different curing ages. (**a**) 3 d UCS; (**b**) 7 d UCS; (**c**) 28 d UCS.

4.2. Effects of Different Factors on Bleeding Rate

Combined with the established regression model (Formula (1)), a comprehensive comparative analysis of the influence of different factors on bleeding rate (Figure 12) found that the amount of furnace bottom slag has a greater impact on the bleeding rate. Appropriately reducing the amount of furnace bottom slag can promote bleeding. The increase in water rate is related to the loose and porous structure of furnace bottom slag, many carbon particles, and the high loss of ignition, which leads to strong water absorption. The amount of desulfurized gypsum has a small effect on the bleeding rate, the concentration and fly ash mixing. The amount negatively affects the bleeding rate, while the gasification slag, furnace bottom slag, and desulfurized gypsum positively affect the bleeding rate, but the effects of the three are not much different.

4.3. Engineering Performance Index Test Analysis

According to the GB/T1346-2011 national standard [35], the test piece was tested and analyzed (Figure 14). The CBSWCB slump under the optimal ratio was 16.6 cm, the bleeding rate was 1.6%, the cohesion was average, and there was no segregation. Water sedimentation problems, good water retention, no water condensate precipitation in the central stack or edge and good segregation resistance show that CBSWCB has good transportation performance. The initial setting time is 5.42 h, and the final setting time is 7 h.

The uniaxial compression test was performed on three specimens after 8 h of curing, and the average value of the three combined test data was taken. The compressive strength reached 0.24 MPa. Although it did not reach 0.29 MPa, the strength of the filling body determined by the empirical formula was relatively large. Emphasis is placed on safety considerations and there is a certain amount of overflow. It can be considered that the CBSWCB under the optimal ratio meets the early filling strength requirements of the target mine, and provides a reference for its subsequent filling and mining, which has certain guiding significance.

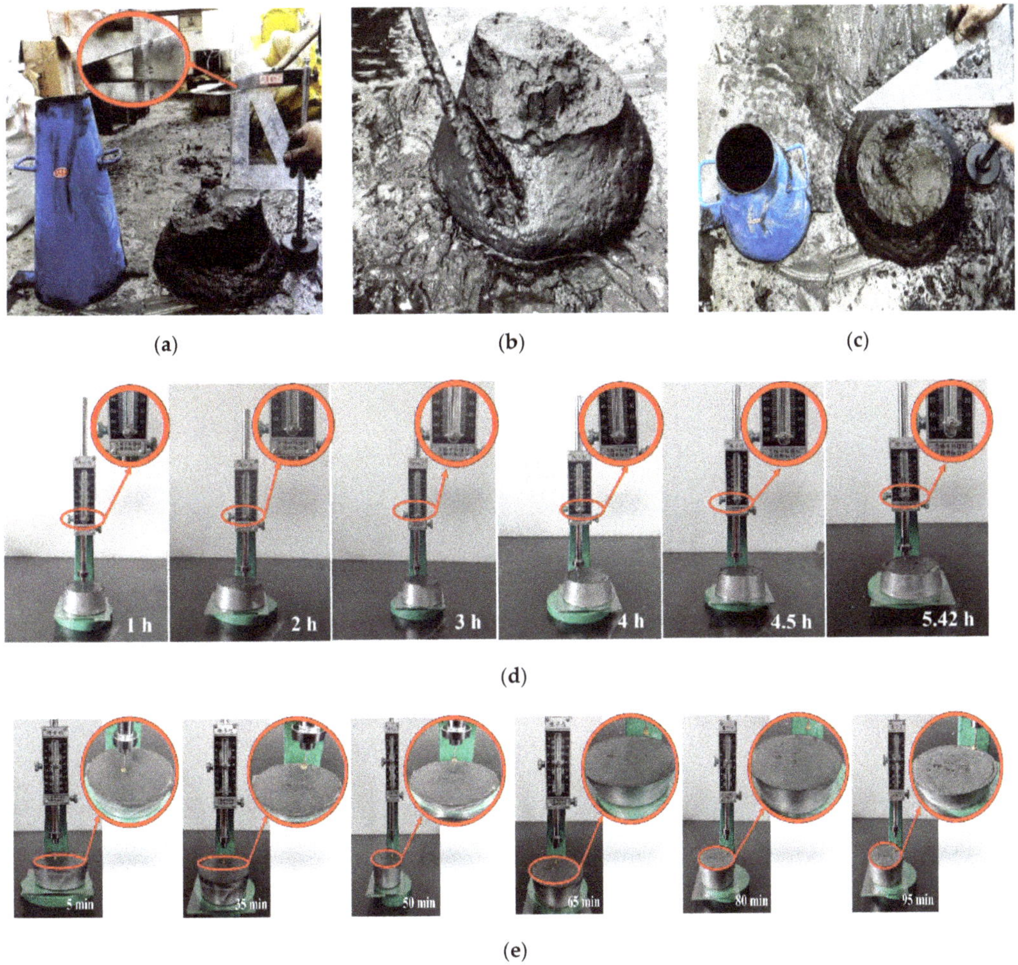

Figure 14. Engineering performance inspection of coal-based solid waste filling body. (**a**) Slump test; (**b**) cohesion test; (**c**) water retention test; (**d**) initial setting time test; (**e**) final setting time test.

5. Conclusions

In this paper, the UCS and bleeding rate of CBSWCB were tested by orthogonal experiment, and the influencing factors of mechanical properties and strength development were analyzed. The multiple generalized linear model of strength and bleeding rate was established, and the optimal filling material ratio was determined. The engineering performance index of CBSWCB with the optimal ratio was tested, and the following conclusions are drawn.

(1) Based on the development characteristics of the industrialized structure of the Coal-Thermal Power-Chemical Industry Base in East Ningxia, a multi-source coal-based solid waste filling material was preliminarily prepared, the strength development law of CBSWCB was analyzed, and the optimal ratio scheme of this test was determined, that is, the concentration was 78%, the ash: gangue ratio was 0.6, the gasification slag: coal gangue was 0.25, the furnace bottom slag: coal gangue was 0.1, and the desulfurization gypsum: coal gangue was 0.4.

(2) Concentration and desulfurized gypsum content had a great influence on the early compressive strength of CBSWCB, while fly ash, gasification slag, and furnace bottom slag had little influence on the early compressive strength, and the influence degree of the three was close. High concentration and high content of fly ash and furnace bottom slag, low content of desulfurization gypsum, and gasification slag can significantly improve the early strength. High concentration and high content of fly ash, low content of gasification slag, furnace bottom slag, and desulfurization gypsum were conducive to the increase in the later strength. At the same time, the compressive strength increases rapidly in the early stage. With the increase in curing age, the impact on the compressive strength gradually decreases and finally tends to be stable.

(3) The content of furnace bottom slag had a greater impact on bleeding rate, while the content of desulfurized gypsum had a small effect on bleeding rate. Concentration and fly ash content were negatively correlated to the bleeding rate, while gasification slag, furnace bottom slag, and desulfurized gypsum had a positive impact on the bleeding rate, and the impact of the three was close.

(4) The engineering performance test of CBSWCB under the optimal proportioning scheme shows that the bleeding rate of the filling body was 1.6%, the slump was 16.6 cm, the cohesion was general, the segregation resistance was good, and the initial setting time was 5.42 h. The final setting time was 7 h, and the early strength after curing for 8 h reached 0.24 MPa, which meets the requirements of the target mine paste filling. The consumption of coal-based solid waste filling materials exceeds 95%, which ensures the resource utilization of coal-based solid waste, reduces the cost of mine filling, and has significant economic and social benefits.

Author Contributions: J.Z. and K.Y. conceived and designed the theoretical framework; J.Z. and X.H. performed the experiments and wrote the manuscript; Z.W., X.Z. and J.F. corrected the tables and figures. All authors participated in the finalization of the written manuscript. J.Z. acted as the supervisor of the project and acquired all necessary funding. All authors have read and agreed to the published version of the manuscript.

Funding: This work was supported by the key projects of National Natural Science Foundation of China (52130402), the National Key Research and Development Program of China (2019YFC1904304), Institute of Energy, Hefei Comprehensive National Science Center under Grant No. 21KZS217 and the 2021 start-up fund for introducing talents and scientific research of Anhui University of Science and Technology (2021yjrc11).

Institutional Review Board Statement: Not applicable.

Informed Consent Statement: Not applicable.

Data Availability Statement: The data presented in this study are available upon request from the corresponding author.

Acknowledgments: The authors would like to thank the project (21KZS217) of the Institute of Energy, Hefei Comprehensive National Science Center (Anhui Energy Laboratory), the key projects of National Natural Science Foundation of China (52130402), the National Key Research and Development Program of China (2019YFC1904304) and the 2021 start-up fund for introducing talents and scientific research of Anhui University of Science and Technology (2021yjrc11) for the support to this research.

Conflicts of Interest: The authors declare that they have no known competing financial interest or personal relationship that could have appeared to influence the work reported in this paper.

References

1. Zhang, J.X.; Ju, F.; Li, M.; Zhou, N.; Zhang, Q. Method of coal gangue separation and coordinated in-situ backfill mining. *J. China Coal Soc.* **2020**, *45*, 131–140. (In Chinese) [CrossRef]
2. Wu, A.X.; Yang, Y.; Chen, H.Y.; Chen, S.M.; Han, Y. Status and prospects of paste technology in China. *Chin. J. Eng.* **2018**, *40*, 517–525. (In Chinese) [CrossRef]
3. Wu, A.X.; Wang, Y.; Wang, H.J. Status and Prospects of the Paste Backfill Technology. *Met. Mine* **2016**, *7*, 1–9. (In Chinese)
4. Li, J.; Wang, J. Comprehensive utilization and environmental risks of coal gangue: A review. *J. Clean. Prod.* **2019**, *239*, 117946. [CrossRef]

5. Stracher, G.B.; Taylor, T.P. Coal fires burning out of control around the world: Thermodynamic recipe for environmental catastrophe. *Int. J. Coal Geol.* **2003**, *59*, 7–17. [CrossRef]
6. Ren, A.; Feng, G.Y.; Guo, Y.X.; Qi, T.Y.; Guo, J.; Zhang, M.; Kang, L.X.; Han, Y.L.; Zhang, P.L. Influence on performance of coal mine filling paste with fly ash. *J. China Coal Soc.* **2014**, *39*, 2374–2380. (In Chinese) [CrossRef]
7. Tang, Y.S.; Zhang, L.F.; Lv, H.Y. Study on proportion optimization of coal-based solid wastes filling materials. *J. Min. Sci. Technol.* **2019**, *4*, 327–336. (In Chinese) [CrossRef]
8. Wu, L.B.; Wang, Y.M.; Chen, W.; Liu, J.M.; Chen, Z.Q. Ratio Optimization of Red Mud-Fly Ash Paste Filling Material Based on Orthogonal Experiment. *Min. Res. Dev.* **2020**, *40*, 45–49. (In Chinese) [CrossRef]
9. Zhao, X.; Yang, K.; He, X.; Wei, Z.; Zhang, J.Q. Study on proportioning experiment and performance of solid waste for underground backfilling. *Mater. Today Commun.* **2022**, *32*, 103863. [CrossRef]
10. Liu, L.; Fang, Z.; Qi, C.; Zhang, B.; Guo, L.; Song, K.-L. Numerical study on the pipe flow characteristics of the cemented paste backfill slurry considering hydration effects. *Powder Technol.* **2018**, *343*, 454–464. [CrossRef]
11. Ma, C.; Zhang, L.; Li, B.; Mao, X. Constitutive Model of Solid Backfill Materials and Numerical Simulation of Overburden Movement and Deformation in Backfill Mining. *Shock Vib.* **2021**, *2021*, 1–14. [CrossRef]
12. Zhao, K.; Huang, M.; Yan, Y.J.; Wan, W.L.; Ning, F.J.; Zhou, J.; Zhang, X.; Wang, J. Study on mechanical properties and synergistic deformation characteristics of tailings cemented filling assembled material body with different lime-sand ratios. *Chin. J. Rock Mech. Eng.* **2021**, *40* (Suppl. S1), 2781–2789. (In Chinese) [CrossRef]
13. Ou, Y.X.; Shi, C.J.; Shi, J.H.; Shan, B.; Wu, Z.M.; Huang, Y.; Xing, Y. Compressive Mechanical Properties and Prediction for Elastic Modulus of Ultra-high Performance Concrete. *J. Chin. Ceram. Soc.* **2021**, *49*, 296–304. (In Chinese) [CrossRef]
14. Zhao, X.Y.; Li, X.W.; Yang, K.; Wei, Z. The Spatiotemporal Characteristics of Coupling Effect between Roof and Backfill Body in Dense Backfill Mining. *Geofluids.* **2021**, *2021*, 1–15. [CrossRef]
15. Wang, J.; Song, W.; Cao, S.; Tan, Y. Mechanical properties and failure modes of stratified backfill under triaxial cyclic loading and unloading. *Int. J. Min. Sci. Technol.* **2019**, *29*, 809–814. [CrossRef]
16. Zhang, C.; Fu, J.; Song, W.; Du, C.; Fu, H. High-volume ultrafine fly ash-cement slurry mechanical properties and strength development model establishment. *Constr. Build. Mater.* **2021**, *277*, 122350. [CrossRef]
17. Gu, Y.; Wang, D.M.; Fang, K.Z.; Yao, G.; Wang, Q.B.; Zhang, M.; Sun, R.; Nan, L. Dissolution Characteristics of Coal Gasification Slag and Its Effect on Cement-Based Materials. *Bull. Chin. Ceram. Soc.* **2021**, *40*, 1579–1585. (In Chinese) [CrossRef]
18. Su, Z.; Chen, Q.; Zhang, Q.; Zhang, D. Recycling Lead–Zinc Tailings for Cemented Paste Backfill and Stabilisation of Excessive Metal. *Minerals.* **2019**, *9*, 710. [CrossRef]
19. Wang, Z.; Yu, W.; Liu, F.; Zhang, H. The Materialization Characteristics and Ratio of a New Soil Paste Filling Material. *Adv. Civ. Eng.* **2020**, *2020*, 6645494. [CrossRef]
20. Wu, J.; Yin, Q.; Gao, Y.; Meng, B.; Jing, H. Particle size distribution of aggregates effects on mesoscopic structural evolution of cemented waste rock backfill. *Environ. Sci. Pollut. Res. Int.* **2021**, *28*, 16589–16601. [CrossRef]
21. Yang, B.G.; Han, Y.M.; Yang, P.F.; Li, Y. Research on Ratio of High Concentration Cementation Stowing Materials in Coal Mine. *Coal Sci. Technol.* **2014**, *42*, 30–33. (In Chinese) [CrossRef]
22. Feng, G.Y.; Jia, X.Q.; Guo, Y.X.; Qi, T.Y.; Li, D.; Li, Z.; Feng, J.; Liu, G.; Song, K.; Kang, L. Influence of the wasted concrete coarse aggregate on the performance of cemented paste backfill. *J. China Coal Soc.* **2015**, *40*, 1320–1325. (In Chinese) [CrossRef]
23. Liang, D.C.; Yang, G.M.; Zhang, X.L.; Liu, D.Q.; Liu, J.C.; Xie, Q. Mechanism of geopolymer synthesis from circulating fluidized bed boiler fly ash and optimization of process parameters by back-propagation artificial neural network. *J. China Coal Soc.* **2021**, *46*, 1–11. (In Chinese) [CrossRef]
24. Sun, Q.; Zhang, X.D.; Zhang, S.K. Experimental Study on the Preparation of High-strength Paste Filling Materials with Mine's Solid Waste. *Non-Met. Mines.* **2015**, *38*, 42–44. (In Chinese)
25. Lan, W.T.; Wu, A.X.; Wang, Y.M.; Li, J.Q. Optimization of filling ratio of hemihydrate phosphogypsum based on orthogonal test. *Chin. J. Nonferrous Met.* **2019**, *29*, 1083–1091. [CrossRef]
26. Coussy, S.; Benzaazoua, M.; Blanc, D.; Moszkowicz, P.; Bussière, B. Arsenic stability in arsenopyrite-rich cemented paste backfills: A leaching test-based assessment. *J. Hazard. Mater.* **2011**, *185*, 1467–1476. [CrossRef]
27. Wang, Y.; Wu, A.; Wang, S.; Wang, H.; Yang, L.; Wang, Y.; Ruan, Z.E. Correlative mechanism of hydraulic-mechanical property in cemented paste backfill. *J. Wuhan Univ. Technol. Mater. Sci. Ed.* **2017**, *32*, 579–585. [CrossRef]
28. Zhang, X.; Lin, J.; Liu, J.; Li, F.; Pang, Z. Investigation of hydraulic-mechanical properties of paste backfill containing coal gangue-fly ash and its application in an underground coal mine. *Energies* **2017**, *10*, 1309. [CrossRef]
29. Ma, D.; Kong, S.; Li, Z.; Zhang, Q.; Wang, Z.; Zhou, Z. Effect of wetting-drying cycle on hydraulic and mechanical properties of cemented paste backfill of the recycled solid wastes. *Chemosphere.* **2021**, *282*, 131163. [CrossRef]
30. Yang, K.; Wei, Z.; Zhao, X.Y.; He, X.; Zhang, J.Q.; Ji, J.S. Theory and technology of green filling mining of solid waste underground in coal power base of Yellow River Basin. *J. China Coal Soc.* **2021**, 1–14. (In Chinese) [CrossRef]
31. Committee, C.N.S. *Standard for Test Method of Performance on Ordinary Fresh Concrete*; Standards Press of China: Beijing, China, 2016.
32. Institute, G.C.R. *Methods for Determination of Physical and Mechanical Properties of Coal and Rock Part 12: Method for Determination of Firmness Coefficient of Coal*; China National Standardization Administration Committee: Beijing, China, 2010; p. 8.
33. Zhou, N.; Zhang, J.X.; Ouyang, S.Y.; Deng, X.J.; Dong, C.W.; Du, E.B. Feasibility study and performance optimization of sand-based cemented paste backfill materials. *J. Clean. Prod.* **2020**, *259*, 120798. [CrossRef]

34. Huang, Q.X.; Li, L. Research on Stowing Material and Its Strength. *Coal Min. Technol.* **2011**, *16*, 38–42. (In Chinese) [CrossRef]
35. Institute, C.B.M.R. *Test Method for Water Consumption, Setting Time and Stability of Cement Standard Consistency*; China National Standardization Management Committee: Beijing, China, 2011; p. 12.

metals

MDPI

Article

The Characteristics of Spiral Pipe Increasing Resistance and Reducing Pressure and the Amendment Equation of Stowing Gradient

Weixiang Wang [1,2], Hongwei Mu [3,*], Guodong Mei [1,2], Lijie Guo [1,2,*], Xinqi Lu [1,2], Anhu Wang [4] and Ran Sun [5]

1 BGRIMM Technology Group, Beijing 102628, China; wangweixiang@bgrimm.com (W.W.); meiguodong@bgrimm.com (G.M.); luxinqi@bgrimm.com (X.L.)
2 National Centre for International Research on Green Metal Mining, Beijing 102628, China
3 Department of Safety Engineering, Qingdao University of Technology, Qingdao 266106, China
4 Technical Support Center for Prevention and Control of Disastrous Accidents in Metal Smelting, University of Science and Technology Beijing, Beijing 100083, China; anhu163@163.com
5 CCCC Water Transportation Consultants Co., Ltd., Beijing 100007, China; sunran112699@163.com
* Correspondence: muhongwei219@163.com (H.M.); guolijie@bgrimm.com (L.G.)

Abstract: To solve the high slurry pressure and severe wear at some sections in backfilling pipelines, this study investigates the solution of using an auxiliary pipe to increase the resistance and reduce the pressure of the mine backfilling pipeline. Using computational fluid dynamics, three auxiliary pipe models, a Z-shaped pipe, a S-shaped pipe and a spiral pipe were constructed and the velocity and pressure distribution characteristics of the filling slurry in the auxiliary pipes were analyzed. The function of friction loss in spiral pipes with different pitches and spiral diameters was established, and the amendment equation for calculating the effective stowing gradient was studied when using spiral pipes to increase resistance and reduce pressure. The results show that, compared with the Z-shaped pipe and the S-shaped pipe, the velocity and pressure in the spiral pipe change continuously and steadily, and there is no obvious sudden change in the local velocity and pressure. Therefore, it is difficult to burst the pipe. When the velocity is 2.5 m/s and the vertical height of the pipe is 2.5 m, the friction loss of the filling slurry in the spiral pipe can reach 3.87~21.26 times that in the vertical pipe, indicating that the spiral pipe can effectively play the role of increasing resistance and reducing pressure. The relationship between the friction loss and spiral diameter is a linear function, and the relationship between the friction loss and pitch is a quadratic function. The three are binary quadratic function relationships. The equation for calculating the effective stowing gradient is obtained, which provides a convenient method for engineering applications and industrial design.

Keywords: backfilling; increasing resistance and reducing pressure; computational fluid dynamics; spiral pipe; stowing gradient

Citation: Wang, W.; Mu, H.; Mei, G.; Guo, L.; Lu, X.; Wang, A.; Sun, R. The Characteristics of Spiral Pipe Increasing Resistance and Reducing Pressure and the Amendment Equation of Stowing Gradient. *Metals* **2022**, *12*, 1105. https://doi.org/10.3390/met12071105

Academic Editors: Srecko Stopic and Reza Ghomashchi

Received: 30 April 2022
Accepted: 23 June 2022
Published: 28 June 2022

1. Introduction

With the increasing depletion of shallow resources, the mining of deep resources has increased at home and abroad [1]. The number of mines over 1000 m is increasing both in China and in other countries [2,3]. Coal mining is already performed at depths greater than 1500 m in China, whereas metal ores are mined at 4350 m [4]. For the reasons of mine safety [5], environmental protection [6,7], deep sustainable mining and national carbon-neutral targets [8,9], backfilling is playing an increasingly important role in mining [10–12], and it has been widely used worldwide [13–16]. Due to the increase in the vertical depth of the backfilling pipe in deep mining, the stowing gradient decreases, which requires effective solutions to the problems of high pressure and severe wear of the pipe [17–19]. Currently, researchers adopt the following methods to reduce the pressure and wear of the slurry on the pipeline [20–26]: reducing the erosion effect of the filling slurry on the pipeline, using the full pipe flow conveying system in the vertical pipeline, using the depressurization conveying system, decreasing the conveying speed of the slurry, changing the properties of the paste by adding binder content, and using neutral water to prepare the filling slurry and flush the pipeline. However, there are few studies on reducing pressure and pipeline wear from the perspective of pipeline structure design. In addressing the problem, this study proposes the method of using an auxiliary pipe to increase resistance and reduce pressure; that is, to add one or more auxiliary pipes at the appropriate position of the vertical backfilling pipe to increase resistance and reduce pressure to decrease the high pressure of the filling slurry in the backfilling pipe, reduce the wear of the local pipe and simultaneously accomplish realize the continuous transportation of the filling slurry.

Computational fluid dynamics (CFD) approximately converts the continuity equation, momentum equation, energy equation and the integral and differential terms in the constitutive equation in fluid mechanics into discrete algebraic forms so that they can be changed into algebraic equations. CFD then uses the algebraic computing ability of the computer to solve the discrete algebraic equations to obtain the numerical solutions of discrete physical quantities such as pressure and velocity at time and space points [27]. CFD Fluent software is currently a relatively popular large-scale commercial software worldwide. It is used to simulate complex flows from incompressible to highly compressible flows. It has rich physical models, advanced numerical fitting methods, powerful processing capacity and physical models confirmed by engineering. Because Fluent adopts a variety of solution methods and multigrid accelerated convergence technology, it can achieve the best convergence speed and solution accuracy. It is widely used in the mine filling pipeline simulation filed all over the world. Kaushal et al. [28] used Ansys Fluent software to analyze the concentration distribution and pressure drop in pipeline flows with high concentrations of fine particles. Nagar et al. [29] conducted CFD simulations for the predication of pressure drop and concentration profiles for better understating of 25 μm fly ash particles mixed with water in the form of solid–liquid slurry in 50 mm pipe diameter with efflux concentrations in the range of 33–47 percent by volume at different flow velocities as 1–3 m/s. Movahedi et al. [30] brought experimental and CFD simulation of slurry flow in the annular flow path using two-fluid model to predict solid–liquid multiphase flow under different flow conditions including static, laminar and turbulence. Zambrano et al. [31] used a three-dimensional, algebraic slip mixture (ASM) model of the CFD software FLUENT 6.3 to obtain the numerical solutions for investigating the pressure-driven slurry flow of heavy oil in a horizontal pipe. Liu et al. [32] employed a CFD method to investigate the flow characteristics of the cemented paste backfill slurry with a mixture model and represent the multi-phase characteristics of pipe flow. Therefore, CFD can be used to study the effect of increasing the resistance and reducing the pressure of different types of auxiliary pipes.

In order to solve the high slurry pressure and severe wear of some sections in the backfilling pipeline, this paper investigates the solution of using an auxiliary pipe to increase the resistance and reduce the pressure of the mine backfilling pipeline by applying computational fluid dynamics to study the flow characteristics of filling slurry in the auxiliary pipe. First, three types of auxiliary pipes, namely the Z-shaped pipe, the S-shaped pipe and the spiral pipe, were designed, and the dimensions of the auxiliary pipes were determined according to the actual conditions in the mine to allow for easy installation. Their computational models were drawn. To verify the feasibility of the CFD method and the matching of simulation parameters, an L-shaped pipe model was established. The reliability of the model and parameters was verified by comparing the similarity between the theoretical and simulated results of pressure loss. Next, the viscosity, density, yield stress and other parameters of the filling slurry were confirmed through laboratory experiments, and the velocity and pressure distribution characteristics of the filling slurry in the three types of auxiliary pipes were simulated using CFD. Then, the friction loss of the filling slurry in spiral pipes with different sizes was calculated, and the functional relationship between friction loss and pitch and spiral diameter was obtained. Finally, the amendment equation for calculating the effective stowing gradient was proposed when using the auxiliary pipes to increase resistance and reduce pressure. This study provides a convenient method for the design of backfilling pipes in engineering applications.

2. Principle of the Increasing Resistance and Reducing Pressure of Auxiliary Pipe

In the backfilling mining method, the cementing material, water and aggregate are evenly mixed in the surface mixing station to form a uniform filling slurry with a solid mass fraction of approximately 60~80%. The filling slurry is transported to the underground goaf by means of gravity or pumping and is hardened for a period of time to form a solid cemented filling body [33,34]. This method can provide support for the rock mass around the goaf, prevent the collapse of the rock mass around the goaf, effectively manage the pressure in the rock mass and improve the safety of underground mining [6,13,35].

In engineering, the friction loss of filling slurry in pipeline transportation directly affects or even determines many core issues, such as whether the filling slurry can be transported by self-flow, the construction of the backfilling system, the selection of equipment (especially pumps) and the layout of the pipeline network [36].

Due to the filling slurry viscosity and the influence of pipe wall roughness, when fluid motion occurs, friction will be produced between the fluid and pipe wall and among fluid particles [37,38], which is called frictional resistance. The energy loss caused by the moving fluid overcoming the frictional resistance along the pipe is called the friction loss. Its calculation equation is shown in Equation (1).

$$P_f = \lambda \frac{l}{D} \frac{v^2}{2g} \gamma \tag{1}$$

where P_f is the friction loss; λ is the resistance coefficient along the pipe, 0.036, which is related to the roughness of the pipe; l represents the pipe length in m; D is pipe inner diameter in m; v represents the velocity of fluid in m/s; g is the acceleration of gravity, 9.81 m/s^2; and γ is the bulk density of filling slurry in N/m^3.

According to Equation (1), the friction loss along the process is proportional to the resistance coefficient, the length and the velocity.

The auxiliary pipe is installed in the vertical pipe section, as shown in Figure 1, and the specific position is adjusted according to the pipe arrangement and pressure distribution in the mine. When the slurry is transported along the pipeline, due to the friction between the slurry particles and the pipe wall and the viscosity of the slurry itself, the slurry will rub against the pipe wall, resulting in energy loss. Therefore, when the backfilling system does not use the auxiliary pipe, the slurry is directly transported along the L-shaped pipe. The energy loss in the vertical section is small, and the slurry pressure is large when it reaches the bottom. When the auxiliary pipe is used, the slurry will be transported along the auxiliary pipe. As the path becomes longer and the friction loss increases, the energy loss of the slurry increases, and the pressure of the slurry when it reaches the bottom is small, which can increase resistance and reduce pressure. In practice, the auxiliary pipe can be connected with the main pipe through the program-controlled electric valve. When the vertical height of the pipe is too large and the stowing gradient is too small, the auxiliary pipe is activated to increase the resistance of the filling slurry, reduce the velocity and pressure of the filling slurry when the filling slurry reaches the bottom of the vertical pipe and relieve wear at the pipe elbow. When the horizontal pipe is long and the stowing gradient is large, the auxiliary pipe is not activated, so that the filling slurry keeps its original transport energy to avoid a blocked pipe.

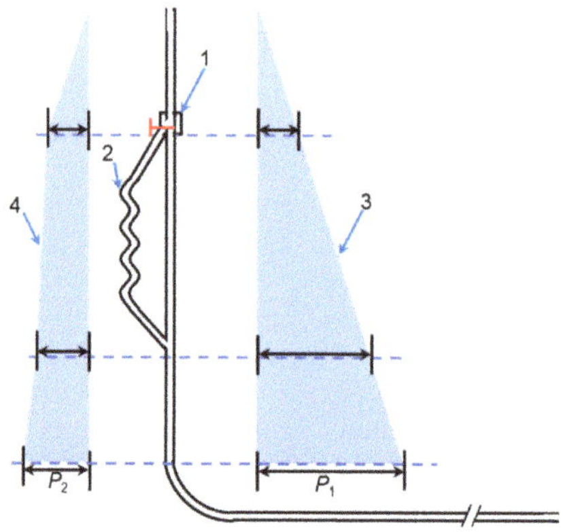

Figure 1. Increasing resistance and reducing pressure schematic of auxiliary pipe (illustration by authors). 1—Program-controlled electric valve, 2—auxiliary pipe, 3—pressure change of vertical pipe of L-shaped pipe, 4—pressure change of vertical pipe after using auxiliary pipe.

3. Numerical Simulation Methods

The mechanical properties of the filling slurry and the complexity of the backfilling pipe system make it impossible to solve the backfilling pipe transport problem accurately using CFD. Therefore, the numerical simulation model needs to be simplified [39,40], i.e., it is assumed that the filling slurry does not undergo phase change during transport and maintains a homogeneous full pipe flow. The effects of mining and temperature on filling slurry transport are not considered during the calculation.

3.1. Mathematical Model and Control Equations

3.1.1. Model Selection

In this study, the mixture multiphase flow model and standard *k-e* turbulence model in Fluent were used to calculate the along-travel friction loss in the flow field and analyze the flow velocity and pressure variation characteristics of the filling slurry.

3.1.2. Boundary Conditions

The *Z*-axis negative direction was set as the gravitational acceleration direction with a magnitude of 9.81 m/s, and the boundary conditions for each type of pipe were the velocity inlet and pressure outlet.

3.1.3. Control Equations

(1) Continuity equation

$$\frac{\partial}{\partial t}\left(\sum_{k=1}^{n} \alpha_k \rho_k\right) + \nabla\left(\sum_{k=1}^{n} \alpha_k \rho_k v_k\right) = 0 \tag{2}$$

where α_k is the volume fraction of the *k*th phase; ρ_k is the density of the *k*th phase in kg/m^3; v_k is the average velocity of the *k*th phase in m/s; and t is time in s.

(2) Momentum equation

$$\frac{\partial}{\partial t}\left(\sum_{k=1}^{n} \alpha_k \rho_k v_k\right) + \nabla\left(\sum_{k=1}^{n} \alpha_k \rho_k v_k v_k\right) = -\alpha_k \nabla p + \nabla\left[\alpha_k\left(\tau_k + \tau'_k\right)\right] + \alpha_k \rho_k g + M_k + (F_{\text{int}})_k + S_{\text{D}} \tag{3}$$

where p is pressure in Pa; τ_k is molecular dynamics in N/m^2; τ'_k is turbulent stress in N/m^3; M_k is interphase momentum transfer per unit volume in N/m^3; $(F_{\text{int}})_k$ is intrinsic force in N/m^3; g is the gravity acceleration in m/s^2; and S_{D} is momentum source term in N/m^3.

3.2. Geometric Modeling and Parameter Setting

3.2.1. Geometric Modeling

The three types of auxiliary pipe models, namely the Z-shaped pipe, the S-shaped pipe and the spiral pipe, were designed as shown in Figure 2, in which Q indicates the pitch, W represents the spiral diameter and D expresses the pipe inner diameter, which is 0.12 m. The height h of the pipe model is 2.5 m. Additionally, 25 dimensional parameters were designed for each of the three types of pipes, as shown in Table 1.

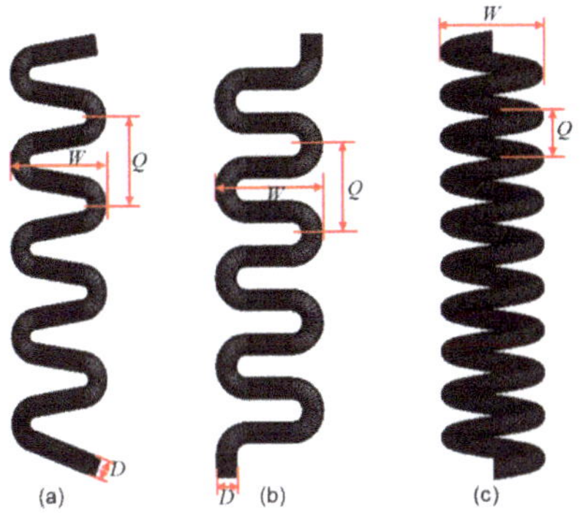

Figure 2. Models of the Z-shaped pipe (**a**), the S-shaped pipe (**b**) and the spiral pipe (**c**).

Table 1. Parameters for different sizes of pipes.

Parameter	Value/m				
Q	0.25	0.5	0.75	1.0	1.25
W	0.5	0.75	1.0	1.25	1.5
D	0.12				
h	2.5				

To verify the feasibility of the CFD method and the matching of simulation parameters, an L-shaped pipe model was established. The specific parameters were vertical pipe height of 100 m, horizontal pipe length of 300 m, pipe inner diameter of 0.12 m, elbow of 90° and radius of curvature of 3 m, as shown in Figure 3.

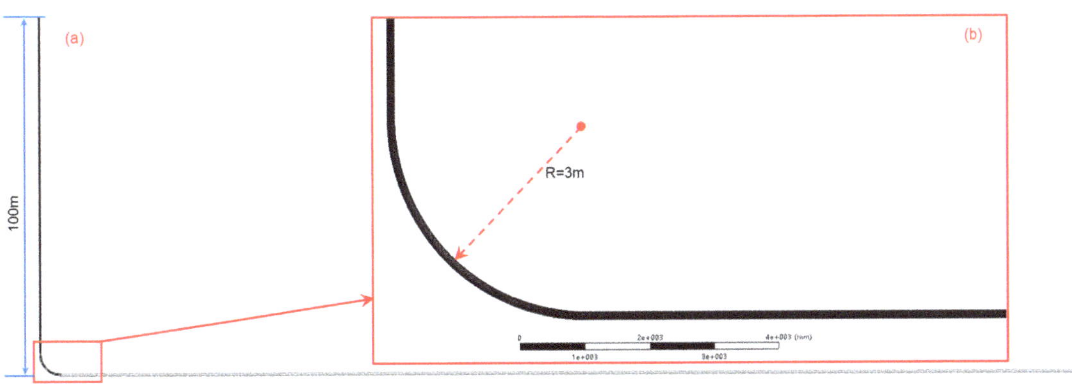

Figure 3. Models of the L-shaped pipe (**a**,**b**) elbow of the L-shaped pipe.

3.2.2. Determination of Filling Slurry Parameters

A gold mine filling slurry in Shandong Province was studied. The filling slurry with a concentration of 72% (wt) was selected for simulation calculations. The same filling slurry sample was prepared in the laboratory using the concentration of filling slurry and the aggregate-to-cement ratio found in the industrial field. The preparation process of the filling slurry sample is shown in Figure 4, and the composition of tailings is shown in Table 2. The prepared filling slurry sample is shown in Figure 5a, and its rheological parameters were tested by a Brookfield R/S Plus rheometer (shown in Figure 5b). The results are shown in Figure 5c.

Figure 4. Preparation process of the filling slurry sample. (**a**) Weigh tailings of a certain mass, (**b**) weigh cementitious materials of a certain mass, (**c**) weigh some water, (**d**) pour the tailings, curing material and water into the hopper of the mixer successively and mix for 10 min at 60 r/min to evenly distribute the components.

Table 2. The composition of tailings.

Composition	SiO_2	Al_2O_3	K_2O	CaO	Fe_2O_3	Na_2O	SO_3	MgO	MnO	Others
Proportion (%)	61.24	22.14	5.10	2.19	2.96	1.82	1.46	1.41	0.17	1.51

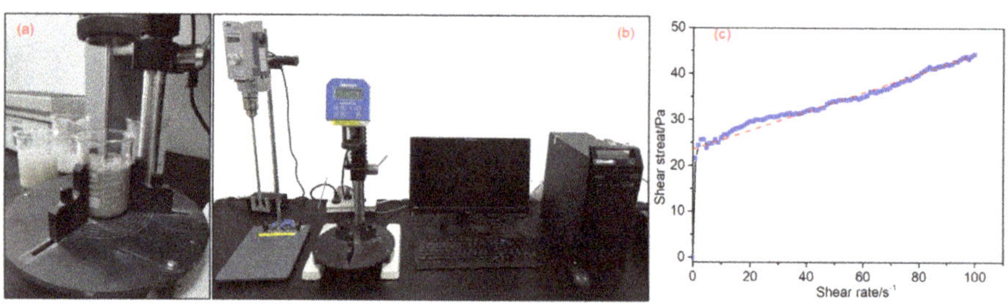

Figure 5. Slurry under test (**a**), Brookfield R/S Plus rheometer (**b**) and rheological curve of filling slurry (**c**).

The regression equation of the slurry rheological curve is shown in Equation (4) [41].

$$\tau_0 = 0.207\eta + 23.53 \tag{4}$$

where τ_0 is the shear stress in Pa and η is the viscosity coefficient of the filling slurry.

Therefore, the yield stress τ is 23.53 Pa and η is 0.207 Pa·s. The filling slurry density ρ is 1700 kg/m^3, which was obtained by weighing 1 L of filling slurry. A 2.5 m/s velocity of the filling slurry in the pipe was selected. The Reynolds number of the filling slurry in the pipe is calculated using Equation (5) as Re = 2463.77 > 2300, and the flow state is judged to be the transition state.

$$\mathrm{Re} = \frac{\rho v D}{\eta} \tag{5}$$

where ρ is the fluid density in kg/m^3; v is the flow velocity in m/s; η is the viscosity coefficient in Pa·s; and D is the pipe inner diameter in m.

3.2.3. Reliability Verification of the Model and Parameters

The reliability of the model and parameters was verified by comparing the similarity ξ between the theoretical and simulated results of pressure loss.

First, the L-shaped pipe model was input into Fluent, and then the physical parameters of the filling slurry were brought into the Fluent solver. After iterative calculation, convergence was achieved, and the pressure difference P_m between the inlet and outlet was obtained as 669,320 Pa.

The total pressure loss of the filling slurry in the pipeline is calculated by Equation (6) [42–44].

$$P_w = \sum P_f + \sum P_j \tag{6}$$

where P_w is the total pressure loss; P_f is the friction loss along the path; and P_j is the local pressure loss.

P_f is calculated according to Equation (1), and P_j is calculated by Equation (7).

$$P_j = \zeta \frac{v^2}{2g} \gamma \tag{7}$$

where ζ is the local resistance coefficient, 1.2; v represents the velocity of the fluid in m/s; g is the acceleration of gravity in m/s^2; and γ is the bulk density of the filling slurry in N/m^3.

The similarity ξ between the theoretical and simulated results of pressure loss can be calculated by Equation (8).

$$\xi = \frac{P_w}{P_m} \times 100\% \tag{8}$$

The total pressure loss of the filling slurry in the pipeline P_w = 643,875 Pa, which was obtained via theoretical calculation. The similarity ξ is 96.2%, which meets the requirements of numerical simulation; that is, the model and parameters are suitable for this study.

4. Results and Analysis

4.1. Velocity and Pressure Fields of the Filling Slurry in the Auxiliary Pipe

Three types of auxiliary pipes were established with $Q = W = 0.5$ m, $D = 0.12$ m and $h = 2.5$ m, and the velocity and pressure fields of the filling slurry were obtained, as shown in Figures 6–8.

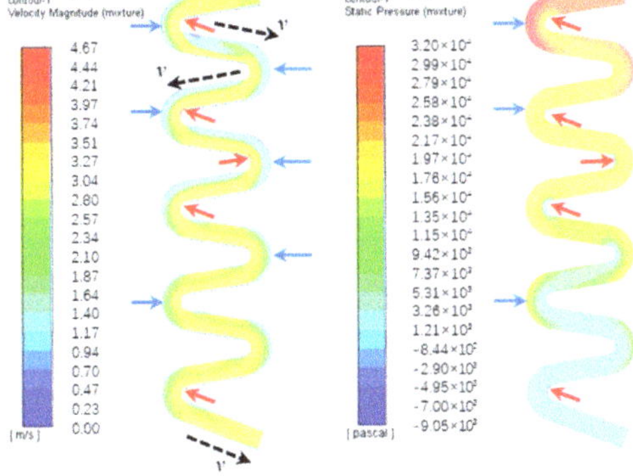

Figure 6. Velocity field (**a**) and pressure field (**b**) of the filling slurry in the Z-shaped pipe.

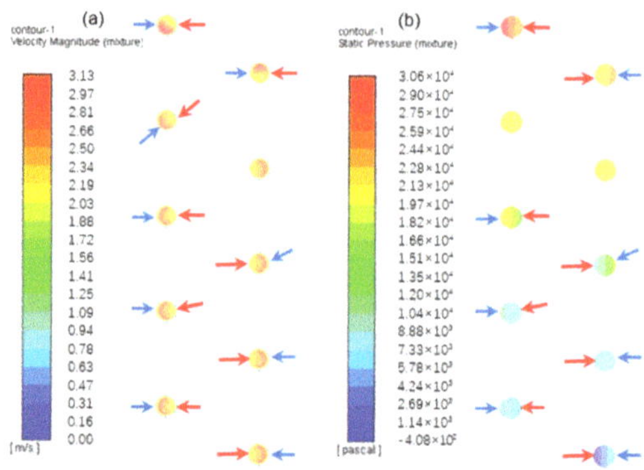

Figure 7. Velocity field (**a**) and pressure field (**b**) of the filling slurry in the S-shaped pipe.

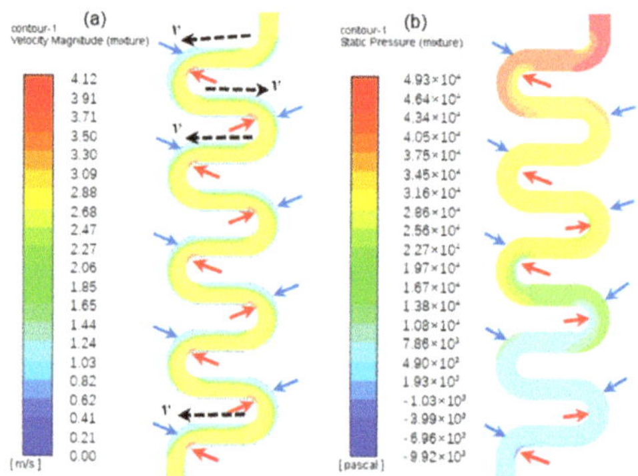

Figure 8. Velocity field (**a**) and pressure field (**b**) of the filling slurry in the spiral pipe.

Figure 6 shows that, for the Z-shaped pipe, the velocity of the filling slurry in the outer Z-shaped pipe is small (as shown by the blue arrow in Figure 6a), and the velocity of the filling slurry in the inner Z-shaped pipe is large (as shown by the red arrow in Figure 6a). At the point shown by the red arrow, there is a sudden increase in velocity, where the velocity of the filling slurry oscillates and the movement direction changes drastically, as shown by the black arrow in the figure. Therefore, the filling slurry energy varies dramatically, and the pipe is liable to burst, while the pressure characteristics of the filling slurry are opposite to the velocity characteristics, that is, the pressure of the filling slurry in the outer Z-shaped pipe is large (as shown by the blue arrow in Figure 6b), the pressure of the filling slurry in the inner Z-shaped pipe is small (as shown by the red arrow in Figure 6b). At the point shown by the red arrow, there is a sudden drop in pressure, where the pipe absorbs the energy of the filling slurry and the local pressure difference is large, so the pipe is liable to burst [45]. This feature is consistent with the results shown in the velocity field. In addition, the more filling slurry that is transported to the lower part of the pipe, the lower the pressure, indicating that the auxiliary pipe has a better effect of increasing resistance and reducing pressure. The characteristics of the S-shaped pipe velocity and pressure fields in Figure 7 are consistent with the characteristics expressed in Figure 6 and will not be repeated here.

Figure 8 clearly shows that the velocity of the filling slurry in the outer spiral pipe is large (as shown by the blue arrow in Figure 8a) and the velocity of the filling slurry in the inner spiral pipe is small (as shown by the red arrow in Figure 8a). There is no obvious velocity change. The pressure characteristics of the filling slurry are similar to the velocity characteristics, i.e., the pressure of the filling slurry in the outer spiral pipe is large (as shown by the blue arrow in Figure 8b) and the pressure of the filling slurry in the inner spiral pipe is small (as shown by the red arrow in Figure 8b). There is no abrupt pressure variation, indicating that the velocity and pressure of the filling slurry in the spiral pipe uniformly change, the impact of filling slurry transport on the pipe is small and the wear is small. Similarly, the further the filling slurry is transported into the lower part of the spiral pipe, the lower its pressure, indicating that the auxiliary pipe has a better effect of increasing resistance and reducing pressure.

Comparing the velocity and pressure fields in the three types of auxiliary pipes shows that the three types of pipes can ensure the continuous transport of filling slurry and achieve the purpose of increasing resistance and reducing pressure. However, the filling slurry velocity and pressure in the Z-shaped pipe and the S-shaped pipe are prone to local sudden changes, and the pipe is liable to burst. However, the spiral pipe slurry velocity and pressure change continuously and steadily, and there are no obvious sudden changes in local velocity and pressure, so it does not easily burst. Hence, in engineering applications, spiral pipes have more obvious advantages than Z-shaped and S-shaped pipes and should be used more often.

The simulation results show that the filling slurry velocity and pressure in the outer spiral pipe are greater than those in the inner spiral pipe, and the wear of the outer spiral pipe is relatively serious. In practical applications, the design and production of the outer spiral pipe need to carry out special treatment, and special materials can be thickened or added to enhance the wear resistance of the local spiral pipe [46,47], so that the pipe structure can maintain overall stability, increasing the service life of the pipe and reducing cost.

4.2. Characteristics of Friction Loss in Spiral Pipes with Different Sizes

The above study shows that spiral pipes are more suitable for engineering applications. Therefore, we only study the effect of increasing resistance and reducing pressure of spiral pipes here and calculate the friction loss (ΔP) of filling slurry in 25 different sized spiral pipes (Figure 9 shows some spiral pipe models with the same height, spiral diameter and different pitch). The results are shown in Table 3. Through the simulation calculation, the friction loss at the center of the inlet and outlet of the vertical pipe with a height of 2.5 m is 5228 Pa at the same parameters of filling slurry. The ratio of friction loss generated by the spiral pipe to that generated by the vertical pipe at the same height is ω, which is shown in Table 4. When the velocity is 2.5 m/s and the vertical height of the pipe is 2.5 m, the friction loss of the spiral pipe can reach 3.87~21.26 times that of the vertical pipe, which indicates that the spiral pipe can effectively play the role of increasing resistance and reducing pressure.

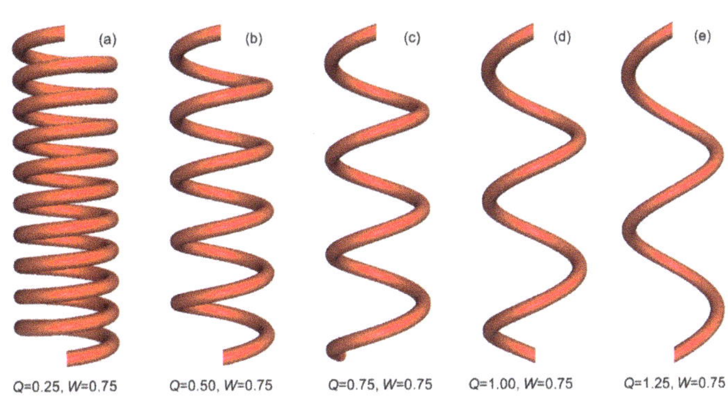

Figure 9. Spiral pipe models with the same height, spiral diameter W and different pitch Q. (**a**) $Q = 0.25$, $W = 0.75$, (**b**) $Q = 0.50$, $W = 0.75$, (**c**) $Q = 0.75$, $W = 0.75$, (**d**) $Q = 1.00$, $W = 0.75$, (**e**) $Q = 1.25$, $W = 0.75$.

Table 3. Friction loss ΔP (Pa) of spiral pipes with different sizes.

Q \ W	0.5	0.75	1	1.25	1.5
0.25	55,393.72	66,217.24	80,239.87	95,401.62	111,148.30
0.5	30,364.67	34,905.20	41,307.39	48,744.80	56,278.46
0.75	24,800.37	25,413.41	29,062.09	33,701.80	38,536.95
1	22,478.22	21,682.92	23,457.82	26,639.43	29,790.07
1.25	22,234.60	20,211.32	20,685.33	22,677.81	24,867.83

Table 4. ω of spiral pipes of different size.

Q \ W	0.5	0.75	1	1.25	1.5
0.25	10.60	12.67	15.35	18.25	21.26
0.5	5.81	6.68	7.90	9.32	10.76
0.75	4.74	4.86	5.56	6.45	7.37
1	4.30	4.15	4.49	5.10	5.70
1.25	4.25	3.87	3.96	4.34	4.76

Using data from Table 3, Figures 10 and 11 are plotted. From Figure 10, when Q is a constant, ΔP increases with increasing W, and the relationship between the two is a linear function. The solid line in the figure is the linear fitting curve for each group of data, and the fitting degree R^2 is above 0.9. From Figure 11, when W is a constant, ΔP decreases with increasing Q, and the relationship between the two is a monotonically decreasing quadratic function. The solid line in the figure is the corresponding quadratic function fitting curve for each group of data, and the fitting degree R^2 is above 0.9. Therefore, to obtain a larger friction loss, a spiral pipe with small Q and large W should be chosen. W is limited by the actual shaft size in the mine, and there is a maximum value, which can be determined in combination with the actual engineering application.

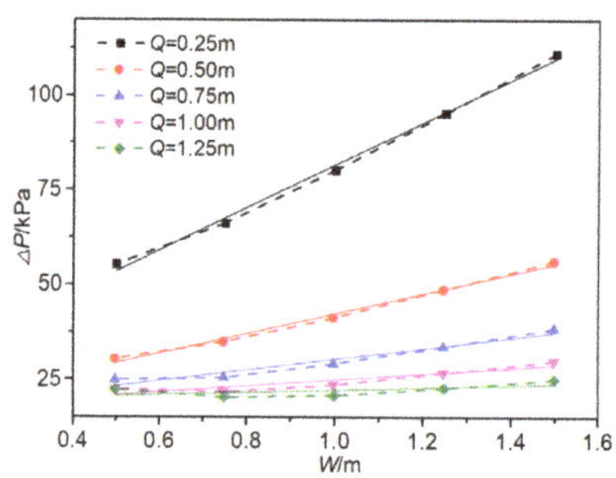

Figure 10. ΔP vs. Q at different W.

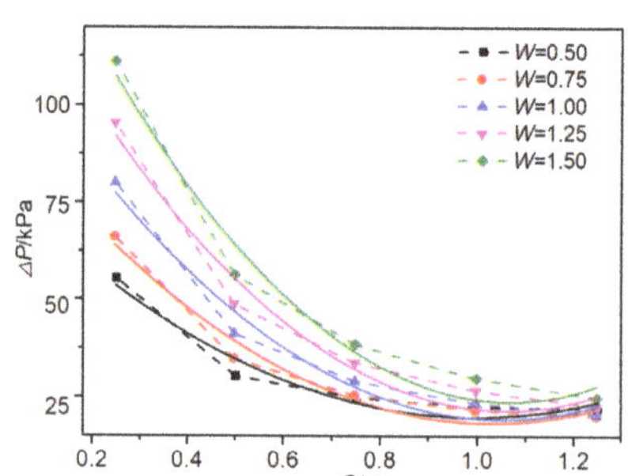

Figure 11. ΔP vs. W at different Q.

4.3. Amendment Equation of the Stowing Gradient

In the backfilling method, the stowing gradient is the most important parameter of the backfilling pipe system, which expresses the engineering characteristics of the mine backfilling pipe and is one of the key factors of whether the mine filling slurry can be transported by self-flow. The calculation equation of the stowing gradient is shown in Equation (9). Since ΔP along the pipe changes after the addition of a spiral pipe to the backfilling pipe, Equation (9) cannot accurately calculate the stowing gradient. Therefore, it is important to propose a calculation method for the effective stowing gradient when using spiral pipes to increase resistance and reduce pressure. The relevant calculation analysis will also become the basis of mine backfilling pipe design. We propose an amendment equation for calculating the stowing gradient based on the friction loss calculation results, which is shown in Equation (10).

$$N = \frac{L}{H} \tag{9}$$

$$N' = \frac{L + (\omega - 1) \cdot h}{H} \tag{10}$$

where N is the stowing gradient; N' is the effective stowing gradient; H is the height difference between the starting point and ending point of the backfilling pipe; L is the total length of the backfilling pipe including the length of the elbows, joints and other fittings; and h is the total height of the spiral pipe.

Because ω is obtained by many simulation calculations, it is not intuitive in practice. To simplify Equation (10), a relationship between ω and Q and W needs to be established.

Figure 12 shows the three-dimensional coordinate of ω and Q and W. According to the above study, when W is a constant, the relationship between ω and Q is a monotonically decreasing quadratic relationship. Because it can determine the relationship between ω and Q and W, the binary quadratic function is chosen to fit the data for Figure 12. The results are shown in Figure 12, and the relationship between ω and Q and W is shown in Equation (11).

$$\omega = 17.47Q^2 + 2.03W^2 - 36.66Q + 0.06W + 20.84 \tag{11}$$

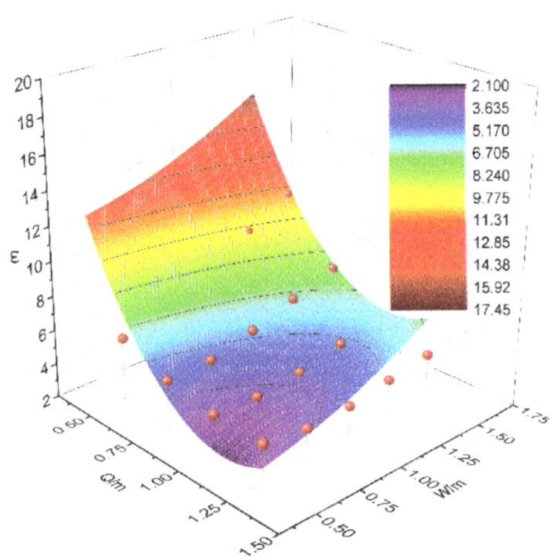

Figure 12. Three-dimensional coordinate diagram of ω with Q and W.

The fitting degree R^2 of the equation in Figure 12 is 0.90. Bringing Equation (11) into Equation (10):

$$N' = \frac{L + \left(17.47Q^2 + 2.03W^2 - 36.66Q + 0.06W + 19.84\right) \cdot h}{H} \tag{12}$$

It can be seen from Equation (12) that when L, Q, W, h and H are known, the effective stowing gradient N' can be found when spiral pipe is used to increase resistance and reduce pressure. Otherwise, when L, h and H are known, the relationship between N' and Q and W can be found, and then suitable backfilling pipe parameters can be designed in combination with the actual industrial application.

5. Conclusions

In this study, the flow state of filling slurry in auxiliary pipes with different sizes was simulated using CFD, and the distribution characteristics of filling slurry velocity and pressure were analyzed. The relationship between friction loss along the pipe and spiral diameter and pitch was established, and the calculation equation of the stowing gradient was revised. The following conclusions were obtained.

(1) The three types of pipes can achieve the purpose of increasing resistance and reducing pressure, but the velocity and pressure of filling slurry in spiral pipe change continuously and steadily, and there are no obvious sudden changes in local velocity and pressure, so it is difficult to burst. Hence, in engineering applications spiral pipes have more obvious advantages than Z-shaped and S-shaped pipes and should be used more often.

(2) In practical applications, the design and production of the outer spiral pipe need to carry out special treatment, and special materials can be thickened or added to enhance the wear resistance of the local spiral pipe, so that the pipe structure can maintain overall stability, increasing the service life of the pipe, and reducing cost.

(3) The relationship between ΔP and Q and W was obtained. ΔP increases with increasing W, and decreases with increasing Q. The relationship between ω and Q and W is a binary quadratic function relationship, and the equation for calculating the effective stowing gradient when using spiral pipe is proposed.

(4) This study provides an idea for industrial design and engineering applications. First, the parameters of the filling slurry are determined through laboratory experimental tests. Second, the parameter combination of the backfilling pipe is determined by the actual industrial situation. Third, the relationship between ω and Q and W is obtained via numerical simulation and analysis. Therefore, the relationship between N' and Q and W can be found by the calculation equation of the effective stowing gradient. Finally, a suitable filling pipe is designed from actual conditions.

Author Contributions: Conceptualization, W.W. and H.M.; Data curation, W.W. and A.W.; Formal analysis, H.M. and R.S.; Funding acquisition, G.M., L.G. and X.L.; Investigation, W.W. and X.L.; Methodology, H.M.; Resources, L.G.; Validation, G.M. and L.G.; Writing—original draft, W.W., A.W. and R.S.; Writing—review and editing, H.M. and L.G. All authors have read and agreed to the published version of the manuscript.

Funding: This research was supported by the National Key R&D Program of China (2018YFC0604605, 2021YFC2900600, 2021YFE0102900); This research was also funded by the Key Research Fund of BGRIMM the Exploration fund of BGRIMM (04-2122, 04-2213).

Institutional Review Board Statement: Not applicable.

Informed Consent Statement: Not applicable.

Data Availability Statement: The original contributions presented in the study are included in the article, and further inquiries can be directed to the corresponding authors.

Conflicts of Interest: The authors declare that the research was conducted in the absence of any commercial or financial relationships that could be construed as a potential conflict of interest.

References

1. Zhang, C.; Jin, G.H.; Liu, C.; Li, S.G.; Xue, J.H.; Cheng, R.H.; Wang, X.L.; Zeng, X.Z. Prediction of rockbursts in a typical island working face of a coal mine through microseismic monitoring technology. *Tunn. Undergr. Space Technol.* **2021**, *113*, 103972. [CrossRef]
2. Aleksakhin, A.; Sala, D.; Golovin, K.; Kovalev, R. Reducing energy costs for pipeline transportation. *Transp. Res. Procedia* **2021**, *57*, 24–32. [CrossRef]
3. Cai, M.F. Prediction and prevention of rockburst in metal mines—A case study of Sanshandao gold mine. *J. Rock Mech. Geotech.* **2016**, *8*, 204–211. [CrossRef]
4. Xie, H.P.; Ju, Y.; Gao, F.; Gao, M.Z.; Zhang, R. Groundbreaking theoretical and technical conceptualization of fluidized mining of deep underground solid mineral resources. *Tunn. Undergr. Space Technol.* **2017**, *67*, 68–70. [CrossRef]
5. Cheng, H.Y.; Wu, S.C.; Li, H.; Zhang, X.Q. Influence of time and temperature on rheology and flow performance of cemented paste backfill. *Constr. Build. Mater.* **2020**, *231*, 117117. [CrossRef]
6. Mashifana, T.; Sithole, T. Clean production of sustainable backfill material from waste gold tailings and slag. *J. Clean. Prod.* **2021**, *308*, 127357. [CrossRef]
7. Rybak, J.; Adigamov, A.; Kongar-Syuryun, C.; Khayrutdinov, M.; Tyulyaeva, Y. Renewable-resource technologies in mining and metallurgical enterprises providing environmental safety. *Minerals* **2021**, *11*, 1145. [CrossRef]
8. Lyu, X.; Yang, K.; Fang, J.J. Utilization of resources in abandoned coal mines for carbon neutrality. *Sci. Total Environ.* **2022**, *822*, 153646. [CrossRef]
9. Zhao, X.; Ma, X.W.; Chen, B.Y.; Shang, Y.P.; Song, M.L. Challenges toward carbon neutrality in China: Strategies and countermeasures. *Resour. Conserv. Recycl.* **2022**, *176*, 105959. [CrossRef]
10. Qi, C.C.; Fourie, A. Cemented paste backfill for mineral tailings management: Review and future perspectives. *Miner. Eng.* **2019**, *144*, 106025. [CrossRef]
11. Chen, Q.S.; Zhou, H.L.; Wang, Y.M.; Li, X.S.; Zhang, Q.L.; Feng, Y.; Qi, C.C. Resistance Loss in Cemented Paste Backfill Pipelines: Effect of Inlet Velocity, Particle Mass Concentration, and Particle Size. *Materials* **2022**, *15*, 3339. [CrossRef] [PubMed]
12. Abdul-Hussain, N.; Fall, M. Thermo-hydro-mechanical behaviour of sodium silicate-cemented paste tailings in column experiments. *Tunn. Undergr. Space Technol.* **2012**, *29*, 85–93. [CrossRef]
13. Yin, S.H.; Shao, Y.J.; Wu, A.X.; Wang, H.J.; Liu, X.H.; Wang, Y. A systematic review of paste technology in metal mines for cleaner production in China. *J. Clean. Prod.* **2020**, *247*, 119590. [CrossRef]
14. Ercikdi, B.; Kuekci, G.; Yilmaz, T. Utilization of granulated marble wastes and waste bricks as mineral admixture in cemented paste backfill of sulphide-rich tailings. *Constr. Build. Mater.* **2015**, *93*, 573–583. [CrossRef]
15. Hefni, M.; Ahmed, H.A.M.; Omar, E.S.; Ali, M.A. The potential re-use of Saudi mine tailings in mine backfill: A path towards sustainable mining in Saudi Arabia. *Sustainability* **2021**, *13*, 6204. [CrossRef]
16. Qi, C.C.; Chen, Q.S.; Fourie, A.; Zhao, J.W.; Zhang, Q.L. Pressure drop in pipe flow of cemented paste backfill: Experimental and modeling study. *Powder Technol.* **2018**, *333*, 9–18. [CrossRef]
17. Wang, Y.Y.; Wang, Y.; Yang, G.F.; Ruan, Z.E.; Wang, Z.Q. Effect of cement-to-tailings ratio and flow rate on the wear performance of filling pipeline. *Powder Technol.* **2022**, *397*, 117027. [CrossRef]
18. Hewitt, D.; Allard, S.; Radziszewski, P. Pipe lining abrasion testing for paste backfill operations. *Miner. Eng.* **2009**, *22*, 1088–1090. [CrossRef]
19. Emad, M.Z.; Mitri, H.; Kelly, C. State-of-the-art review of backfill practices for sublevel stoping system. *Int. J. Min. Reclam. Environ.* **2015**, *29*, 544–556. [CrossRef]
20. Javaheri, V.; Porter, D.; Kuokkala, V.T. Slurry erosion of steel—Review of tests, mechanisms and materials. *Wear* **2018**, *408*, 248–273. [CrossRef]
21. Mc Guinness, M. Wear Profile of the Kidd Mine Pastefill Distribution System. Master's Thesis, McGill University, Montreal, QC, Canada, 2014.
22. Steward, N.R.; Spearing, A. The performance of backfill pipelines. *South. Afr. Inst. Min. Metall.* **1992**, *92*, 27–34.
23. Buchan, A.J.; Spearing, A.J.S. The effect of corrosion on the wear rate of steel pipelines conveying backfill slurry. *South. Afr. Inst. Min. Metall.* **1994**, *94*, 37–45.
24. Calderon-Hernandez, J.W.; Sinatora, A.; de Melo, H.G.; Chaves, A.P.; Mano, E.S.; Leal Filho, L.S.; Paiva, J.L.; Braga, A.S.; Pinto, T.C.S. Hydraulic convey of iron ore slurry: Pipeline wear and ore particle degradation in function of pumping time. *Wear* **2020**, *450*, 203272. [CrossRef]
25. Zhou, K.P.; Gao, R.; Gao, F. Particle flow characteristics and transportation optimization of superfine unclassified backfilling. *Minerals* **2017**, *7*, 6. [CrossRef]
26. Wang, X.L.; Wang, H.J.; Wu, A.X.; Kang, G.W. Wear law of Q345 steel under the abrasion–corrosion synergistic effect of cemented paste backfill. *Constr. Build. Mater.* **2022**, *332*, 127283. [CrossRef]
27. Garrido, P.; Burgos, R.; Concha, F.; Bürger, R. Software for the Design and Simulation of Gravity Thickeners. *Miner. Eng.* **2003**, *16*, 85–92. [CrossRef]
28. Kaushal, D.R.; Thinglas, T.; Tomita, Y.; Kuchii, S.; Tsukamoto, H. CFD modeling for pipeline flow of fine particles at high concentration. *Int. J. Multiph. Flow* **2012**, *43*, 85–100. [CrossRef]

29. Nagar, H.; Gupta, T.; Kumar, N. CFD modeling for the flow of fly ash slurry in straight pipeline. *Mater. Today.* **2022**, *57*, 2223–2227. [CrossRef]
30. Movahedi, H.; Jamshidi, S. Experimental and CFD simulation of slurry flow in the annular flow path using two-fluid model. *J. Pet. Sci. Eng.* **2021**, *198*, 108224. [CrossRef]
31. Zambrano, H.; Sigalotti, L.D.G.; Klapp, J.; Peña-Polo, F.; Bencomo, A. Heavy oil slurry transportation through horizontal pipelines: Experiments and CFD simulations. *Int. J. Multiph. Flow* **2017**, *91*, 130–141. [CrossRef]
32. Liu, L.; Fang, Z.Y.; Qi, C.C.; Zhang, B.; Guo, L.J.; Song, K.I.-I.L. Numerical study on the pipe flow characteristics of the cemented paste backfill slurry considering hydration effects. *Powder Technol.* **2019**, *343*, 454–464. [CrossRef]
33. Peng, X.P.; Guo, L.J.; Liu, G.S.; Yang, X.C.; Chen, X.Z. Experimental Study on Factors Influencing the Strength Distribution of In Situ Cemented Tailings Backfill. *Metals* **2021**, *11*, 2059. [CrossRef]
34. Zhu, J.Q.; Wu, S.C.; Cheng, H.Y.; Geng, X.J.; Liu, J. Response of Floc Networks in Cemented Paste Backfill to a Pumping Agent. *Metals* **2021**, *11*, 1906. [CrossRef]
35. Sun, Q.; Tian, S.; Sun, Q.W.; Li, B.; Cai, C.; Xia, Y.J.; Wei, X.; Mu, Q.W. Preparation and microstructure of fly ash geopolymer paste backfill material. *J. Clean. Prod.* **2019**, *225*, 376–390. [CrossRef]
36. Füreder, K.; Svardal, K.; Krampe, J.; Kroiss, H. Rheology and friction loss of raw and digested sewage sludge with high TSS concentrations: A case study. *Water Sci. Technol.* **2018**, *2017*, 276–286. [CrossRef]
37. Azizi, N.; Homayoon, R.; Hojjati, M.R. Predicting the Colebrook–White friction factor in the pipe flow by new explicit correlations. *J. Fluids Eng.* **2019**, *141*, 051201. [CrossRef]
38. Li, M.Z.; He, Y.P.; Liu, Y.D.; Huang, C. Effect of interaction of particles with different sizes on particle kinetics in multi-sized slurry transport by pipeline. *Powder Technol.* **2018**, *338*, 915–930. [CrossRef]
39. Messa, G.V.; Malin, M.; Malavasi, S. Numerical prediction of fully-suspended slurry flow in horizontal pipes. *Powder Technol.* **2014**, *256*, 61–70. [CrossRef]
40. Rathore, R.K.; Gupta, P.K.; Kumar, N. Numerical investigation of zinc tailings slurry flow field in a horizontal pipeline. *Mater. Today* **2021**, *45*, 2702–2706. [CrossRef]
41. Shao, X.P.; Wang, L.; Li, X.; Fang, Z.Y.; Zhao, B.C.; Tao, Y.Q.; Liu, L.; Sun, W.L.; Sun, J.P. Study on Rheological and Mechanical Properties of Aeolian Sand-Fly Ash-Based Filling Slurry. *Energies* **2020**, *13*, 1266. [CrossRef]
42. Sinha, S.L.; Dewangan, S.K.; Sharma, A. A review on particulate slurry erosive wear of industrial materials: In context with pipeline transportation of mineral—Slurry. *Part. Sci. Technol.* **2017**, *35*, 103–118. [CrossRef]
43. Bharathan, B.; McGuinness, M.; Kuhar, S.; Kermani, M.; Hassani, F.P.; Sasmito, A.P. Pressure loss and friction factor in non-newtonian mine paste backfill: Modelling, loop test and mine field data. *Powder Technol.* **2019**, *344*, 443–453. [CrossRef]
44. Yang, X.B.; Xiao, B.L.; Gao, Q.; He, J.Y. Determining the pressure drop of cemented Gobi sand and tailings paste backfill in a pipe flow. *Constr. Build. Mater.* **2020**, *255*, 119371. [CrossRef]
45. Zhang, Q.L.; Cui, J.Q.; Zheng, J.J.; Wang, X.M.; Wang, X.L. Wear mechanism and serious wear position of casing pipe in vertical backfill drill-hole. *Trans. Nonferrous Met. Soc. China* **2011**, *21*, 2503–2507. [CrossRef]
46. Sanin, V.N.; Andreev, D.E.; Yukhvid, V.I. Self-propagating high-temperature synthesis metallurgy of pipes with wear-resistant protective coating with the use of industrial wastes of metallurgy production. *Russ. J. Non-Ferr. Met.* **2013**, *54*, 274–279. [CrossRef]
47. Kumar, K.; Kumar, S.; Singh, G.; Singh, J.P.; Singh, J. Erosion Wear investigation of HVOF sprayed WC-10Co4Cr coating on slurry pipeline materials. *Coatings* **2017**, *7*, 54. [CrossRef]

metals

Article

Regional Distribution and Causes of Global Mine Tailings Dam Failures

Shui-Quan Lin [1], Guang-Jin Wang [1,2,*], Wen-Lian Liu [1,3], Bing Zhao [1], Ying-Ming Shen [1], Meng-Lai Wang [3] and Xiao-Shuan Li [4]

1 Faculty of Land Resource Engineering, Kunming University of Science and Technology, Kunming 650093, China; linshuiquan@stu.kust.edu.cn (S.-Q.L.); lwenl702@sina.com (W.-L.L.); zhaobing1995@stu.kust.edu.cn (B.Z.); xinyingming@stu.kust.edu.cn (Y.-M.S.)
2 Yunnan Key Laboratory of Sino-German Blue Mining and Utilization of Special Underground Space, Kunming 650093, China
3 National Engineering and Technology Center for Development & Utilization of Phosphate Resources, Yunnan Phosphate Group Co., Ltd., Kunming 650113, China; wml26382022@163.com
4 School of Civil Engineering, Shaoxing University, Shaoxing 312000, China; xsli2011@126.com
* Correspondence: wangguangjin2005@163.com

Abstract: Tailings ponds are one of the three major production facilities in metal mines. The volume of tailings increases year by year, but the storage capacity of existing tailings ponds is limited. Therefore, tailings dams must become more fine-grained and larger. The potential hazard they represent should not be underestimated. This paper reveals the causes and regional distribution patterns of 342 tailings dam failures globally from 1915 to 2021 through statistical analysis. It was found that tailings pond failures occur almost every year, with an average of 4.4 accidents/year (1947–2021). The frequency has been gradually increasing in recent years, and most tailings pond failures are directly related to heavy rainfall or earthquakes. The frequency of tailings pond failures was significantly higher in Asia (21.3%) and the Americas (57.9%), especially in China (n = 43) and the United States (n = 107). Causes of tailings pond failures differed among regions. Most tailings pond failures in Asia and Europe were related to hydroclimate, while those in South America were mainly triggered by earthquakes. This study will provide theoretical data for the pre-design as well as the safe and stable operation of global tailings ponds, which will help to prevent global tailings pond failures.

Keywords: tailings pond; regional distribution; dam break; accident statistics; causation analysis

Citation: Lin, S.-Q.; Wang, G.-J.; Liu, W.-L.; Zhao, B.; Shen, Y.-M.; Wang, M.-L.; Li, X.-S. Regional Distribution and Causes of Global Mine Tailings Dam Failures. *Metals* **2022**, *12*, 905. https://doi.org/10.3390/met12060905

Academic Editor: Lijie Guo

Received: 8 April 2022
Accepted: 21 May 2022
Published: 26 May 2022

1. Introduction

With the continued refinement of tailings particles and the increase in tailings emissions year by year, tailings dams will inevitably become more fine-grained and larger. However, extreme weather events and seismic activity will lead to frequent tailings pond failures.

From 1993 to 2009, the global population grew by 23%, while the global economy grew by 153% [1]. Rapid economic growth and population increase require the supply of more mineral resources. In this context, even low-grade minerals are now being extracted. Mineral extraction harms the environment [2,3] and generates large amounts of byproducts [4,5]. The fine particulate matter produced after beneficiation is called tailings and is usually stored in the form of a slurry in large, man-made dikes, i.e., tailings dams that are intended to protect the natural environment from pollution [6–8]. The global mining industry has produced enormous amounts of tailings, e.g., an estimated 14 billion tons in 2010 alone [4]. Ore grades are showing a decreasing trend, which means that more tailings will have to be stored in tailings impoundments in the future [9–11]. However, once a tailings pond leaks, it will severely damage downstream industrial and agricultural production and the wellbeing of local residents, as well as polluting the environment [12–14]. These negative effects will be exacerbated by pre-existing environmental threats such as land use change (LUC), water pollution, acid mine drainage (AMD) and the loss of biodiversity due to mining activities [15–17]. The failure rate of tailings dams in the last 100 years was reported at about 1.2% [18,19]. However, it should be noted that this figure is not accurate due to information loss, and it is certain that the failure rate of tailings dams is significantly higher than that of water storage dams [20,21].

Since the beginning of the 20th century, the frequency of tailings dam failures has been high worldwide. Dam failures that caused extremely severe damage are as follows [22–44]. On 28 March 1965, 16 tailings ponds in Chile collapsed almost simultaneously due to a 7.25 magnitude earthquake, resulting in 270 deaths. On 26 February 1972, the Buffalo tailings ponds in West Virginia, United States, collapsed due to dam instability, resulting in 125 deaths, destroying 500 homes, and causing more than $65 million of economic damage. On 19 July 1985, the Prealpi Mineraia tailings dam in Stava, Italy, failed due to a frozen drainage system, resulting in 268 fatalities; approximately 180,000 cubic meters of semi-fluid tailings were released, burying the downstream villages of Stava and Tesero. On 22 February 1994, the Merriespruit tailings dam in South Africa failed due to a gap in the dam caused by heavy rainfall, resulting in the loss of 600,000 m^3 of tailings which affected infrastructure up to 4 km downstream and caused 17 fatalities. On 30 January 2000, the Baia Mare tailings pond in Romania collapsed, seriously contaminating water sources; as a result, more than 2 million people had limited access to drinking water. On 8 September 2008, the Xinta mine in Xianfen County, Shanxi Province, collapsed due to illegal construction and local seepage damage, resulting in 277 deaths. On 4 August 2014 at the Mount Polley mine tailings pond in Canada, failure to take into account the ice layer led to the collapse of the glacial lake layer at the base of the dam, resulting in the discharge of about 17 million m^3 of wastewater and 8 million m^3 of tailings into the lake, causing extremely serious pollution to water sources. On 5 November 2015, the Samarco iron ore tailings pond in Brazil collapsed due to a small earthquake that triggered dam liquefaction; about 32 million m^3 of tailings gushed out, flooding 158 houses in downstream villages, killing at least 17 people and polluting 650 km of the river that flows into the Atlantic Ocean. On 25 January 2019, Minas Gerais Cérrego do Feijío Vale's iron ore waste dam in Brazil, that has operated at a high-water level for years, experienced a momentary dam failure due to dam slope instability, leaving 325 people dead or missing and causing 12 million m^3 of tailings to flow out, polluting 650 km of rivers. This final accident is recognized as the most catastrophic mine dam failure of all time. The evolution of this event is shown in Figure 1 [42].

Currently, information on tailings pond dam failures is commonly obtained from the International Commission on Large Dams (ICOLD), the United Nations Environment Programme (UNEP) and the World Energy Information Service (WISE). Kamrul Islam et al. (2021) analyzed the water pollution caused by dam failures including information that is lacking in the aforementioned databases, such as mine location and production data, and using the gray water footprint (WF) as a proxy [43]. Paola Dutto et al. (2017) numerically reproduced a landslide using a pure viscous model with a pure frictional model, taking the Aberfan and gypsum tailings impoundment flow slide as an example [44]. Darve and Laouafa (2002) studied dam damage patterns, referring to the damage pattern of loose sand or normally consolidated clay as "diffuse failure" to distinguish it from the "localised mode". In tailings ponds where tailings are used to build the dam body, "diffuse failure"-type events tend to be more common [45]. Ledesma et al. (2022) proposed a method to assess the damage of tailings dams due to flow liquefaction, and analyzed and validated the Fundão dam to derive three factors that may lead to dam damage: (i) a surface load applied to the crest of the upstream raise; (ii) horizontal deformation at the toe of the setback; and (iii) an increase in the water level within the tailings [46]. Such damage likely imposes a static load impact on the dam, while a dam failure caused by an earthquake is referred to as dynamic ruin. The importance of these databases and the published literature is unquestionable. This study aims to analyze the impact of mine tailings dam failures worldwide over the past 100 years. To this end, we updated the tailings dam failure database, analyzed tailings dam failures from the perspectives of dam height, mine type, and geographic location, summarized the main causal factors (type of failure and regional analyses), and revealed regional distribution patterns of tailings dam failures. It is expected that this study will help to reduce tailings dam failures in the future.

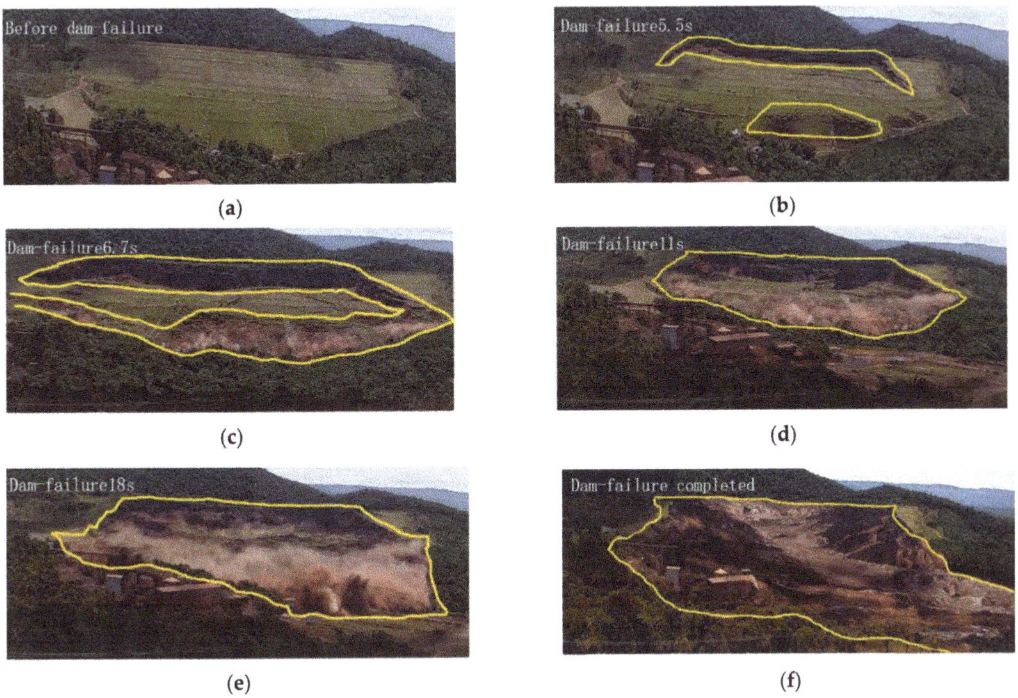

Figure 1. Feijío tailings pond dam failure, Brazil, 2019. (**a**) Before dam-failure; (**b**) Dam-failure 5.5s; (**c**) Dam-failure 6.7s; (**d**) Dam-failure 11s; (**e**) Dam-failure 18s; (**f**) Dam-failure completed.

2. Causal Analysis of Tailings Pond Dam Failures

2.1. Statistics of Tailings Pond Dam Failures

The years 1947–2021 may be divided into three phases (the period of 1915–1946 is excluded due to data loss), with every 25 years being classified as a distinct phase. In the first phase (1947–1971), 73 tailings pond dam failures occurred. In the second phase (1972–1996), 143 tailings pond dam failures took place. During the third phase (1997–2021), 115 tailings pond failures occurred. Figure 2 shows a statistical analysis of the frequency of tailings pond failures; as can be seen, the annual average of global tailings pond failures was at least 4.4 from 1947 to 2021, i.e., 5.7 from 1972 to 1996 (red line), 4.6 from 1997 to 2021 (green line), and 2.9 from 1947 to 1971 (blue line). The frequency of tailings pond failures from 1972 to 1996 was the highest. The probability of tailings pond failures was found to be more than 10 times higher than that of reservoir failures [47]. The yellow spheres in Figure 3 indicate tailings pond failures. When the Z-axis value (indicating the number of tailings pond failures) is 0–5, the projection points (green dots) on the XZ surface are the densest, indicating the highest chance of 0–5 dam failure accidents in tailings ponds worldwide each year. As the Z-axis value increases, the projection points become more and more sparse. The yellow sphere at the highest point indicates the largest number of tailings pond accidents worldwide in 1965 ($n = 23$), when 18 such events occurred almost simultaneously due to an earthquake in Chile. The projection points (red dots) on the XY surface are the densest ($n = 62$) when the X-axis value is 1974–1983, which means that there were more dam failure accidents in tailings ponds during this period. When the *Y*-axis value is 5–10, the number of projection points on the XY surface is the largest.

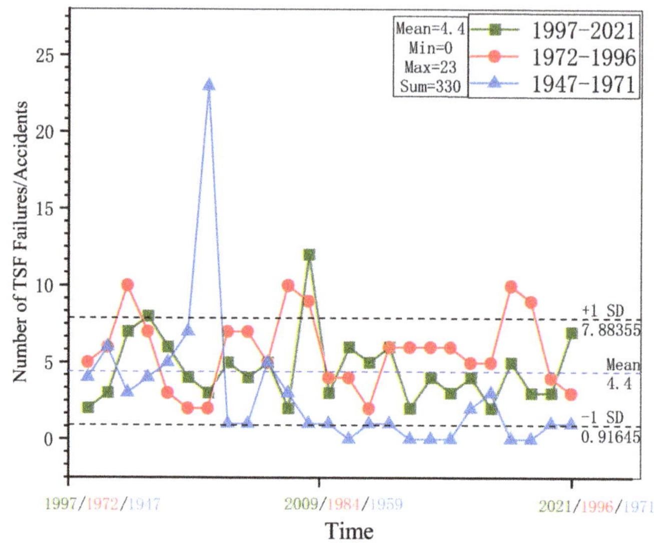

Figure 2. Temporal distribution of tailings pond failures (TSF—Tailings Storage Facility).

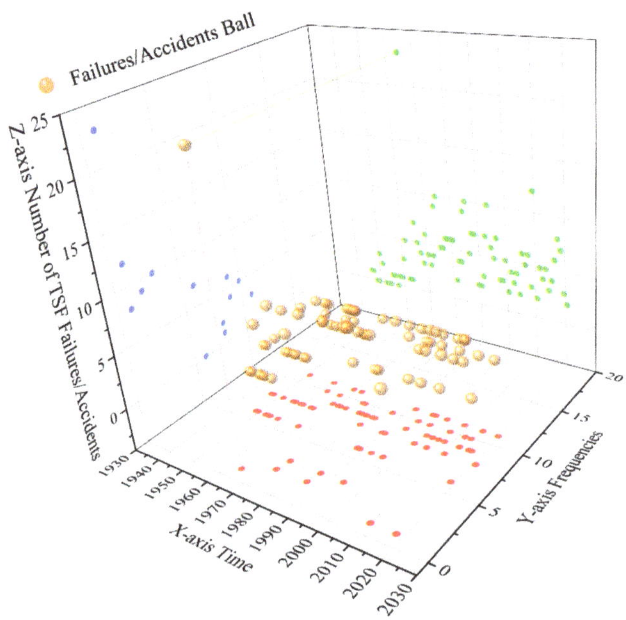

Figure 3. Three-dimensional distribution of tailings pond failures.

The magnitude of tailings pond failures is classified according to the number of human casualties, the degree of pollution to the downstream environment, and the quantity of tailings discharged. The classification is as follows:

1. Very serious tailings dam failures: multiple loss of life (>20) and/or release of 1,000,000 m^3 totals discharge, and/or release travel distance of 20 km or more.
2. Serious tailings dam failures: loss of life and/or release of $\geq$100,000 m^3 semi-solids.
3. Other tailings dam failures: engineering/facility failures other than those classified as very serious or serious, no loss of life.
4. Other tailings-related accidents: accidents are other than those classified as type 1, 2, or 3.

Figure 4 demonstrates the relationship between the severity of tailings pond failures and time. Other failures occupy the largest proportion, followed by serious failures; the number of very serious failures ranks third. In terms of a linear regression analysis, both the red and black lines show an increasing trend, indicating that tailings pond failures are becoming more and more serious, which may be related to the growth of tailings pond dam height.

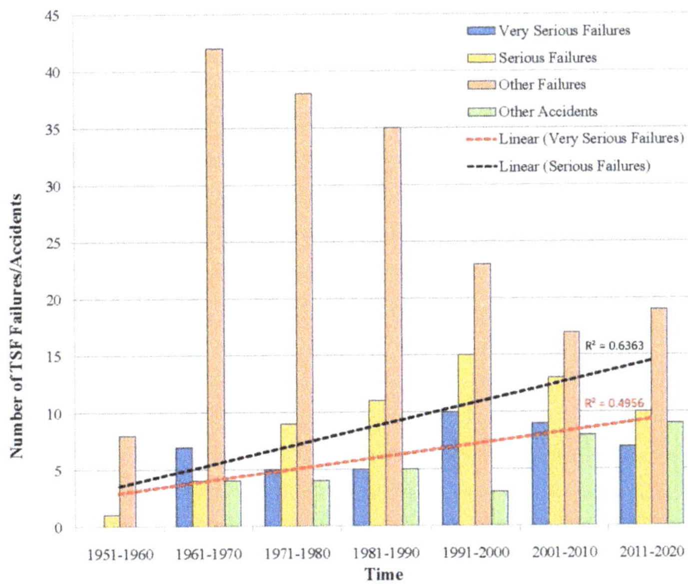

Figure 4. Magnitude of tailings pond failures over time.

From existing dam failure cases, it was found that there are numerous causes of tailings pond failures, and the corresponding patterns and mechanisms are very complex. Tailings pond failures are not brought about by a single factor, but rather, are often the result of multiple factors acting together. In essence, they are due to the influence of the external environment (earthquakes, rainfall, flooding, dam foundation subsidence, etc.), which causes changes in the stress and seepage fields, leading to the destabilization of the structure. The causes of ICOLD tailings pond failures are classified as follows:

- Structure and condition of the dam foundation (FN)

Foundations that have been subjected to great pressure and horizontal thrust may cause damage to the stability of the dam body due to deformation or sliding. The main two causes of this phenomenon are as follows; 1. Low strength of the dam foundation, especially when a karst foundation exists. 2. Poor slip resistance of the dam foundation. Foundation fracture, rock softening, or modifications may cause the foundation slip resistance to decline.

- Earthquakes (EQ)

The mechanisms by which earthquakes cause tailings dam failure are as follows: 1. liquefaction of tailings sand; 2. weakening of tailings dam material; and 3. destabilization of the dam itself. The main factors affecting the seismic liquefaction of the dam are: 1. dynamic load conditions; 2. the physical properties of the tailings conditions; and 3. The burial conditions [48]. Most of the tailings ponds damaged due to earthquakes were upstream-type dams; such ponds have problems, including generally high infiltration lines and poor seismic resistance.

- Mine subsidence (MS)

Generally speaking, the ground cannot withstand the huge pressure of tailings pond settlement or collapse, leading to dam failure. Such tailings ponds are generally built above an underground mine.

- Structural inadequacies, inadequate or failed decants (ST)

Design errors or the failure of a component to function as designed. Failed decants (which drain water from the impoundments) are a common cause.

- External erosion (ER)

At the scale of the structure, erosion by water flow can be divided into two types: if the water erodes the visible part of the structure, it is called external erosion; conversely, if the water erodes an invisible part of the structure or its foundation, it is called internal erosion [49]. On a large scale, rainfall can damage the soil and wash it away by runoff, potentially causing tailings dam failures [50].

- Seepage and internal erosion (SE)

Under the action of the seepage field, tailings sand material may undergo infiltration deformation. For tailings dams, when the infiltration deformation conditions are met, a pipe surge will occur within the dam body, causing cracks and local collapse within the pond and local instability of the dam slope, eventually leading to the failure of the tailings pond.

- Overtopping (OT)

Flood topping triggered-dam failure can be summarized as follows. Excessive wind speed or rainfall, the blockage or destruction of flood discharge and drainage structures, or reduced flood discharge capacity may cause the water to impact the dam or the water level to rise, eventually leading to the destruction of the dam. Flooding in tailings ponds is often the result of multi-factor coupling, with rainfall being the main factor and impermeable dams, aging flood relief facilities, scale blockage, damage, and other phenomena being secondary factors. As such, it can be said that strong rainfall is a necessary condition for flooding dam failure.

- Slope instability (SI)

Dam destabilization refers to damage due to tailing sand extrusion or construction, heavy rainfall, or other factors (non-seepage disturbance). Dam cracks or dam body slip resistance are usually insufficient to cause tailings pond failure. The usual signs of damage are cracking and bulging of the dam face, protrusion of the slope, sinking of the top of the dam, and deformation of the dike edge.

The causes of 258 failures were identified, while those of the remaining 84 failures remain unknown. Among the 258 tailings pond failures with known causes, 10.1% were caused by ST (26), 22.1% by SI (57), 17.1% by EQ (44), 5% by ER (13), 11.2% by SE (29), more than 24.4% by OT (63), 0.4% by MS (1), 9.7% by FN (25). Most tailings pond failures were directly related to heavy rainfall or earthquakes, as shown in Table 1.

Table 1. Statistics on the causes of tailings pond failures.

Reason	Number of Accidents	Region
SI	n = 57 (22.1%)	n = 1 (Bulgaria, Italy, Ukraine, Russia, Yugoslavia, Romania, Zambia, Spain, Australia, South Africa); n = 2 (Brazil); n = 3 (South Africa); n = 4 (UK); n = 5 (Canada); n = 11 (China); n = 21 (USA)
MS	n = 1 (0.4%)	n = 1 (China)
SE	n = 29 (11.2%)	n = 1 (Finland, Hungary, Peru, South Africa, France, UK, Australia); n = 3 (Canada); n = 4 (China); n = 15 (USA)
ST	n = 26 (10.1%)	n = 1 (UK, Ecuador, India, Canada, Macedonia, Romania, Hungary, Bulgaria Mexico); n = 2 (Brazil); n = 3 (China, Philippines); n = 8 (USA)
FN	n = 25 (9.7%)	n = 1 (China, Australia, New Zealand, Russia, Spain, China); n = 2 (UK); n = 3 (Philippines); n = 5 (Canada); n = 10 (USA)
OT	n = 63 (24.4%)	n = 1 (Zambia, Portugal, Peru, Zimbabwe, South Africa, Spain, Brazil); n = 2 (Mexico, Canada); n = 3 (Australia); n = 5 (UK); n = 6 (Chile, Philippines); n = 10 (China); n = 14 (USA); n = 1 (Region unknown)
EQ	n = 44 (17.1%)	n = 1 (USA); n = 2 (Peru Argentina); n = 3 (China); n = 5 (Japan); n = 29 (Chile)
ER	13 (5%)	n = 1 (Philippines, China, Chile, Sweden, Guyana, Montenegro, Bulgaria, Brazil); n = 2(Canada, USA).

Figure 5 represents the relationships among damming methods (X), causes of dam failure (Y), and the number of dam failures (Z). Tailings pond damming methods are generally divided into the following types: upstream (US), water retention (WR), downstream (DS), and centerline (CL). It can be seen that the US damming method is the most common, while the CL method is the least widely used. Among tailings ponds using the US method, 30 failed due to EQ and 28 due to SI. Both EQ and SI cause far more dam failures in tailings ponds applying the US method than other factors. It is worth noting that EQ (yellow cone) and SI (dark green cone) also account for a much larger proportion of total dam failures than other factors (FN, MS, ER, ST, SE, OT); MS (cyan cone) causes the fewest failures.

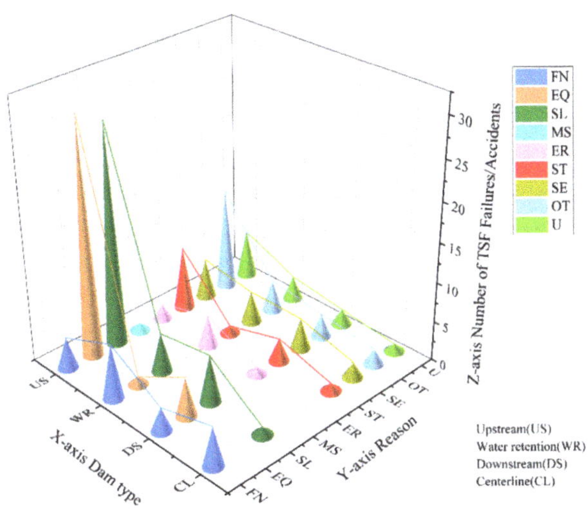

Figure 5. Three-dimensional diagram of the relationship between dam construction method and dam failure causes.

2.2. Analysis of Tailings Pond Dam Failures

Figure 6 shows the global distribution of tailings pond dam failures. The United States (n = 107) had the highest number of tailings pond dam failure accidents, followed by China (n = 43). Figure 7 displays the causes of dam failures in seven key regions; it can be seen that the number of dam failures with unknown causes (yellow block, n = 62) is high, which indicates that the information recording system is not perfect. In the United States, 29.9% (n = 32) of tailings pond failures have unknown causes, and most tailings pond failures were caused by SI (19.6%). In China, 25.5% of tailings pond failures were caused by SI and 20.9% by OT. In Chile, 78% of tailings pond failures were caused by EQ, but the 16 accidents in 1965 may result in the true value being less than 78%. In Canada, 19.2% of tailings pond failures were caused by SI (n = 5) and FN (n = 5), while the causes of seven dam failures remain unknown. OT was the main trigger for tailings pond failures in the Philippines (33.3%) and the United Kingdom (33.3%). Among the seven countries, Brazil had the most significant loss of data (46.1%) regarding the cause of dam failures. It is likely that climate and earthquakes were the main causes of these tailings pond failures, as discussed in detail below.

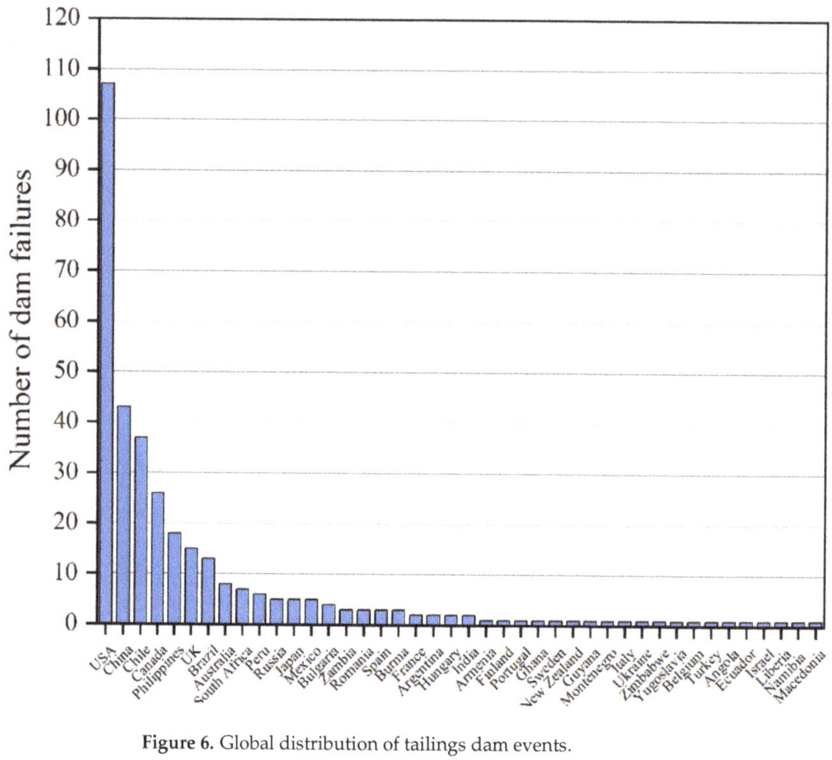

Figure 6. Global distribution of tailings dam events.

Although the causes of 258 mine failures have been determined, the causes of 84 tailings pond failures (24.6%) remain unknown. Figure 8 shows the dam failure causes for different mine types and dam construction methods, where the size of the blue circle is proportional to the number of tailings pond failures. Dam failures in copper mine tailings ponds were mainly caused by EQ and OT, accounting for 24 and 13 cases, respectively. In addition, copper tailings ponds (n = 38) are mainly applied with the US damming method. It is indicated that most US tailings pond failures were triggered by EQ (consistent with the results in Figure 5). Twenty dam failures in Chilean copper mine tailings ponds were caused by earthquakes. The number of gold tailings ponds using the US method was also the highest (n = 13), and ST (n = 9) and OT (n = 11) were the two most common causes of failures. Five (45.5%) gold tailings pond failures in the Philippines were attributed to OT. For the 23 Pb mine tailings pond failures, the main causes were SI (n = 6) and OT (n = 8). In summary, rainfall and extreme weather events are two important factors leading to tailings dam failures.

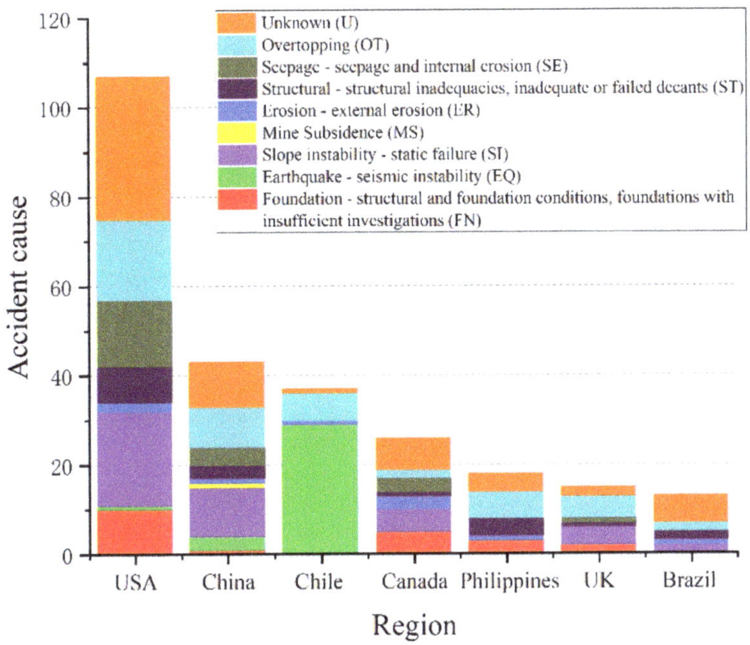

Figure 7. Causes in regions with a high frequency of tailings dam failures.

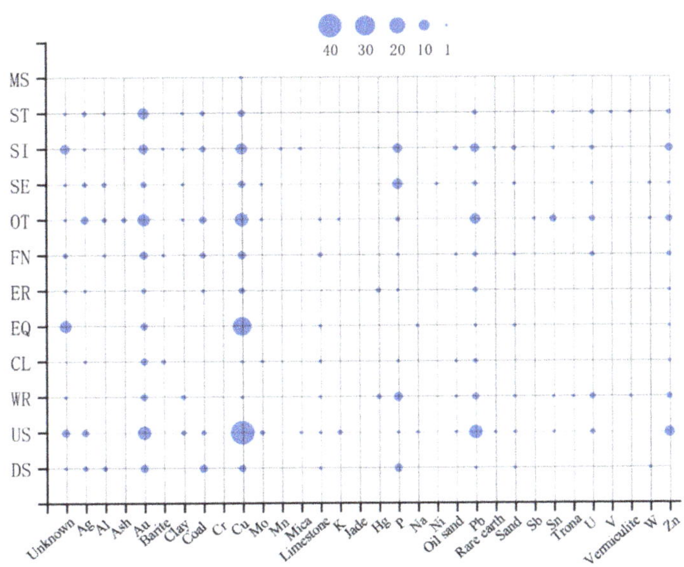

Figure 8. Causes of tailings pond failures in different mines.

The height of a tailings dam is another important factor affecting its safety. The tailings dam failure database contains information, including the dam height, of 159 failure events.

Approximately 89.9% (n = 143) of these events occurred with dam heights between 0 and 50 m; only 8.8% (n = 14) occurred with dam heights between 50 and 100 m, and 1.3% (n = 2) were with dam heights greater than 100 m. However, increasing the height of tailings dams does not prevent dam failures. There is a moderate correlation (r^2 = 0.54) between tailings dam height and mine size (in terms of production). Tailings dam heights between 0 and 100 m correspond to an average production of about 104 million tons, while heights greater than 100 m correspond to an average production of about 643 million tons [43]. The box line in Figure 9 shows the distribution of tailings dam heights in different mines. The blue dots represent the height of the tailings dam and the hollow squares inside the yellow blocks represent the average height of the dam. We found that the chance of dam failure is higher in copper (n = 36), gold (n = 19) and lead (n = 15) mines. As can be seen in Figure 9, the dam heights in instances of copper tailings dam failures were 5–140 m, with the average being 29.8 m and the median being 22.5 m. The dam heights in instances of gold tailings dam failures were 5–94 m, with the average being 24.9 m and the median being 24 m. The dam heights in instances of lead mine tailings pond failures ranged from 5 to 45 m, with the average being 19.2 m and the median being 15 m. Both the average and median height of lead mine tailings pond failures were lower than those of the copper and gold mine pond failures. Figure 10 shows the time distribution relative to the height of tailings dam failures. The linear regression equation ($y = (0.28 \pm 0.1)\,x - (528.74 \pm 200.94)$) indicates that tailings dam heights slowly increased with time.

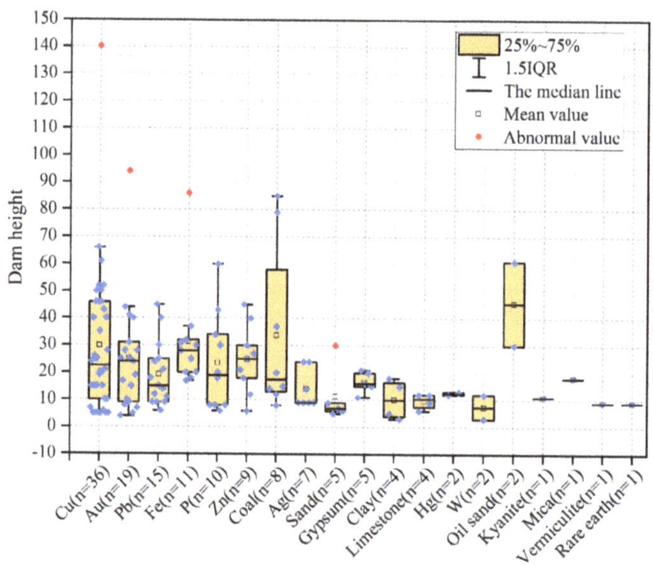

Figure 9. Dam height distributions of tailings dam failures for different mines.

Figure 11 shows the distribution of released tailings (percentage of storage) after a tailings pond break. The black dots represent the proportion of released tailings to the storage volume after the tailings pond break, and the hollow square represents the average height of the tailings dam. The average proportion of sand burst was 0.3455, indicating that after the dam break, the released tailings accounted for 34.6% of the storage volume. The mean and median values of the proportion of released tailings due to FN were 0.561 and 0.463, respectively, i.e., larger than those of the proportion of released tailings caused by other factors. Figure 12 shows the relationship between the proportion of released tailings after a tailings dam failure and the severity of the accident. The mean and median values of the proportion of released tailings after very serious failures were 0.419 and 0.447, respectively, i.e., larger than those after serious failures and other failures.

Figure 13 shows the relationship between the tailings pond storage volume and the tailings release volume after a dam failure. The x-axis of the blue point in Figure 13a represents the tailings storage volume and the y-axis represents the tailings release volume after dam failure. These data are logarithmically transformed in Figure 13b. The linear regression equation in Figure 13b shows a positive correlation between the tailings release volume and the tailings storage volume after dam failure. $y = (0.79 \pm 0.54) + (0.75 \pm 0.08)\,x$, where y is Log (Release volume) (Mm^3), and x is Log (Storage volume) (Mm^3). In general, as the storage volume increases, so does the release of tailings sand.

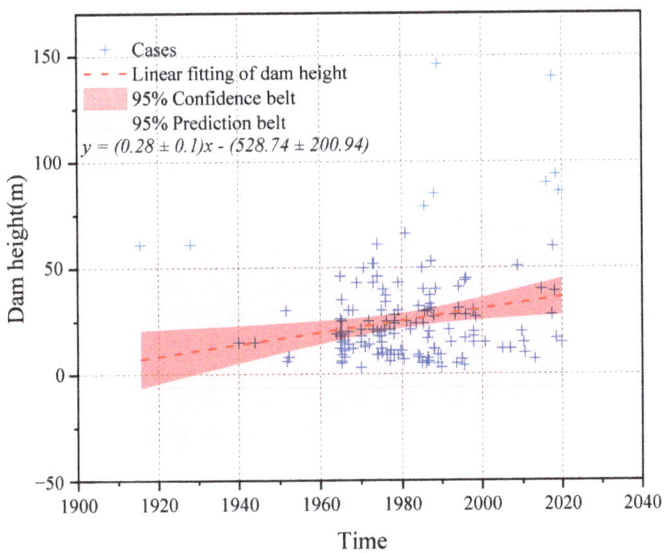

Figure 10. Linear regression analysis of dam height.

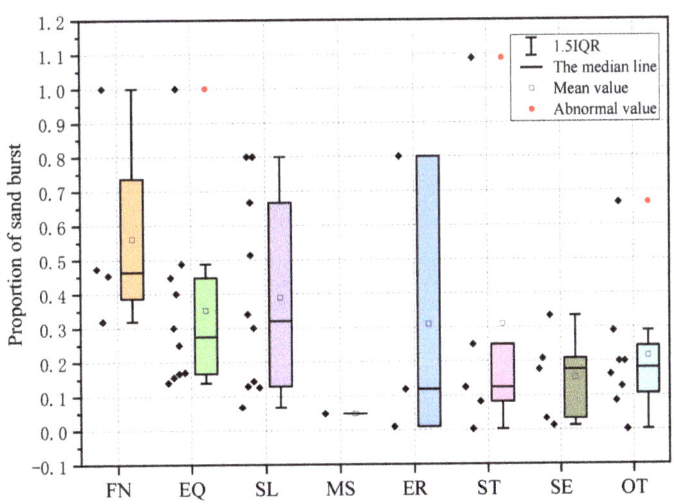

Figure 11. Relationship between the proportion of released tailings and the dam construction method after dam failure.

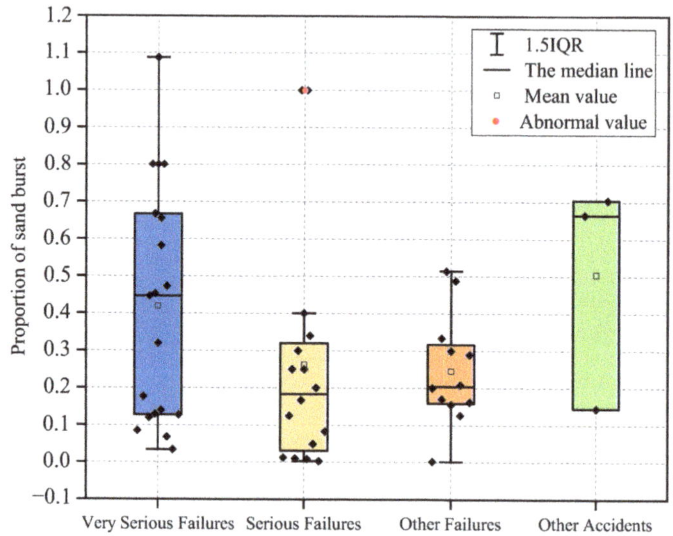

Figure 12. Relationship between the proportion of released tailings and the hazard level after dam failure.

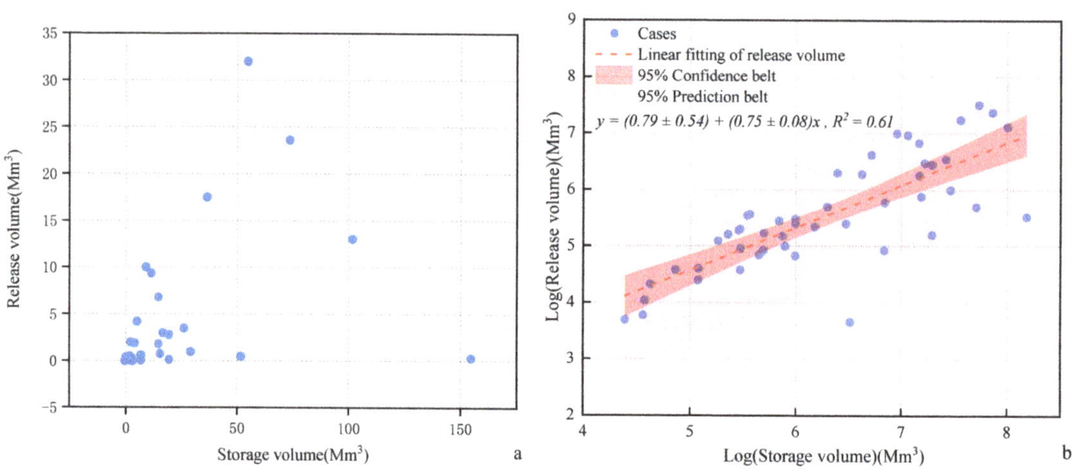

Figure 13. Relationship between tailings storage volume and release volume. (**a**) Before logarithmic transformation; (**b**) after logarithmic transformation.

3. Regional Analysis of Tailings Pond Dam Failure Accidents

Table 2 provides statistics about tailings pond dam failures in different regions. It can be seen that 21.3% of tailings dam failures occurred in Asia, 12.6% in Europe, 57.9% in the Americas, 4.8% in Africa, 2.6% in Oceania, and 0.8% in unknown regions. Figure 14 shows the distribution of tailings pond failures in different regions. The frequency of tailings pond failures in Asia and the Americas, especially in China (n = 43) and the United States (n = 107), was significantly higher than in other regions.

Table 2. Statistics of global tailings dam failures.

Region		Number	Causes of Dam Failure	
Oceania	New Zealand	1	1 FN	9 (3 OT, 1 ER, 1 SI, 2 Unknown, 2 FN)
	Australia	8	3 OT, 1 ER, 1 SI, 2 Unknown, 1 FN	
Asia	Turkey	1	Unknown	73 (4 FN, 8 EQ, 12 SI, 1 MS, 2 ER, 8 ST, 4 SE, 15 OT, 19 Unknown)
	Israel	1	SI	
	India	2	1 ST, 1 Unknown	
	Myanmar	3	3 Unknown	
	Japan	5	5 EQ	
	Philippines	18	6 OT, 4 ST, 3 FN,1 ER, 4 Unknown	
	China	43	1 FN, 3 EQ, 11 SI, 1 MS, 1 ER, 3 ST, 4 SE, 9 OT, 10 Unknown	
Europe	Finland	1	SE	43 (7 OT,4 SE,5 ST,12 SI,4 FN,3 ER, 8 Unknown)
	Portugal	1	OT	
	Swedish	1	ER	
	Montenegro	1	ER	
	Italy	1	SI	
	Ukraine	1	SI	
	Yugoslavia	1	SI	
	Belgium	1	Unknown	
	Macedonia	1	ST	
	French	2	1 SE, 1 Unknown	
	Hungary	2	1 SE, 1 ST	
	Romania	3	1 ST, 1 SI, 1 Unknown	
	Spain	3	1 OT, 1 FN, 1 SI	
	Bulgaria	4	2 SI, 1 ER, 1 ST	
	Russia	5	1 SI, 1 FN, 3 Unknown	
	UK	15	5 OT, 1 SE, 1 ST, 4 SI, 2 FN, 2 Unknown	
Africa	Armenia	1	OT	16 (4 SI, 1 SE, 4 OT, 1 ST, 6 Unknown)
	Ghana	1	Unknown	
	Zimbabwe	1	OT	
	Angola	1	Unknown	
	Liberia	1	ST	
	Namibia	1	Unknown	
	Zambia	3	1 OT, 1 SI, 1 Unknown	
	South Africa	7	3 SI, 1 SE, 1 OT, 2 Unknown	
America	Guyana	1	ER	198 (31 OT, 19 SE, 13 ST, 8 ER, 29 SI, 34 EQ, 15 FN, 49 Unknown)
	Ecuador	1	ST	
	Argentina	2	2 EQ	
	Mexico	5	2 OT, 1 SI, 1 ST, 1 Unknown	
	Peru	6	2 EQ, 1 SE, 1 OT, 2 Unknown	
	Brazil	13	2 OT, 2 ST, 1 ER, 2 SI, 6 Unknown	
	Canada	26	2 OT, 3 SE, 1 ST, 3 ER, 5 SI, 5 FN, 7 Unknown	
	Chile	37	1 ER, 6 OT, 29 EQ, 1 Unknown	
	USA	107	18 OT, 15 SE, 8 ST, 2 ER, 21 SI, 1 EQ, 10 FN, 32 Unknown	
Unknown	Unknown	3	1 OT, 2 EQ	3 (1 OT, 2 EQ)
		Total		342 (61 OT, 27 ST, 28 SE, 44 EQ, 25 FN, 58 SI, 1 MS,14 ER, 84 Unknown)

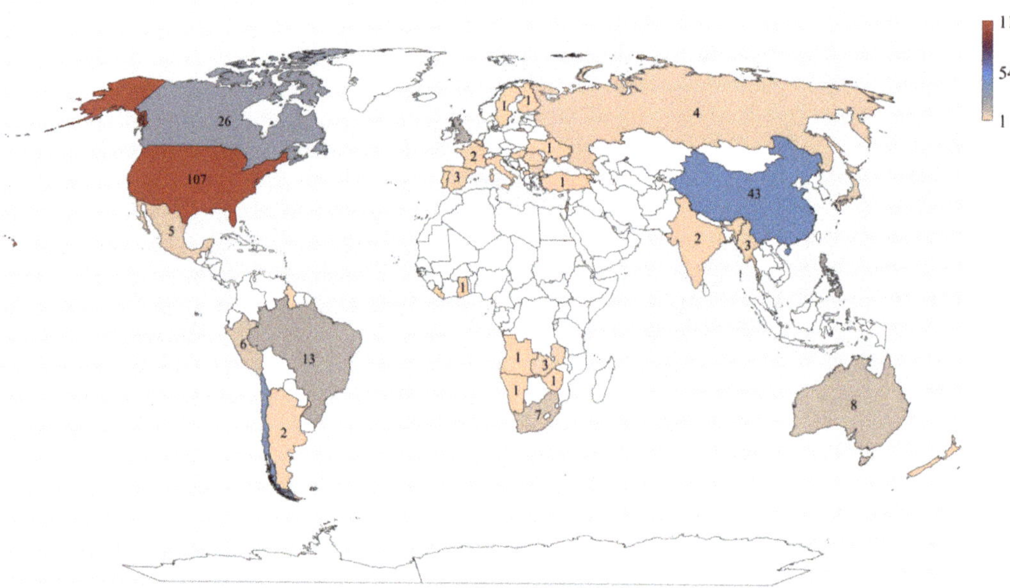

Figure 14. Global geographic distribution of tailings pond dam failures.

Figure 14 illustrates the global distribution of tailings pond failures. Figure 15 shows the locations of some tailings ponds in seismically active zones [43,51]; it can be seen that most of the tailings ponds in South America are located in areas with frequent earthquakes. The Global Seismic Hazard Map depicts the geographic distribution of the Peak Ground Acceleration (PGA) with a 10% probability of being exceeded in 50 years, computed for reference rock conditions (shear wave velocity of 760–800 m/s). Figure 16 shows the distribution of climate types and seismic zones around the world [52–54]. Figure 17 shows the global distribution of rainfall intensity. Seventy-three tailings pond failures occurred in Asia, with 43 in China and 18 in the Philippines. Fifteen of the cases in Asia were caused by OT. As shown in Figure 16, China is located at the intersection of the world's two major seismic zones, i.e., the circum-Pacific seismic belt and the Eurasian seismic belt, and is mainly extruded by the Indian plate in the Cenozoic, resulting in frequent earthquakes in five regions, namely, southwest, northwest, north, Taiwan, and southeast coastal areas [55–57]. However, the tailings ponds in China and seismic zones show a roughly staggered distribution, leading to a small impact of earthquakes on tailings ponds; the number of tailings pond failures triggered by EQ (n = 3) in the study period was much less than that caused by OT (n = 9). It is noteworthy that tailings pond failures in China are mainly concentrated in summer (May, June, July) (n = 13). This pattern is related to climatic characteristics. Summer winds from the Pacific Ocean in the southeast and the Indian Ocean in the southwest are warm and moist, bringing more precipitation. However, the wind in China, especially in the north of the country, is cold and dry because of winter winds from inland Asia [58]. Other tailings pond failures in Asia were also mainly related to the hydroclimate and were concentrated in summer. There were fewer EQ-induced failures in China than in South America. There were 198 tailings dam failures in the Americas, mainly in the United States (n = 107) in North America (n = 138) and in Chile (n = 37) in South America (n = 60); such events were primarily caused by EQ (n = 34) and OT (n = 31), with 29 EQ-induced tailings pond failures in Chile and 18 OT-induced failures in the United States. It is noteworthy that Chile is in a highly seismically active zone and had 17 tailings pond failures due to EQ in 1965, as shown in Figure 16 [59]. The situation in North America is the opposite to that in South America, where 22 tailings pond failures were caused by OT.

In North America, when warm and cold currents meet, heavy rainfall can occur. As such, in North America, the effect of rainfall on the stability of tailings ponds is greater, while the effect of earthquakes is weaker.

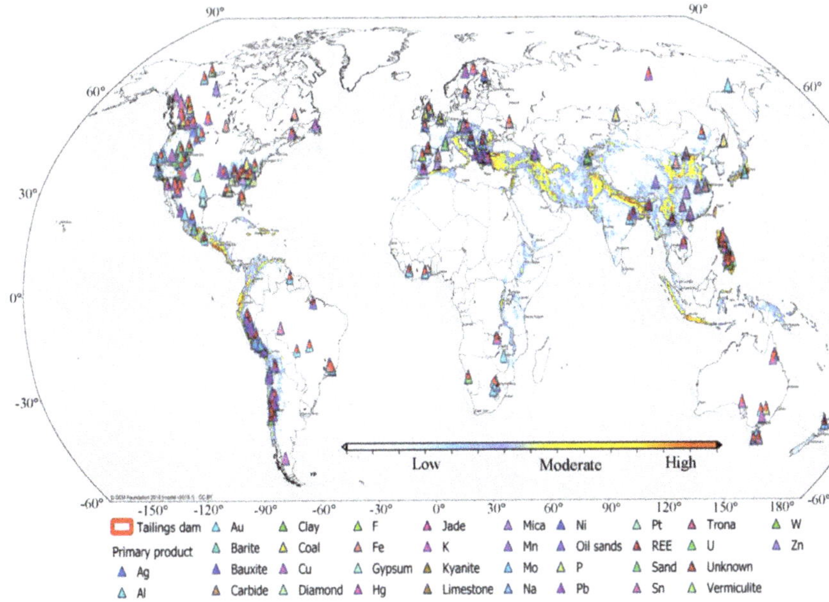

Figure 15. Distribution of global mines in seismic zones.

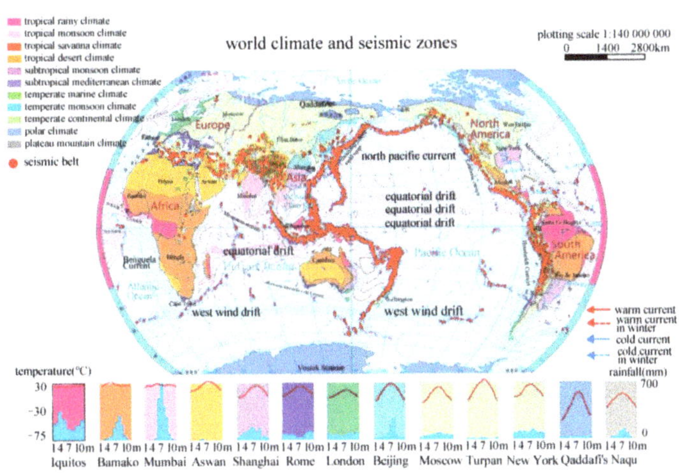

Figure 16. Global climate and seismic zone distribution.

Figure 17. Global average rainfall distribution.

A total of 44 tailings pond failures occurred in Europe, mainly in the United Kingdom (n = 15), which is related to the distribution of seismic zones, as shown in Figure 16. No tailings pond failures due to earthquakes in Europe were found in our statistical analysis. The distribution of failures in Europe is more dispersed than in Asia and America. Tailings pond dam failures in Europe were mainly caused by OT (n = 7) and were concentrated in winter (November, December, January). Stable oxygen (δ18O) and hydrogen (δD) isotopes in Europe show a positive correlation with temperature. They have a negative correlation with precipitation in summer and a positive correlation with precipitation in winter. Both δ18O and δD are significantly heavier in summer and lighter in winter, implying that rainfall in Europe decreases with increasing temperature as well as increasing δ18O and δD [60]. Rainfall in Europe is concentrated in winter, and summers are dry, resulting in a higher incidence of tailings dam failures in the winter. Therefore, factors related to freezing should be considered when damming tailings ponds in Europe [61,62].

Among global tailings pond failures, 4.8% and 2.6% were in Africa and Oceania, respectively. These lower frequencies compared to the other three continents are related to the location of minerals, regional climates and the distribution of seismic zones. Figures 15 and 16 show that earthquakes have little impact on Africa and Oceania, and tailings pond failures in those regions are mainly triggered by OT.

4. Conclusions

A tailings pond failure can cause irrevocable changes in the surrounding ecosystem [63,64], and if the toxic material pollutes the groundwater, the degree of harm will be immense. Following a tailings pond dam break, a tiny number of harmful compounds will be spilled, and these substances will remain in the environment for a long time. In this study, we collected and classified the information on 342 tailings pond failures and used these data to graph tailings pond failures in terms of time and regional distributions, mine type, dam construction methods, etc. Our analysis will provide appropriate data to help future tailings pond researchers. The following conclusions can be drawn:

- We found that the average frequency of tailings pond accidents from 1947 to 2021 was 4.4 per year, and the frequency of tailings pond failures in Asia and the Americas, especially in China (n = 43) and the United States (n = 107), was significantly higher than in other regions. With the increase of large and high tailings ponds, the number of very serious tailings pond accidents is also increasing.

- The causes of 258 failures were identified, while those of 84 failures remain unknown. Among the 258 tailings pond failures with known causes, 10.1% were caused by ST (26), 22.1% by SI (57), 17.1% by EQ (44), 5% by ER (13), 11.2% by SE (29), more than 24.4% by OT (63), 0.4% by MS (1), and 9.7% by FN (25). Most tailings pond failures were directly related to heavy rainfall or earthquakes. Since much of the information regarding tailings pond failures is not disclosed or was undetected, we need to build a better database to capture and document this information.
- Economic development cannot be achieved without the exploitation of mineral resources. We found that the increase in the number of tailings dam failures in developing countries is closely associated with damming methods, climate and earthquakes. The US method is used by most developing countries due to the low construction difficulty and low cost, but tailings dams constructed with the US method have the highest risk of failure. Developing countries need to improve their tailings pond construction, maintenance and monitoring capabilities.

Author Contributions: Conceptualization, S.-Q.L.; methodology, S.-Q.L. and G.-J.W.; writing—original draft preparation, S.-Q.L., W.-L.L. and B.Z.; writing—review and editing, S.-Q.L., M.-L.W. and Y.-M.S.; supervision, S.-Q.L. and X.-S.L.; funding acquisition, G.-J.W. All authors have read and agreed to the published version of the manuscript.

Funding: This study was partially funded by the National Natural Science Foundation of China (No. 52174114).

Institutional Review Board Statement: Not applicable.

Informed Consent Statement: Not applicable.

Data Availability Statement: All data and models of this study are available from the corresponding author upon reasonable request.

Conflicts of Interest: The authors declare no conflict of interest.

References

1. Venter, O.; Sanderson, E.W.; Magrach, A.; Allan, J.R.; Beher, J.; Jones, K.R.; Possingham, H.P.; Laurance, W.F.; Wood, P.; Fekete, B.M.; et al. Sixteen years of change in the global terrestrial human footprint and implications for biodiversity conservation. *Nat. Commun.* **2016**, *7*, 12558. [CrossRef] [PubMed]
2. Wang, G.J.; Tian, S.; Hu, B.; Kong, X.Y.; Chen, J. An experimental study on tailings deposition characteristics and variation of tailings dam saturation line. *Geomech. Eng.* **2020**, *23*, 85–92.
3. Edraki, M.; Baumgartl, T.; Manlapig, E.; Bradshaw, D.; Franks, D.M.; Moran, C.J. Designing mine tailings for better environmental, social and economic outcomes: A review of alternative approaches. *J. Clean. Prod.* **2014**, *84*, 411–420. [CrossRef]
4. Adiansyah, J.S.; Rosano, M.; Vink, S.; Keir, G. A framework for a sustainable approach to mine tailings management: Disposal strategies. *J. Clean. Prod.* **2015**, *108*, 1050–1062. [CrossRef]
5. Wang, S.; Mei, G.; Xie, X.; Guo, L. The Influence of the Instantaneous Collapse of Tailings Pond on Downstream Facilities. *Adv. Civ. Eng.* **2021**, *2021*, 4253315. [CrossRef]
6. Clarkson, L.; Williams, D. Critical review of tailings dam monitoring best practice. *Int. J. Min. Reclam. Environ.* **2020**, *34*, 119–148. [CrossRef]
7. Mei, G.D.; Wu, Z.Z. Research on the dam-break hazard vulnerability assessment index system and methods of tailings pond. In *Applied Mechanics and Materials*; Trans Tech Publications Ltd.: Freinbach, Switzerland, 2012; Volume 204, pp. 3450–3456. [CrossRef]
8. Kheirkhah Gildeh, H.; Halliday, A.; Arenas, A.; Zhang, H. Tailings dam breach analysis: A review of methods, practices, and uncertainties. *Mine Water Environ.* **2021**, *40*, 128–150. [CrossRef]
9. Kossoff, D.; Dubbin, W.E.; Alfredsson, M.; Edwards, S.J.; Macklin, M.G.; Hudson-Edwards, K.A. Mine tailings dams: Characteristics, failure, environmental impacts, and remediation. *Appl. Geochem.* **2014**, *51*, 229–245. [CrossRef]
10. Lv, S.R. Discussion on stability analysis method for tailings dam. *Appl. Mech. Mater.* **2012**, *204*, 93–96. [CrossRef]
11. Burritt, R.L.; Christ, K.L. Full cost accounting: A missing consideration in global tailings dam management. *J. Clean. Prod.* **2021**, *321*, 129016. [CrossRef]
12. Hudson-Edwards, K.A.; Macklin, M.G.; Miller, J.R.; Lechler, P.J. Sources, distribution and storage of heavy metals in the Rıo Pilcomayo, Bolivia. *J. Geochem. Explor.* **2001**, *72*, 229–250. [CrossRef]

13. Chryss, A.; Fourie, A.B.; Monch, A.; Nairn, D.; Seddon, K.D. Towards an integrated approach to tailings management. *J. South. Afr. Inst. Min. Metall.* **2012**, *112*, 965–969. [CrossRef]
14. Caldwell, J.A.; Oboni, F.; Oboni, C. Tailings facility failures in 2014 and an update on failure statistics. In *Tailings and Mine Waste*; University of British Columbia: Kelowna, BC, Canada, 2015; pp. 25–28. [CrossRef]
15. Sovacool, B.K.; Ali, S.H.; Bazilian, M.; Radley, B.; Nemery, B.; Okatz, J.; Mulvaney, D. Sustainable minerals and metals for a low-carbon future. *Science* **2020**, *367*, 30–33. [CrossRef] [PubMed]
16. Islam, K.; Vilaysouk, X.; Murakami, S. Integrating remote sensing and life cycle assessment to quantify the environmental impacts of copper-silver-gold mining: A case study from Laos. *Resour. Conserv. Recycl.* **2020**, *154*, 104630. [CrossRef]
17. Glotov, V.E.; Chlachula, J.; Glotova, L.P.; Little, E. Causes and environmental impact of the gold-tailings dam failure at Karamken, the Russian Far East. *Eng. Geol.* **2018**, *245*, 236–247. [CrossRef]
18. Azam, S.; Li, Q. Tailings dam failures: A review of the last one hundred years. *Geotech. News* **2010**, *28*, 50–54.
19. Rana, N.M.; Ghahramani, N.; Evans, S.G.; McDougall, S.; Small, A.; Take, W.A. Catastrophic mass flows resulting from tailings impoundment failures. *Eng. Geol.* **2021**, *292*, 106262. [CrossRef]
20. Bowker, L.N.; Chambers, D.M. In the dark shadow of the supercycle tailings failure risk & public liability reach all time highs. *Environments* **2017**, *4*, 75. [CrossRef]
21. Wang, G.; Tian, S.; Hu, B.; Xu, Z.; Chen, J.; Kong, X. Evolution pattern of tailings flow from dam failure and the buffering effect of debris blocking dams. *Water* **2019**, *11*, 2388. [CrossRef]
22. de Paiva, C.A.; da Fonseca Santiago, A.; do Prado Filho, J.F. Content analysis of dam break studies for tailings dams with high damage potential in the Quadrilátero Ferrífero, Minas Gerais: Technical weaknesses and proposals for improvements. *Nat. Hazards* **2020**, *104*, 1141–1156. [CrossRef]
23. Rico, M.; Benito, G.; Salgueiro, A.R.; Díez-Herrero, A.; Pereira, H.G. Reported tailings dam failures: A review of the European incidents in the worldwide context. *J. Hazard. Mater.* **2008**, *152*, 846–852. [CrossRef] [PubMed]
24. Cuervo, V.; Burge, L.; Beaugrand, H.; Hendershot, M.; Evans, S.G. Downstream geomorphic response of the 2014 Mount Polley tailings dam failure, British columbia. In *Workshop on World Landslide Forum*; Springer: Berlin/Heidelberg, Germany, 2017; pp. 281–289. [CrossRef]
25. Vanden Berghe, J.F.; Ballard, J.C.; Pirson, M.; Reh, U. Risks of Tailings Dams Failure. In Proceedings of the 3rd International Symposium on Geotechnical Safety and Risk, ISGSR, Munich, Germany, 2–3 June 2011; pp. 209–216.
26. Thompson, F.; de Oliveira, B.C.; Cordeiro, M.C.; Masi, B.P.; Rangel, T.P.; Paz, P.; Freitas, T.; Lopes, G.; Silva, B.S.; Cabral, A.S.; et al. Severe impacts of the Brumadinho dam failure (Minas Gerais, Brazil) on the water quality of the Paraopeba River. *Sci. Total Environ.* **2020**, *705*, 135914. [CrossRef] [PubMed]
27. Liu, S.; Chai, B.; Du, J.; Luo, F.; Xiao, L. Risk post-assessment and management of a waste slag site under extreme scenarios. *Bull. Eng. Geol. Environ.* **2020**, *79*, 2659–2677. [CrossRef]
28. Zheng, X.; Xu, X.H.; Xu, K.L. Study on the risk assessment of the tailings dam break. *Procedia Eng.* **2011**, *26*, 2261–2269. [CrossRef]
29. Quaresma, V.S.; Bastos, A.C.; Leite, M.D.; Costa Jr, A.; Cagnin, R.C.; Grilo, C.F.; Zogheib, L.F.; Oliveira, K.S.S. The effects of a tailing dam failure on the sedimentation of the eastern Brazilian inner shelf. *Cont. Shelf Res.* **2020**, *205*, 104172. [CrossRef]
30. Wang, K.; Yang, P.; Hudson-Edwards, K.; Lu, W.S.; Bu, L. Status and development for the prevention and management of tailings dam failure accidents. *Chin. J. Eng.* **2018**, *40*, 526–539. [CrossRef]
31. Zobrist, J.; Giger, W. Mining and the environment. *Environ. Sci. Pollut. Res.* **2013**, *20*, 7487–7489. [CrossRef]
32. Liang, L.; Liu, Q.; Li, M. Dam-break risk assessment model of tailings reservoir based on variable weight synthesis and analytic hierarchy process. *J. Northeast. Univ. (Nat. Sci.)* **2017**, *38*, 1790. [CrossRef]
33. Ishihara, K.; Ueno, K.; Yamada, S.; Yasuda, S.; Yoneoka, T. Breach of a tailings dam in the 2011 earthquake in Japan. *Soil Dyn. Earthq. Eng.* **2015**, *68*, 3–22. [CrossRef]
34. Cleary, P.W.; Prakash, M.; Mead, S.; Tang, X.; Wang, H.; Ouyang, S. Dynamic simulation of dam-break scenarios for risk analysis and disaster management. *Int. J. Image Data Fusion* **2012**, *3*, 333–363. [CrossRef]
35. Armstrong, M.; Petter, R.; Petter, C. Why have so many tailings dams failed in recent years? *Resour. Policy* **2019**, *63*, 101412. [CrossRef]
36. Rotta, L.H.S.; Alcantara, E.; Park, E.; Negri, R.G.; Lin, Y.N.; Bernardo, N.; Mendes, T.S.G.; Souza Filho, C.R. The 2019 Brumadinho tailings dam collapse: Possible cause and impacts of the worst human and environmental disaster in Brazil. *Int. J. Appl. Earth Obs. Geoinf.* **2020**, *90*, 102119. [CrossRef]
37. Lumbroso, D.; Collell, M.R.; Petkovsek, G.; Davison, M.; Liu, Y.; Goff, C.; Wetton, M. DAMSAT: An eye in the sky for monitoring tailings dams. *Mine Water Environ.* **2021**, *40*, 113–127. [CrossRef]
38. Villavicencio, G.; Espinace, R.; Palma, J.; Fourie, A.; Valenzuela, P. Failures of sand tailings dams in a highly seismic country. *Can. Geotech. J.* **2014**, *51*, 449–464. [CrossRef]
39. Shen, L.; Luo, S.; Zeng, X.; Wang, H. Review on anti-seepage technology development of tailings pond in China. *Procedia Eng.* **2011**, *26*, 1803–1809. [CrossRef]
40. Wu, Z.Z.; Mei, G.D. Statistical analysis of tailings pond accidents and cause analysis of dam failure. *J. Saf. Sci. Technol.* **2014**, *24*, 70–76. [CrossRef]
41. Zardari, M.A.; Mattsson, H.; Knutsson, S.; Khalid, M.S.; Ask, M.V.; Lund, B. Numerical analyses of earthquake induced liquefaction and deformation behaviour of an upstream tailings dam. *Adv. Mater. Sci. Eng.* **2017**, 1–12. [CrossRef]

42. Teramoto, E.H.; Gemeiner, H.; Zanatta, M.B.; Menegário, A.A.; Chang, H.K. Metal speciation of the Paraopeba river after the Brumadinho dam failure. *Sci. Total Environ.* **2021**, *757*, 143917. [CrossRef]
43. Islam, K.; Murakami, S. Global-scale impact analysis of mine tailings dam failures: 1915–2020. *Glob. Environ. Chang.* **2021**, *70*, 102361. [CrossRef]
44. Dutto, P.; Stickle, M.M.; Pastor, M.; Manzanal, D.; Yague, A.; Tayyebi, S.M.; Lin, C.; Elizalde, M.D. Modelling of Fluidized Geomaterials: The Case of the Aberfan and the Gypsum Tailings Impoundment Flowslides. *Materials* **2017**, *10*, 562. [CrossRef]
45. Laouafa, F.; Darve, F. Modelling of slope failure by a material instability mechanism. *Comput. Geotech.* **2002**, *29*, 301–325. [CrossRef]
46. Ledesma, O.; Sfriso, A.; Manzanal, D. Procedure for assessing the liquefaction vulnerability of tailings dams. *Comput. Geotech.* **2022**, *144*, 104632. [CrossRef]
47. Chen, C.C.; Zhao, Y.Q.; Jiang, L.J. Review on research of dam break of tailings pond. *Min. Res. Dev.* **2019**, *39*, 103–108.
48. Wang, W.; Yin, G.; Wei, Z.; Zhang, Q.; Jin, X. Study of the dynamic stability of tailings dam based on time-history analysis method. *J. China Univ. Min. Technol.* **2018**, *47*, 271–279. [CrossRef]
49. Bonelli, S. (Ed.) *Erosion of Geomaterials*; John Wiley & Sons: Hoboken, NJ, USA, 2012. [CrossRef]
50. Coulibaly, Y.; Belem, T.; Cheng, L. Numerical analysis and geophysical monitoring for stability assessment of the Northwest tailings dam at Westwood Mine. *Int. J. Min. Sci. Technol.* **2017**, *27*, 701–710. [CrossRef]
51. Silva, V.; Amo-Oduro, D.; Calderon, A.; Dabbeek, J.; Despotaki, V.; Martins, L.; Rao, A.; Simionato, M.; Viganò, D.; Yepes-Estrada, C.; et al. *Global Earthquake Model (GEM) Seismic Risk Map (Version 2018.1)*; GEM Foundation: Pavia, Italy, 2018. [CrossRef]
52. Song, Z.P.; Yin, J.Y.; Xue, Y.; Zhang, G.M.; Liu, J.; Zhu, Y.Q.; Zhang, Y.X. The global and sub-zone period characteristics for large earthquakes. *Chin. J. Geophys.* **2013**, *56*, 1868–1876. [CrossRef]
53. Kossobokov, V.G.; Nekrasova, A.K. Global seismic hazard assessment program maps are erroneous. *Seism. Instrum.* **2012**, *48*, 162–170. [CrossRef]
54. Rastogi, B.K.; Sharma, J. Global seismic temporal pattern and enhanced seismicity since 2000. *J. Ind. Geophys* **2016**, *20*, 316–324.
55. Liang, G.H. Distribution of seismic zones in China. *Encycl. Knowl.* **2016**, *14*, 1–2.
56. Du, G.; Wu, Q.; Zhang, X.; He, J.; Zou, L.; Feng, Y.; Liu, J.; Romanelli, F. Pn wave velocity and anisotropy underneath the central segment of the North-South Seismic Belt in China. *J. Asian Earth Sci.* **2019**, *184*, 103941. [CrossRef]
57. You, T.; Wu, R.; Liu, G.; Chai, Z. Contribution of precipitation events with different consecutive days to summer rainfall change over China. *Theor. Appl. Climatol.* **2020**, *141*, 1493–1510. [CrossRef]
58. Wang, Z.Y.; Ding, Y.H. Climatic characteristics of rainy season in China. *Chin. J. Atmos. Sci.* **2008**, *32*, 1–13. [CrossRef]
59. Lu, P.; Zhang, H.; Gao, L.; Comte, D. Seismic imaging of the double seismic zone in the subducting slab in Northern Chile. *Earthq. Res. Adv.* **2021**, *1*, 100003. [CrossRef]
60. Celle-Jeanton, H.; Travi, Y.; Blavoux, B. Isotopic typology of the precipitation in the Western Mediterranean region at three different time scales. *Geophys. Res. Lett.* **2001**, *28*, 1215–1218. [CrossRef]
61. Zhong, J.; Li, S.L.; Ibarra, D.E.; Ding, H.; Liu, C.Q. Solute production and transport processes in Chinese monsoonal rivers: Implications for global climate change. *Glob. Biogeochem. Cycles* **2020**, *34*, e2020GB006541. [CrossRef]
62. Kong, F.; Wang, Y.F.; Lv, L.L. Spatial and temporal variations in global, continental and regional scale rainfall over the past 100 years (1900–2010). *J. Catastrophology* **2018**, *33*, 81–88+95.
63. Clarkson, L.; Williams, D. An overview of conventional tailings dam geotechnical failure mechanisms. *Min. Metall. Explor.* **2021**, *38*, 1305–1328. [CrossRef]
64. Omachi, C.Y.; Siani, S.M.; Chagas, F.M.; Mascagni, M.L.; Cordeiro, M.; Garcia, G.D.; Thompson, C.C.; Siegle, E.; Thompson, F.L. Atlantic Forest loss caused by the world' s largest tailing dam collapse (Fundão Dam, Mariana, Brazil). *Remote Sens. Appl. Soc. Environ.* **2018**, *12*, 30–34. [CrossRef]

metals

Article

A Particle Size Distribution Model for Tailings in Mine Backfill

Zongnan Li [1,2,3], Lijie Guo [1,3,*], Yue Zhao [1,3], Xiaopeng Peng [1,3] and Khavalbolot Kyegyenbai [4]

1 Beijing General Research Institute of Mining and Metallurgy, Beijing 100160, China; lizongnan@bgrimm.com (Z.L.); zhaoyue@bgrimm.com (Y.Z.); pengxiaopeng@bgrimm.com (X.P.)
2 School of Civil and Resources Engineering, University of Science and Technology Beijing, Beijing 100083, China
3 National Centre for International Research on Green Metal Mining, Beijing 102628, China
4 School of Geology and Mining Engineering, Mongolian University of Science and Technology, Ulaanbaatar 210646, Mongolia; khavalbolot@must.edu.mn
* Correspondence: guolijie@bgrimm.com

Abstract: With the increasing awareness of sustainable mining, the cement tailings backfill (CTB) method has been developed rapidly over the past decades. In the CTB technique, the two main mechanical properties engineers were concerned with are the rheological properties of CTB slurry and the resulting CTB strength after curing. Particle size distribution (PSD) of tailings material or PSD of the slurry is a significant factor that highly influences the rheological of CTB slurry and the strength performance of CTB. However, the concentrically partial size distribution curve and existing mathematical model could not represent the PSD of tailings material. In this study, a mathematical model for the particle size distribution of mine tailings was established using three model coefficients *A B* and *K*, which mainly reflect the characteristics of particles from three aspects respectively, the average size of particles, the proportion of the coarse or the fine parts of particles, and the distribution width of particles; meanwhile, an optimal coefficient solution method based on error analysis is given. Twelve tailing materials sourced from metal mines around China were used for the model establishment and validation. The determination coefficient of error analysis (R^2) for all twelve modeled PSD lognormal curves was more significant than 0.99, and the modeled PSD lognormal curves are highly consistent with the determined particle size distribution curve.

Keywords: backfill; tailings; particle size distribution; metal mine; log-sigmoid

Citation: Li, Z.; Guo, L.; Zhao, Y.; Peng, X.; Kyegyenbai, K. A Particle Size Distribution Model for Tailings in Mine Backfill. *Metals* **2022**, *12*, 594. https://doi.org/10.3390/met12040594

Academic Editor: Antoni Roca

Received: 24 February 2022
Accepted: 29 March 2022
Published: 30 March 2022

1. Introduction

With the increasing awareness of sustainable mining, the cement tailings backfill (CTB) method has been developed rapidly over the past decades [1–4]. Cement tailings backfill is a technology that assists waste management and mitigates the mine environment from being hazardous by utilizing tailings (or other waste materials) to underground mined voids resulting from underground mine operations [5–8]. It somehow performs as both a support system or an underground working platform to improve the underground mine stability and promote ore extraction [9–12]. The cement tailings backfill is normally mixed to a high-density slurry with a non-settling character, consisting of a low cementitious material content, mine tailings as aggregate, and processed mine water, which could be gravity-transferred or facile pumping into mined cavities [13–16]. After placing CTB slurries in mined cavities, it could then be consolidated and cured to a designed period to achieve particular strength for further mine exaction [17–20].

Tailings used for filling in mines are usually obtained through the beneficiation process, and their particle size range varies according to different beneficiation processes [21–23]. Generally, the tailings produced by the flotation process for copper, lead-zinc, gold, and other raw ores can reach about 80% below 37 μm [24]. The tailings produced by the magnetic separation and gravity separation process are relatively coarse for iron ore, tin ore, etc. According to the 74 μm boundary, the coarse particle part above can account up for about 60% [25,26].

Over the past few decades, due to the low operating costs and well-performed mechanical performance compared with other backfilling methods, CTB technology has been increasingly applied in the mining industry [27–29]. In the CTB technique, the two main mechanical property engineers were concerned with are the rheological properties of CTB slurry and the resulting CTB strength after curing [30–34]. The governing factors of CTB rheological and strength performance have been well documented in the literature [35–38] (e.g., physical and chemical properties of the tailings, chemical composition and content of the mixing water, binder type, and content, the CTB mix design, and in situ curing conditions for strength performance only).

Particle size distribution (PSD) of tailings material or PSD of the slurry is a significant factor that highly influences the rheological and strength performance of CTB [29–31]. Conventionally, there are two forms: interval distribution and cumulative distribution, representing material PSD. Interval distribution, also known as differential distribution or frequency distribution, represents the percentage content of particles in a series of particle size ranges; cumulative distribution represents the percentage content of particles less than specific particle size [37,39]. As the cumulative distribution curve can easily make the cumulative proportion of particles smaller than a specific size, it is widely used in the mine backfill [36,37,39–41]. On the PSD curve, the coarse and fine characteristics of a specific backfill tailing material can be approximately reflected by choosing some points. Such common representative points are usually d_{30}, d_{50}, d_{60}. Here, the symbol d represents the particle size, and the number subscript represents the proportion smaller than the particle size. For example, d_{10} represents a particle size with a cumulative volume fraction less than 10%.

Conventionally, researchers often use these representative points to represent tailing PSD in investigating the relationship between the tailings' PSD with the rheological and strength performance of CTB to solve the problems encountered in CTB slurry transportation underground support [42–44]. Therefore, it is of great significance to study the particle distribution characteristics of tailings. However, the conventional PSD curve is not easy enough to describe the characteristics of particle distribution entirely because the curve is only a collection of scattered points and the selected representative points are random to some extent, i.e., there is no certain equivalent size that can represent the whole particle group features in a conventional PSD curve. Thus, it is of great significance to study the full-size description method of tailings.

Fredlund et al. [45–47] established a mathematical equation representing soil particle size distribution. However, Fredlund's model mainly focuses on naturally grained soils and could not represent the particle size distribution of artificial mine tailings. In addition, in Fredlund's model, five model coefficients are required to represent the PSD, including the initial breaking point of the PSD curve, the steepest slope of the curve, the shape of the fines portion of the curve, the amount of fine, and the diameter of the minimum allowable size particle. These model coefficients are difficult to obtain and lead to difficulties in the study of PSD. Hence, a mathematical model for tailings material PSD with fewer model coefficients will benefit CTB research.

The present study aims to build a mathematical model of tailings material using twelve different tailings sources from various mines in China. Loop iteration was used to obtain a more reliable model function to express the characteristics of particle size compositions with three coefficients. The model could then be validated using the twelve different tailings materials and was further applied in industrial applications.

2. Materials

Twelve different tailings sources from various mines in China were used in this study and each of them are conform to Non-hazardous industrial solid waste standard [48]. The tailings include coarse-grained tailings to extremely fine-grained tailings, representing the typical particle size distribution (PSD) range of tailings materials in underground metal mines. After sampling, the particle size distribution (PSD) of the used tailings were determined by Laser Particle Size Analyzer (PSA) (Malvern Mastersizer 2000: Malvern Instruments Ltd., Malvern, UK) and the specific gravity (ρ_s) of each tailing was measured [49]. As shown in Table 1, the measured particle size distribution and specific gravity for all twelve tailings were listed.

Table 1. The particle size distribution and specific gravity of tailings.

Samples	ρ_s	PSD Measured Curve, μm					
		d_{10} (1)	d_{30} (2)	d_{50} (3)	d_{60} (4)	d_{70} (5)	d_{90} (6)
Classified fine Copper tailing: S1	3.02	1.76	6.41	14.42	20.43	28.71	64.62
Unclassified Copper tailing: S2	2.64	2.25	9.34	36.27	56.83	81.42	172.7
Unclassified Copper-Nickel tailing: S3	2.94	2.62	10.56	27.94	42.53	62.52	132.48
Unclassified Polymetallic tailing: S4	3.19	2.75	12.86	33.71	53.15	82.34	203.57
Unclassified Copper tailing: S5	2.87	2.75	13.15	39.58	68.51	116.14	251.02
Unclassified Copper tailing: S6	2.75	2.95	11.78	28.97	43.56	65.03	151.48
Unclassified Copper tailing: S7	2.98	3.31	23.54	78.86	119.77	179.22	393.43
Unclassified Copper-Gold tailing: S8	2.95	4.44	11.1	21.88	31.21	45.9	118.76
Unclassified Copper-Gold tailing: S9	2.94	7.24	40.4	76.42	99.45	130.37	268.87
Unclassified Copper tailing: S10	2.96	9.31	46.57	82.46	105.45	137.32	284.11
Unclassified Iron tailing: S11	2.84	10.22	42.7	79.81	104.39	137.8	296.31
Classified coarse Copper tailing: S12	2.94	13.62	60.82	106.92	137.79	179.24	345.65

(1) The portion of particles with diameters smaller than this value is 10%. (2) The portion of particles with diameters smaller than this value is 30%. (3) The portion of particles with diameters smaller than this value is 50%. (4) The portion of particles with diameters smaller than this value is 60%. (5) The portion of particles with diameters smaller than this value is 70%. (6) The portion of particles with diameters smaller than this value is 90%.

Figure 1 illustrates the particle size distribution of all twelve tailings materials in semi-logarithmic coordinate space. S1 is the finest material used in this study in the twelve tailings, which is the classified fine part, followed by unclassified tailings S2 to S11 sourced from different metal mines and classified coarse tailing S12. Hence, the tailings materials from S1 to S12 are gradually coarsened, and the average grain size increases.

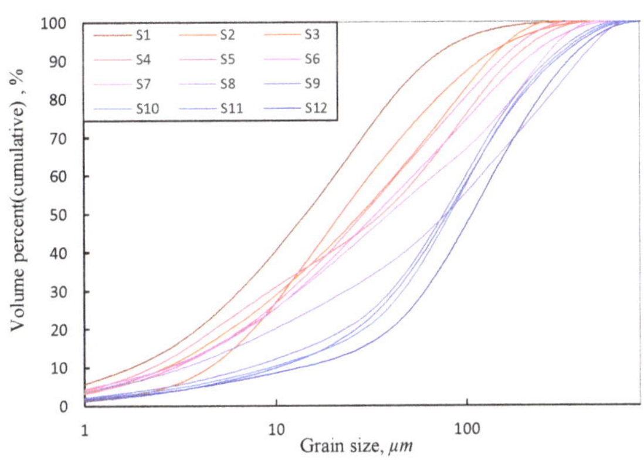

Figure 1. The particle size distribution of twelve different tailings.

3. Mathematical Model

3.1. Definition of Coefficients

As shown in Figure 1, the PSD of tailings on the logarithmic curve has S-shaped characteristics. The ordinate is the cumulative percentage value passing a specific particle size, and the abscissa is the logarithm of the particle diameter. Therefore, to establish a Mathematical Model for PSD of the tailings material, a Sigmoid function can be used to simulate the tailing's grain size distribution characteristics, as shown in Equation (1). The following contents will discuss further analysis by the goodness-of-fit test for its reliability.

$$N_i = \frac{K}{1 + AX_i{}^B} \tag{1}$$

where N_i refers to the cumulative percentage value of particles less than specific particle size, X_i, and A, B, K are the model coefficients.

The equivalent form of Equation (1) could be written as follows:

$$\ln(A) + B\ln(X_i) = \ln(\frac{K}{N_i} - 1) \tag{2}$$

The independent variable and dependent variable of Equation (2) can be equivalent, and then the equation could be modified as follows:

$$\overline{A} + B\overline{X} = \overline{Y} \tag{3}$$

where

$$\begin{cases} \overline{A} = \ln(A) \\ \overline{X_i} = \ln(X_i) \\ \overline{Y_i} = \ln(\frac{K}{N_i} - 1) \end{cases}$$

The model coefficients A, B and K in Equation (1) then can be obtained in the following steps:

Step 1: Three methods to determine Coefficient K

(a). Method 1: The meaning of K value is the cumulative fraction of particles when the particle size reaches infinity. Therefore, the approximate value is Approach to 100%. It is means $K = 100$ (Excluding percent sign, the same below).

(b). Method 2: According to Equation (2), three equidistant points are selected to eliminate the coefficients A and B. The value K can be calculated by solving the Equation (4). The equidistant points can be 37 μm, 74 μm, and 150 μm.

$$K = \frac{N_1[2N_0N_2 - N_1(N_0 + N_2)]}{N_0N_2 - N_1^2} \tag{4}$$

where N_0, N_1, and N_2 are the cumulative percentage values of particles passing 37 μm, 74 μm, and 150 μm. It should be pointed out that the K value can be calculated for N_0, N_1 and N_2 of any equidistant points. The above value method can cover most of the particle size range of tailings for common tailings, and the value is relatively reasonable.

(c). Method 3: The K value is optimal fitting solved by loop iterative calculation, which will be discussed in Section 3.2.

Step 2: Take points and linear regression to obtain coefficients A and B

The coefficients A and B can be obtained by linear regression of the measured tailing's particle size distribution scatters by Equation (3). A series of representative points are taken for regression analysis. In the present work, d_{10}, d_{30}, d_{50}, d_{60}, d_{70}, d_{90} are proposed. The linear regression equation could be written as follows:

$$\begin{cases} \overline{A} = \frac{\sum X_iY_i - \frac{1}{N}\sum X_i \sum Y_i}{\sum X_i^2 - \frac{1}{N}(\sum X_i)^2} \\ \overline{B} = \frac{\sum Y_i - \overline{A}\sum X_i}{N} \end{cases} \tag{5}$$

where (X_i, Y_i) is the sample point, i.e., $(d_{10},10)$, $(d_{30},30)$, $(d_{50},50)$, $(d_{60},60)$, $(d_{70},70)$, $(d_{90},90)$, and N is the number of samples ($N = 6$ in this study). The coefficients A and B can be obtained by substitution with $\overline{A}$ and $\overline{B}$ in Equation (3).

Step 3: Error analysis

To further analyze the reliability of this regression model, the following goodness-of-fit test error analysis method is used:

$$R^2 = 1 - \frac{SSR}{SST} = 1 - \frac{\sum(\hat{Y}_i - Y_i)^2}{\sum(\overline{Y}_i - Y_i)^2} \tag{6}$$

where R^2 reflects the goodness-of-fitting, also known as Determinants of coefficients, the maximum value of R^2 is 1, and the closer the value to 1, the better the fitting is. Generally, it should not be less than 0.8. *SST* is the square sum of total deviations; *SSR* is the square sum of errors; $\hat{Y}_i$ is the predictive value of the model; and $\overline{Y}_i$ is the average value of samples.

Through the above steps, the S-curve model can be obtained, but the reliability of the model is greatly affected by the measuring points, and the reliability index of error analysis could be low. In order to solve this problem, a simple loop iteration calculation is carried out based on the goodness-of-fit to find the most suitable model coefficients.

3.2. Iterative Analysis for the Optimal Fitting Coefficient

In order to obtain a more reliable model function to express the characteristics of particle size composition, the three model coefficients (*A*, *B* and *K*) are obtained by loop iteration. Figure 2 illustrates the structure of the loop iterative control flow chart. As shown in Figure 2, Equation (6) is used as the discrimination function of the iterative loop, where the initial value of *K* is taken as 100, and the loop step is 1 ($K = K + 1$). The Coefficient *A* and *B* in each cycle step are obtained by Equation (5), the corresponding model calculation value is calculated by Equation (1), and the corresponding goodness-of-fit is calculated by Equation (6).

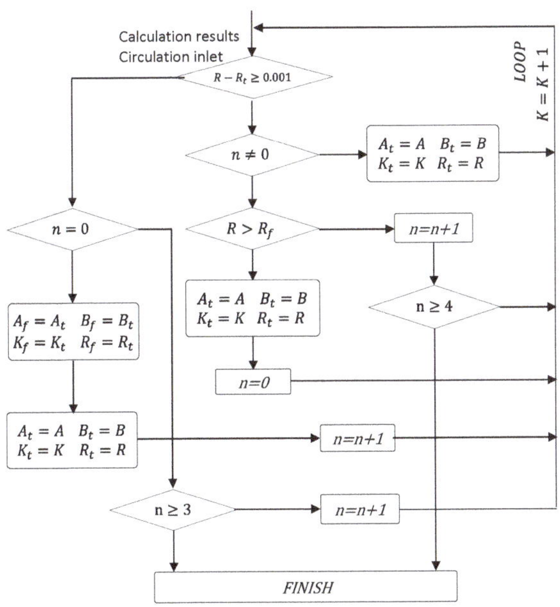

Figure 2. Loop iteration calculation method chart (Main loop idea: Set the circulation step of *K* value as 1.0, and calculate the *A* and *B* values according to the measured points. If the fitting coefficient R^2 of the model is greater than 0.99, it is regarded as the potential solution, and the potential solution appears three times in a row is the optimal solution, otherwise the output failures.).

At the starring of loop interactive, the initial value of K was taken as 100, the characteristic points (d_{10},10), (d_{30},30), (d_{50},50), (d_{60},60), (d_{70},70), (d_{90},90) as the sample points for linear regression were obtained, and the model coefficients A and B were calculated according to Equation (3). When the corresponding model calculation value $\hat{N}_i$ of the cumulative proportion of particles is calculated by Equation (1), R^2 can then be obtained by Equation (6). If R^2 satisfies the condition of loop termination, the calculation ends by increasing K by 1 for a new loop until the R^2 fits the loop termination requirements. The final values of model coefficients could finally be outputted. In this study, the condition for cycle terminations is $R^2 \leq 0.99$. Generally, the PSD model with enough goodness-of-fit can be obtained through a few loop-steps

3.3. Coefficients Interpretation

The three coefficients of the model determine the distribution characteristics of the particle size, each of them is analyzed as follows.

(a) Coefficient A reflects the average particle size

We plot the model curves with various A values for 10 to 100 when K = 100 and $B = -1.0$ as shown in Figure 3. The PSD curves move towards a coarse particle area with the increase, indicating that the Coefficient A is positively correlated with the overall particle size. Hence, the larger the value of A is, the larger the average particle size is, and vice versa.

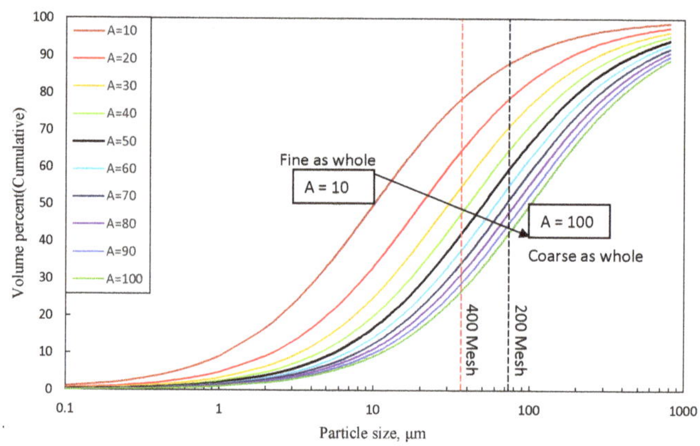

Figure 3. Model curve under different coefficient A.

(b) Coefficient B represents the proportion of coarse and fine tailings

Similarly, Figure 4 illustrates the model curves under different coefficient B values for -1.25 increasing to -0.8 when A = 50, K = 100. As shown in Figure 4, the fine fraction content increased in the particle size distribution with an increase in the B value. Therefore, the Coefficient B can reflect the proportion of the fine part to the coarse. The larger that B is, the smaller the fine particles contained, and vice versa.

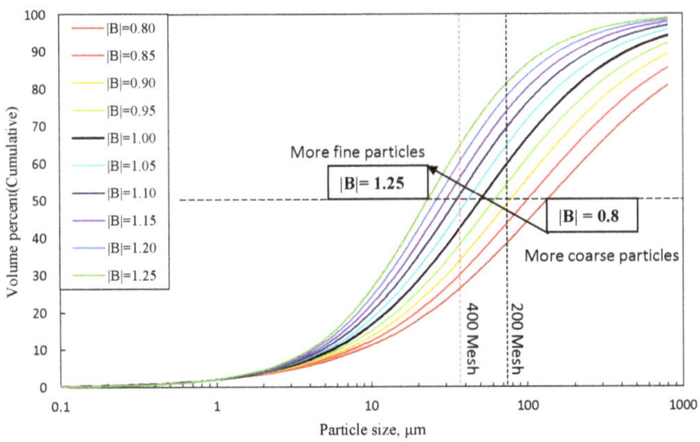

Figure 4. Model curve under different coefficient *B*.

(c) Coefficient *K* represents the width of particle distribution

Similarly, Figure 5 illustrates the modeled curve under different coefficients *K* increasing from 100 to 145 when $A = 50$ and $B = -1.0$. As shown in Figure 5, with the increase of *K*, the cumulative volume fraction of particles reaches 100% rapidly, and the corresponding particle size decreases significantly, which indicates that the Coefficient *K* can represent the maximum particle size and the distribution width of particles. The higher the *K* value is, the smaller the maximum particle size is and the narrower the particle distribution width, and vice versa.

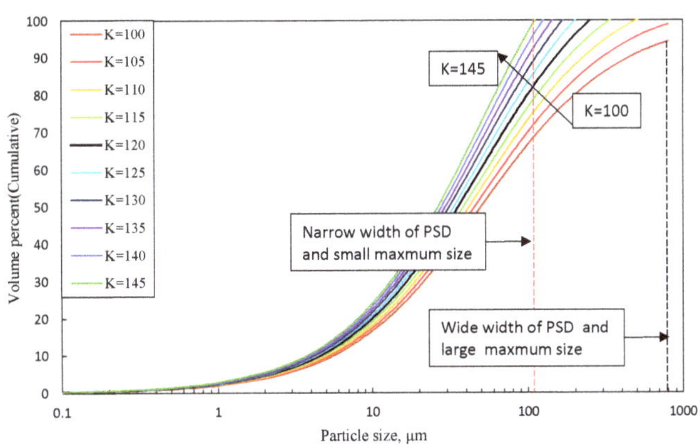

Figure 5. Model curve under different Coefficient *K*.

4. Validation and Discussion

Twelve kinds of tailings are used for verification. The coefficients of sample materials are shown in Table 1. The model is established by cyclic iteration, as described in Section 3. The results are shown in Table 2.

Coefficient *A* reflects the average fineness of tailings particles, as shown in Table 1, with the increase of samples number, the average particles sizes show an increasing trend; $A = 19.32$ for S1 sample, which is the smallest one, indicating that the sample is the finest tailings, $A = 234.78$ for S12, which is the largest one, indicating that the sample is the coarsest tailings. Coefficient *B* reflects the portion of coarse and fine tailings, as shown in Table 1, $B = -0.69$ of S8 sample is the largest, indicating the fine part proportion is the least, $B = -1.29$ of S2 sample is the smallest, indicating the fine part proportion is the most. Coefficient K reflects the width of PSD, as shown in Table 1, $K = 101$ of S2 sample is the lowest, indicating the distribution of PSD is the broadest, $K = 126$ of S8 sample is the highest, indicating the distribution of PSD is the narrowest.

Table 2. Tailings sample PSD model and coefficients.

Samples	Model	Coefficients			
		A	B	K	R^2
S1	$N_i = \frac{104}{1+19.32e^{-1.12}}$	19.32	−1.12	104	0.999
S2	$N_i = \frac{101}{1+57.78e^{-1.29}}$	57.78	−1.29	101	0.999
S3	$N_i = \frac{120}{1+25.15e^{-0.87}}$	25.15	−0.87	120	0.999
S4	$N_i = \frac{115}{1+19.57e^{-0.79}}$	19.57	−0.79	115	0.994
S5	$N_i = \frac{109}{1+28.79e^{-0.96}}$	28.79	−0.96	109	0.999
S6	$N_i = \frac{108}{1+25.18e^{-0.89}}$	25.18	−0.89	108	0.999
S7	$N_i = \frac{116}{1+22.6e^{-0.77}}$	22.60	−0.77	116	0.996
S8	$N_i = \frac{126}{1+27.98e^{-0.69}}$	27.98	−0.69	126	0.995
S9	$N_i = \frac{115}{1+90.63e^{-1.0}}$	90.63	−1.00	115	0.995
S10	$N_i = \frac{115}{1-133.67e^{-1.07}}$	133.67	−1.07	115	0.995
S11	$N_i = \frac{108}{1+167.68e^{-1.16}}$	167.68	−1.16	108	0.997
S12	$N_i = \frac{104}{1-19.32e^{-1.12}}$	234.78	−1.13	114	0.993

Overall, the model coefficients A, B, and K could well describe the average particle sizes, portion of coarse and fine tailings and the range for the particle size distribution. Figure 6 illustrates the semi-logarithmic PSD of all twelve samples modeled using the three coefficients, and their PSD characteristics are highly consistent with the measured PSD graph shown in Figure 1.

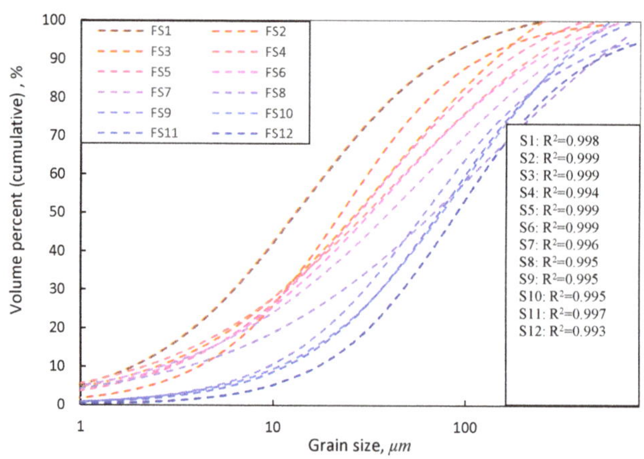

Figure 6. Model curve of tailings samples.

We compared the model curve with the measured curve and selected the particle size characteristic curves of two samples (S1 and S2) for a clear representation, in which the solid line is the measured PSD, and the dotted line is the model calculated PSD.

It can be seen from Figure 7, although there is a slight deviation in some local places of the curve, that the overall modeled curve is highly consistent with the measured one, which vividly reflects the fact of a high goodness-of-fit (R^2). Similarly, other samples also have the same regular characteristics, but due to the limited space, it will not be shown one by one.

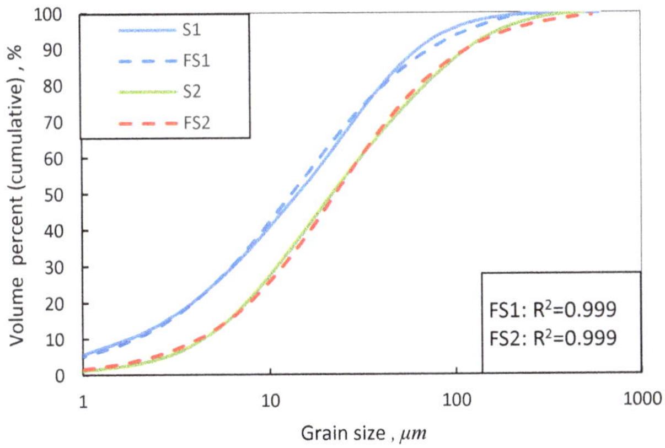

Figure 7. Comparison between modeled PSD curve and measured PSD curve.

5. Conclusions

This study firstly presents a mathematical model for the particle size distribution of mine tailings. The model contains three model coefficients (A, B and K) which mainly reflect the characteristics of particles from three aspects respectively, the average size of particles can be reflected by coefficient A, the proportion of the coarse or the fine parts of particles can be reflected by coefficient B, and the distribution width of particles can be reflected by coefficient K. Loop iteration was used to obtain a more reliable model function to express the characteristics of particle size composition with the three coefficients. Twelve tailing materials sourced from metal mines around China were used for the model establishment and validation. The goodness-of-fitting was given by R^2 for all twelve samples, each of them was greater than 0.99, showing a highly consistent between the test values and the model calculate values.

Compared with other particle characterization methods, using the proposed model to research the PSD features of tailings can intuitively obtain the overall particle size, the proportion characteristics tailings of coarse parts to fine parts, the distribution width of the particle size, which can provide a reference for studying the PSD features and its influence on other physical quantities, such as the strength characteristics of cemented backfill and the flow pattern characteristics of tailings slurry. This model is mainly focused on artificially grinded tailing materials in mineral procession, its applicability for natural formed particles such as sand or soil need be verified and the application on artificial sand such as construction sand, slag powder could be further studied.

Author Contributions: Conceptualization, Z.L.; Data curation, Y.Z.; Formal analysis, Z.L. and Y.Z.; Funding acquisition, L.G.; Investigation, Y.Z. and X.P.; Methodology, Z.L. and K.K.; Resources, L.G.; Validation, X.P.; Writing—original draft, Z.L.; Writing—review and editing, L.G. and K.K. All authors have read and agreed to the published version of the manuscript.

Funding: This research was supported by the National Key R&D Program of China (No. 2021YFC2900600, No. 2021YFE0102900); This research was also funded by the Key Research Fund of BGRIMM (No. 02-1911, 04-2208).

Data Availability Statement: The original contributions presented in the study are included in the article, and further inquiries can be directed to the corresponding authors.

Conflicts of Interest: The authors declare that the research was conducted in the absence of any commercial or financial relationships that could be construed as a potential conflict of interest.

References

1. Jones, H.; Boger, D.V. Sustainability and Waste Management in the Resource Industries. *Ind. Eng. Chem. Res.* **2012**, *51*, 10057–10065. [CrossRef]
2. Franks, D.M.; Boger, D.V.; Côte, C.M.; Mulligan, D.R. Sustainable development principles for the disposal of mining and mineral processing wastes. *Resour. Policy* **2011**, *36*, 114–122. [CrossRef]
3. Sivakugan, N.; Veenstra, R.; Naguleswaran, N. Underground Mine Backfilling in Australia Using Paste Fills and Hydraulic Fills. *Int. J. Geosynth. Ground Eng.* **2015**, *1*, 18. [CrossRef]
4. Birat, J.-P. Society, Materials, and the Environment: The Case of Steel. *Metals* **2020**, *10*, 331. [CrossRef]
5. El-Batsh, H.M.; Doheim, M.A.; Hassan, A.F. On the application of mixture model for two-phase flow induced corrosion in a complex pipeline configuration. *Appl. Math. Model.* **2012**, *36*, 5686–5699. [CrossRef]
6. Behera, S.K.; Mishra, D.; Singh, P.; Mishra, K.; Mandal, S.K.; Ghosh, C.; Kumar, R.; Mandal, P.K. Utilization of mill tailings, fly ash and slag as mine paste backfill material: Review and future perspective. *Constr. Build. Mater.* **2021**, *309*, 125120. [CrossRef]
7. Ting, X.; Xinzhuo, Z.; Miedema, S.A.; Xiuhan, C. Study of the characteristics of the flow regimes and dynamics of coarse particles in pipeline transportation. *Powder Technol.* **2019**, *347*, 148–158. [CrossRef]
8. Mourinha, C.; Palma, P.; Alexandre, C.; Cruz, N.; Rodrigues, S.M.; Alvarenga, P. Potentially Toxic Elements' Contamination of Soils Affected by Mining Activities in the Portuguese Sector of the Iberian Pyrite Belt and Optional Remediation Actions: A Review. *Environments* **2022**, *9*, 11. [CrossRef]
9. Liu, Q.; Liu, D.; Tian, Y.; Liu, X. Numerical simulation of stress-strain behaviour of cemented paste backfill in triaxial compression. *Eng. Geol.* **2017**, *231*, 165–175. [CrossRef]
10. Zhao, Y.; Taheri, A.; Karakus, M.; Deng, A.; Guo, L. The Effect of Curing under Applied Stress on the Mechanical Performance of Cement Paste Backfill. *Minerals* **2021**, *11*, 1107. [CrossRef]
11. Xu, W.; Cao, P.; Tian, M. Strength Development and Microstructure Evolution of Cemented Tailings Backfill Containing Different Binder Types and Contents. *Minerals* **2018**, *8*, 167. [CrossRef]
12. Le, Z.-H.; Yu, Q.-L.; Pu, J.-Y.; Cao, Y.-S.; Liu, K. A Numerical Model for the Compressive Behavior of Granular Backfill Based on Experimental Data and Application in Surface Subsidence. *Metals* **2022**, *12*, 202. [CrossRef]
13. Öhlander, B.; Chatwin, T.; Alakangas, L. Management of Sulfide-Bearing Waste, a Challenge for the Mining Industry. *Minerals* **2012**, *2*, 1–10. [CrossRef]
14. Wu, D.; Yang, B.; Liu, Y. Pressure drop in loop pipe flow of fresh cemented coal gangue–fly ash slurry: Experiment and simulation. *Adv. Powder Technol.* **2015**, *26*, 920–927. [CrossRef]
15. Li, M.-Z.; He, Y.-P.; Liu, Y.-D.; Huang, C. Pressure drop model of high-concentration graded particle transport in pipelines. *Ocean Eng.* **2018**, *163*, 630–640. [CrossRef]
16. Wilson, K.C. Deposition limit nomograms for particles of various densities in pipeline flow. *Hydrotransport* **1979**, *1*, 1–12.
17. Kesimal, A.; Yilmaz, E.; Ercikdi, B.; Alp, I.; Deveci, H. Effect of properties of tailings and binder on the short-and long-term strength and stability of cemented paste backfill. *Mater. Lett.* **2005**, *59*, 3703–3709. [CrossRef]
18. Libos, I.L.S.; Cui, L. Effects of curing time, cement content, and saturation state on mode-I fracture toughness of cemented paste backfill. *Eng. Fract. Mech.* **2020**, *235*, 107174. [CrossRef]
19. Zhao, Y.; Taheri, A.; Karakus, M.; Chen, Z.; Deng, A. Effects of water content, water type and temperature on the rheological behaviour of slag-cement and fly ash-cement paste backfill. *Int. J. Min. Sci. Technol.* **2020**, *30*, 271–278. [CrossRef]
20. Peng, X.; Guo, L.; Liu, G.; Yang, X.; Chen, X. Experimental Study on Factors Influencing the Strength Distribution of In Situ Cemented Tailings Backfill. *Metals* **2021**, *11*, 2059. [CrossRef]
21. Wilson, K.C.; Clift, R.; Addie, G.R.; Maffett, J. Effect of broad particle grading on slurry stratification ratio and scale-up-ScienceDirect. *Powder Technol.* **1990**, *61*, 165–172. [CrossRef]
22. Guo, W.; Han, Y.; Gao, P.; Tang, Z. Effect of operating conditions on the particle size distribution and specific energy input of fine grinding in a stirred mill: Modeling cumulative undersize distribution. *Miner. Eng.* **2022**, *176*, 107347. [CrossRef]
23. Quezada, G.R.; Ramos, J.; Jeldres, R.I.; Robles, P.; Toledo, P.G. Analysis of the flocculation process of fine tailings particles in saltwater through a population balance model. *Sep. Purif. Technol.* **2020**, *237*, 116319. [CrossRef]
24. Guo, Z.; Qiu, J.; Jiang, H.; Zhang, S.; Ding, H. Improving the performance of superfine-tailings cemented paste backfill with a new blended binder. *Powder Technol.* **2021**, *394*, 149–160. [CrossRef]
25. Zanko, L.M.; Niles, H.B.; Oreskovich, J.A. Mineralogical and microscopic evaluation of coarse taconite tailings from Minnesota taconite operations. *Regul. Toxicol. Pharmacol.* **2008**, *52* (Suppl. 1), S51–S65. [CrossRef]
26. Uzi, A.; Levy, A. Flow characteristics of coarse particles in horizontal hydraulic conveying. *Powder Technol.* **2018**, *326*, 302–321. [CrossRef]
27. Zhang, B.; Xin, J.; Liu, L.; Guo, L.; Song, K.-I. An Experimental Study on the Microstructures of Cemented Paste Backfill during Its Developing Process. *Adv. Civ. Eng.* **2018**, *2018*, 1–10. [CrossRef]
28. Singh, J.P.; Kumar, S.; Mohapatra, S.K. Modelling of two phase solid-liquid flow in horizontal pipe using computational fluid dynamics technique. *Int. J. Hydrogen Energy* **2017**, *42*, 20133–20137. [CrossRef]
29. Qi, C.; Chen, Q.; Fourie, A.; Zhao, J.; Zhang, Q. Pressure drop in pipe flow of cemented paste backfill: Experimental and modeling study. *Powder Technol.* **2018**, *333*, 9–18. [CrossRef]

30. Wu, D.; Yang, B.; Liu, Y. Transportability and pressure drop of fresh cemented coal gangue-fly ash backfill (CGFB) slurry in pipe loop. *Powder Technol.* **2015**, *284*, 218–224. [CrossRef]
31. Li, M.; He, Y.; Liu, Y.; Huang, C. Effect of interaction of particles with different sizes on particle kinetics in multi-sized slurry transport by pipeline. *Powder Technol.* **2018**, *338*, 915–930. [CrossRef]
32. Lu, G.; Fall, M.; Yang, Z. An evolutive bounding surface plasticity model for early-age cemented tailings backfill under cyclic loading. *Soil Dyn. Earthq. Eng.* **2019**, *117*, 339–356. [CrossRef]
33. Liu, L.; Fang, Z.; Wu, Y.; Lai, X.; Wang, P.; Song, K.-I. Experimental investigation of solid-liquid two-phase flow in cemented rock-tailings backfill using Electrical Resistance Tomography. *Constr. Build. Mater.* **2018**, *175*, 267–276. [CrossRef]
34. Matoušek, V.; Chára, Z.; Konfršt, J.; Novotný, J. Experimental investigation on effect of stratification of bimodal settling slurry on slurry flow friction in pipe. *Exp. Therm. Fluid Sci.* **2022**, *132*, 110561. [CrossRef]
35. Zhang, J.; Cheng, L.; Yuan, H.; Mei, N.; Yan, Z. Identification of viscosity and solid fraction in slurry pipeline transportation based on the inverse heat transfer theory. *Appl. Therm. Eng.* **2019**, *163*, 114328. [CrossRef]
36. Adiguzel, D.; Bascetin, A. The investigation of effect of particle size distribution on flow behavior of paste tailings. *J. Environ. Manag.* **2019**, *243*, 393–401. [CrossRef]
37. US Department of Commerce. *NIST Recommended Practice Guide: Particle Size Characterization*; National Institute of Standards and Technology: Gaithersburg, MD, USA, 2001.
38. Zhu, J.; Wu, S.; Cheng, H.; Geng, X.; Liu, J. Response of Floc Networks in Cemented Paste Backfill to a Pumping Agent. *Metals* **2021**, *11*, 1906. [CrossRef]
39. Kaushal, D.R.; Sato, K.; Toyota, T.; Funatsu, K.; Tomita, Y. Effect of particle size distribution on pressure drop and concentration profile in pipeline flow of highly concentrated slurry. *Int. J. Multiph. Flow* **2005**, *31*, 809–823. [CrossRef]
40. Kaushal, D.R.; Thinglas, T.; Tomita, Y.; Kuchii, S.; Tsukamoto, H. CFD modeling for pipeline flow of fine particles at high concentration. *Int. J. Multiph. Flow* **2012**, *43*, 85–100. [CrossRef]
41. Lang, L.; Song, K.-I.; Lao, D.; Kwon, T.-H. Rheological Properties of Cemented Tailing Backfill and the Construction of a Prediction Model. *Materials* **2015**, *8*, 2076–2092. [CrossRef]
42. Pang, B.; Wang, S.; Liu, G.; Jiang, X.; Lu, H.; Li, Z. Numerical prediction of flow behavior of cuttings carried by Herschel-Bulkley fluids in horizontal well using kinetic theory of granular flow. *Powder Technol.* **2018**, *329*, 386–398. [CrossRef]
43. Jiang, F.; Zhao, P.; Qi, G.; Li, N.; Bian, Y.; Li, H.; Jiang, T.; Li, X.; Yu, C. Pressure drop in horizontal multi-tube liquid–solid circulating fluidized bed. *Powder Technol.* **2018**, *333*, 60–70. [CrossRef]
44. Arolla, S.K.; Desjardins, O. Transport modeling of sedimenting particles in a turbulent pipe flow using Euler–Lagrange large eddy simulation. *Int. J. Multiph. Flow* **2015**, *75*, 1–11. [CrossRef]
45. Fredlund, M.D.; Wilson, G.W.; Fredlund, D.G. Use of the grain-size distribution for estimation of the soil-water characteristic curve. *Can. Geotech. J.* **2002**, *39*, 1103–1117. [CrossRef]
46. Fredlund, M.D.; Fredlund, D.G. Prédiction of the soil-water characteristic curve from the grain-size distribution curve. In Proceedings of the 3rd Symposium on Unsaturated Soil, Rio de Janeiro, Brazil, 22–25 April 1997.
47. Fredlund, M.; Wilson, G.W.; Fredlund, D.G. Use of Grain-Size Functions in Unsaturated Soil Mechanics. In *Advances in Unsaturated Geotechnics*; American Society of Civil Engineers: Reston, VA, USA, 2000.
48. Standardization Administration. *Standard for Pollution Control on the Non-Hazardous Industrial Solid Waste Storage and Landfill;* State Administration for Market Regulation, Ed.; China Environmental Publishing Group: Beijing, China, 2021.
49. *D02.04.0D*; Standard Test Method for Density, Relative Density, and API Gravity of Liquids by Digital Density Meter (ASTM D4052-2009). ASTM: West Conshohocken, PA, USA, 2009.

metals

MDPI

Article

Numerical Modelling of Blasting Dust Concentration and Particle Size Distribution during Tunnel Construction by Drilling and Blasting

Jianjun Shi [1], Wei Zhang [1], Shucheng Guo [1] and Huaming An [2,*]

[1] Beijing Key Laboratory of Urban Underground Space Engineering, School of Civil and Resource Engineering, University of Science and Technology Beijing, Beijing 100083, China; keyan@ces.ustb.edu.cn (J.S.); g20198051@xs.ustb.edu.cn (W.Z.); g20198012@xs.ustb.edu.cn (S.G.)

[2] Faulty of Public Security and Emergency Management, Kunming University of Science and Technology, Kunming 650093, China

* Correspondence: huaming.an@kust.edu.cn

Abstract: In order to reduce the blasting dust concentration in the tunnel during the drilling and blasting, accelerate the tunnel excavation process, and improve the working environment for the construction workers, a three-dimensional geometric model of dust transport was established based on the gas-solid two-phase flow model using the DesginModeler software, and the discrete phase model (DPM) in the FLUENT software was used to simulate the variation of dust concentration and the distribution of dust particle size at different locations along the tunnel route within 1200 s after tunnel blasting. The results showed that the concentration of blasting dust gradually decreased over time, with the fastest decrease in the range of 2 s to 120 s, and after 900 s, the dust concentration stabilized. The overall spatial distribution of the dust concentration showed a trend of decreasing from the palm face to the tunnel entrance and from the bottom plate to the upper part. The distribution pattern of dust with different particle sizes was not the same along the length of the tunnel. The large particles settled in the area of 25 m from the palm face under the action of gravity. With the increases of distance, the mass flow rate decreased, and the dust particle size became smaller, but the proportion of small particles gradually increased, while the R-R distribution index increased. The results in this study were confirmed to be reliable by comparing the measured data to provide guidance for the dust reduction technology in tunnel blasting, so as to quickly remove the dust generated during the blasting process and improve the engineering construction efficiency.

Keywords: blasting dust movement; dust concentration; particle size distribution; blasting dust reduction; numerical simulation

Citation: Shi, J.; Zhang, W.; Guo, S.; An, H. Numerical Modelling of Blasting Dust Concentration and Particle Size Distribution during Tunnel Construction by Drilling and Blasting. *Metals* **2022**, *12*, 547. https://doi.org/10.3390/met12040547

Academic Editors: Lijie Guo and Antoni Roca

Received: 24 January 2022
Accepted: 19 March 2022
Published: 23 March 2022

1. Introduction

The green low-carbon technology of metal mining process has become a research hotspot for scholars at home and abroad in recent years. Green low-carbon technology is vital to promote the development of metalliferous mineral resources shifting from extensive destructive mining to clean and energy-saving mining in future decades. The drilling and blasting method is one of the most widely used and cost-effective excavation construction methods in tunneling projects, but a large amount of dust with extremely high mass concentrations would be generated in the process of blasting [1,2], which is not in line with green low-carbon mining. The high dust concentration not only greatly affects the safety and progress of tunnel construction, but also causes a serious threat to the health of the operators. By the end of 2020, a total of 17,064 new cases of various occupational diseases were reported nationwide, of which occupational pneumoconiosis alone accounted for 14,367 cases [3]. Therefore, the study of the tunnel blasting dust transport law and particle size distribution characteristics can be used to explore a green low-carbon mining method to reduce the dust generated by blasting, improve the tunnel construction environment, and protect the health of operators.

At present, the research on dust migration properties by international and domestic scholars is mainly focused on the field of coal mining. Its roadway cross-sectional area is small, much smaller than the current cross-sectional area of the new road tunnel. Therefore, the dust generated by blasting mainly accumulates near the blasting excavation surface, which produces more serious pollution [4,5]. Yucheng Li [6] studied the non-stationary transport law of blasting dust in the roadway space and derived the distribution trend of the settling amount of different particle sizes of dust along the longitudinal direction of the roadway. Guoliang Zhang [7] simulated the dust motion trajectory of a plateau mine based on FLUENT, and the results can be used to determine the minimum dust exhaust air velocity of a mine in a plateau environment. Lichao Zhang [8] analyzed the spatial distribution of the dust concentration and particle size at the fully mechanized coal mining face. Sa [9] derived reasonable ventilation parameters by studying the changes of the dust concentration by ventilation in 20 min after blasting in a cavern stope and the dust trajectory at different wind speeds. In terms of tunnel dust migration, Kanaoka [10] performed numerical simulations of wind flow, dust mass concentration distribution, and particle movement in construction tunnels. Yapeng Wang [11] simulated the drilling dust migration in a long-distance single-ended tunnel and established the ventilation parameters and dust removal methods. Zhongqiang Sun [12] analyzed the stress condition of dust particles in the tunnel space and simulated the transport law of dust in different processes. Mengmeng Quan [13] conducted a simulation analysis of blasting dust transport and obtained the time required for the dust concentration to fall to the permissible value. Cao [14] conducted a numerical simulation to analyze the variation pattern of the dust concentration in tunnels during the construction period at different altitudes, and the results showed that the dust concentration tends to decrease linearly with increasing altitude.

In summary, in terms of researching the transport characteristics of tunnel blasting dust, previous studies have focused on the variation pattern of the dust concentration. However, at the microscopic level, the distribution pattern of the dust particle size in the tunnel space has been rarely investigated. At the same time, due to the influence of the confinement of the tunnel space, as well as the large amount of blasting dust and the characteristics of a wide distribution of the particle size, the previous research results are not applicable to exploring the effect of the particle size of tunnel blasting dust on dust reduction and transport patterns. In view of this, on the basis of the above research results, this paper used numerical simulation to study not only the law of dust concentration changes with time after tunnel blasting by the drilling and blasting method, but also analyzed the distribution characteristics of the dust particle size at different locations along the tunnel and explains the law of dust transport from the microscopic level. The results of this study were compared with the experimental data obtained by previous authors to optimize blasting methods for green and low-carbon mining, so as to provide guidance for blasting dust reduction at tunnel sites.

2. Mathematical Model of Air Flow and Dust Flow

2.1. Overview of Gas–Solid Two-Phase Flow Model

The mathematical models for simulating gas–solid two-phase flow in the FLUENT software (ANSYS Inc., Canonsburg, PA, USA) include the discrete phase model (DPM), the mixture model, the Eulerian model, and the VOF model. These models can be grouped into two categories: the Eulerian-Eulerian method and the Eulerian-Lagrange method. The discrete phase model belongs to the Euler–Lagrange method. It requires that the volume of the particle phase should not be too large and uniformly distributed in the continuous phase, which means the local volume concentration of particles should be less than 10% [15]. By calculating the velocity and turbulent kinetic energy of the continuous flow field, the trajectories of individual particles are obtained, and the trajectories of a large number of particles are calculated, so that the flow field distribution of the discrete phase can be derived. At present, the study of the trajectory of tunnel blasting dust particles in a continuous flow field usually uses the DPM model, which essentially belongs to the Euler–Lagrange method [16].

In this paper, the air flow field in the tunnel space was considered as a continuous phase and the blasting dust particles were considered as a discrete phase, while the variation characteristics of the tunnel blasting dust concentration and particle size distribution were simulated and analyzed using the DPM model. First, the quality of the grid division was verified. Second, the transient pressure-based solver was used, and the wind flow velocity and turbulent kinetic energy of the flow field were calculated using the SIMPLE algorithm. Finally, the random orbit model was opened, and the trajectories of individual particles were integrated in the Lagrangian coordinate system to obtain all particle transport trajectories.

2.2. Mathematical Model of Airflow

The mathematical model of gas flow is mainly used to determine the velocity field and pressure distribution of the air in the tunnel. As Re is generally larger than 1×10^6 in the tunnel space and the airflow state is usually turbulent flow, so the turbulence can be simulated numerically by solving the transient 3D equations in the grid size of the turbulence scale. Considering the gas flow in the tunnel as a constant incompressible adiabatic flow, the 3D Euler-Euler model was used to describe the continuous gas phase in the tunnel, and the continuity and momentum equations for the gas phase in the tunnel are as follows [17,18].

$$\frac{\partial \rho}{\partial t} + \frac{\partial(\rho u_i)}{\partial x_i} = 0 \tag{1}$$

$$\frac{\partial(\rho u_i)}{\partial t} + \frac{\partial(\rho u_i u_j)}{\partial x_j} = -\frac{\partial p}{\partial x_i} + \frac{\partial \tau_{ij}}{\partial x_j} + \rho g_i - F_i \tag{2}$$

where ρ is the density of air, kg/m^3; τ_{ij} is the Reynolds stress tensor; g_i is the acceleration of gravity, kg/m^2; F_i is the particle flow resistance, N.

The Reynold's time-averaged equation was used to model turbulence, the basic idea of which is that the transient pulsations in the time-averaged equation are represented by a k-ε two-path turbulence model [19,20].

The kinetic energy equation for turbulent pulsations (also known as the k equation) can be expressed as:

$$\frac{\partial(\rho k)}{\partial t} + \frac{\partial(\rho k u_i)}{\partial x_i} = \frac{\partial}{\partial x_j}\left[(\mu + \frac{\mu_t}{\sigma_k})\frac{\partial k}{\partial x_j}\right] + G_k - \rho\varepsilon \tag{3}$$

The equation for the energy dissipation rate of turbulent pulsation kinetic energy (also known as the ε equation) can be expressed as:

$$\frac{\partial(\rho \varepsilon)}{\partial t} + \frac{\partial(\rho \varepsilon u_i)}{\partial x_i} = \frac{\partial}{\partial x_j}\left[(\mu + \frac{\mu_t}{\sigma_\varepsilon})\frac{\partial \varepsilon}{\partial x_j}\right] + \frac{C_{1\varepsilon}\varepsilon}{k}G_k - C_{2\varepsilon}\rho\frac{\varepsilon^2}{k} \tag{4}$$

where, $G_k = \mu_t\left[\frac{\partial u_i}{\partial x_j} + \frac{\partial u_j}{\partial x_i}\right]\frac{\partial u_i}{\partial x_j}$, $\mu_t = \rho C_\mu \frac{k^2}{\varepsilon}$.

In the above equation sets, k is the turbulent kinetic energy, m^2/s^2; ε is the turbulent kinetic energy dissipation rate, m^2/s^3; μ is the laminar viscosity coefficient; μ_t is the viscosity coefficient of turbulent flow, Pa·s; u_i, u_j are the velocities in the x, y directions, respectively, m/s; G_k is the generation term of turbulent kinetic energy generated by the average velocity gradient, kg/(s^3.m); $C_{1\varepsilon}$, $C_{2\varepsilon}$, C_μ, σ_ε, and σ_k are constants with values of 1.44, 1.92, 0.09, 1.00, and 1.30, respectively [21,22].

2.3. Control Equation of the Movement of Dust

In the process of calculating the trajectory of tunnel blasting dust movement, only the effect of gravity and resistance was considered, and other forces can be neglected. The equilibrium equation for the motion of dust particles is [23–25]:

$$\frac{du_p}{dt} = F_{D(u-u_p)} + \frac{g_x(\rho_p - \rho)}{\rho_p} \tag{5}$$

$$F_D = 0.75\frac{C_D\rho|u_p - u|}{\rho_p d_p} \tag{6}$$

where $F_{D(u-up)}$ is the particle traction force per unit mass, N; C_D is the traction coefficient; μ is the fluid phase velocity, m/s; μ_p is the particle motion velocity, m/s; ρ is the fluid density, kg/m^3; ρ_p is the particle density, kg/m^3; d_p is the particle diameter, m.

The controlling equation for the particle trajectory is [14]:

$$\frac{du_p}{dt} = \frac{1}{\tau_p}\left[\overline{u} + u'(t) - u_p\right] \tag{7}$$

where τ_p is the particle relaxation time, s; $\overline{\mu}$ is the average velocity, m/s; $\mu'(t)$ is the pulsation velocity, m/s. here, $u' = \zeta\sqrt{\overline{(u')^2}}$, and ζ is a random number obeying normal distribution.

For the k-ε model, assuming that the local turbulence is isotropic, the following equation is satisfied [26].

$$\sqrt{\overline{(u')^2}} = \sqrt{\overline{(v')^2}} = \sqrt{\overline{(w')^2}} = \sqrt{\frac{2}{3}k} \tag{8}$$

In order to obtain the turbulent diffusion law and dust concentration distribution of particles, the instantaneous velocity was integrated within segmented time, i.e., performing the integration in discrete time steps to derive the equation of the particle trajectory.

2.4. Dust Particle Size Distribution Function

One of the more widely used models and methods for studying the dust particle size distribution is the Rosin–Rammler distribution function (R-R distribution) [27,28]. This paper assumed that the tunnel face is the injection source of dust particles, and the particle size distribution of dust follows the R-R distribution, which was used to describe the particle size distribution characteristics and settlement law of dust in the tunnel space. The R-R distribution in FLUENT describes the function relationship between the particle size and mass fraction in the following equation.

$$Y_d = e^{-(d/\overline{d})^n} \tag{9}$$

where $\overline{d}$ is the average particle size, m; n is the size distribution index.

3. Geometric Model Establishment and Solution

3.1. Geometric Modeling and Meshing

Due to the enclosed nature of the tunnel space, the variety of operating equipment, and the complexity of the environment, it is impossible to realistically and accurately reproduce the tunnel model. Therefore, with little influence on the numerical simulation results and for more convenient physical modeling, the dust transport space was appropriately simplified as follows: (1) The tunnel was simplified as a semi-circular arch space with a section radius of 8m and a length of 200 m. (2) Considering the effect of the air flow of the duct on the dust, the effect of other equipment in the tunnel was not considered.

(3) The dust generated by other processes and the lifting of the bottom plate was not considered [12].

Based on the above simplifications and assumptions, a 3D geometric model was created using the DesginModeler software at 1:1 based on the actual tunnel blasting space. Among them, the tunnel duct was suspended on the upper right side of the tunnel, with a diameter of 1.4 m, a distance of 3 m from the bottom plate, and the distance from the outlet of the duct to the palm face was 50 m. Finally, the model was imported into the MESH software for meshing, and the number of divided mesh cells was 119,940. The results are shown in Figure 1.

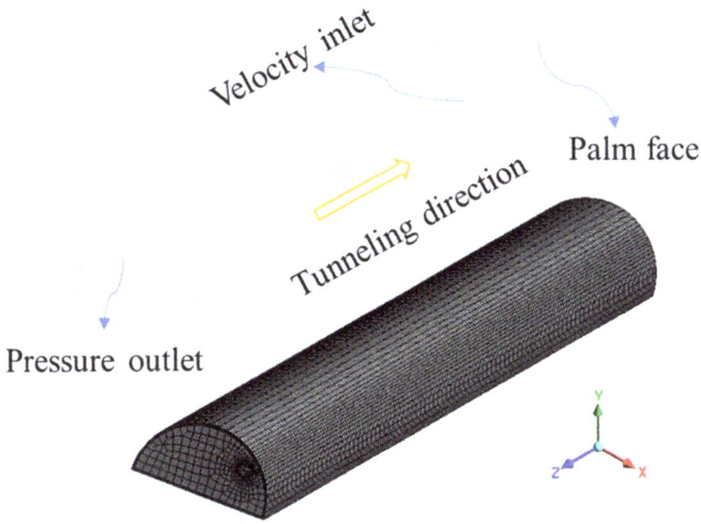

Figure 1. Tunnel geometry model and meshing.

3.2. The Setting of the Simulation Parameters and Boundary Conditions

By extensively reviewing the relevant literature on tunnel blasting dust transport characteristics and particle size distribution, combined with relevant studies in the field of coal mine roadways [29,30], the discrete phase model parameters, injection source parameters, and boundary conditions were taken according to the requirements of the discrete phase model setting in the FLUENT software. The numerical simulation parameters are listed in Table 1.

Table 1. Calculation model parameters.

Parameter Name	Parameter Setting
Solver	Pressure-based solver
Viscous model	k-epsilon model
Inlet boundary type	Velocity inlet
Inlet velocity	20 m/s
Outlet boundary type	Pressure outlet
Discrete phase model	Open
Material	Dolomite
Particle size distribution type	R-R distribution
Min. particle diameter	2 μm
Max. particle diameter	100 μm
Median diameter	12 μm
Distribution index	1.93
Mass flow rate	0.03 kg/s
Turbulent diffusion mode	Stochastic trajectory model

4. Numerical Simulation Results and Analysis

4.1. Analysis of the Distribution of the Air Flow Field in the Tunnel

The movement process of dust particles generated by blasting occurs in the continuous phase airflow field. In order to study the movement pattern of blasting dust in the tunnel space, the distribution of the flow field in the tunnel was firstly analyzed. The outlet of the duct was set as the velocity inlet, and the velocity of the inlet was 20 m/s. The distribution of the air flow field in the tunnel was obtained as shown in Figures 2 and 3:

(1) The wind flow was shot out from the air duct, along the right side of the tunnel towards the palm face, and then returned out along the left side after colliding with the palm face, forming a return flow area between the duct outlet and the palm face. In this area, part of the return wind flow will be impacted, mixed, and integrated with the outlet wind flow, and the original flow field is strongly disturbed, forming an extremely turbulent vortex area;

(2) Most of the wind flows out of the vortex area after turning back from the palm surface and continues to flow forward. Behind 50 m from the palm surface, the flow velocity gradually decreases, and the flow field distribution tends to be stable and uniform. The wind speed is basically stable at about 0.5 m/s;

(3) The initial velocity of the wind flow is 20 m/s. After the impulse and momentum exchange with the surrounding boundary layer, the wind velocity gradually decreases, and after returning back from the palm surface, the wind velocity decreases sharply again. The wind velocity of the flow field along the tunnel shows a trend of maximum wind velocity in the vortex area, followed by the transition area, and small and uniform wind velocity is in the stable area.

4.2. Dust Concentration Variation Law with Time

We set the palm face as the dust injection source and gave the dust a 2 s release time. The horizontal plane with a height of Y = 1.5 m (the height of respiratory zone) was cut as the concentration peak monitoring surface, and the distribution pattern of dust concentration peak at 10 s, 100 s, 300 s, 600 s, and 900 s on this plane was observed, as shown in Figure 4.

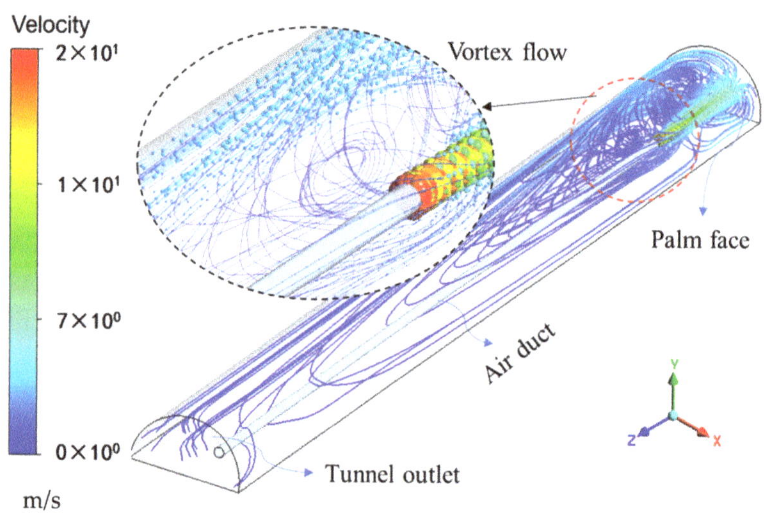

Figure 2. Distribution of airflow streamlines in the tunnel.

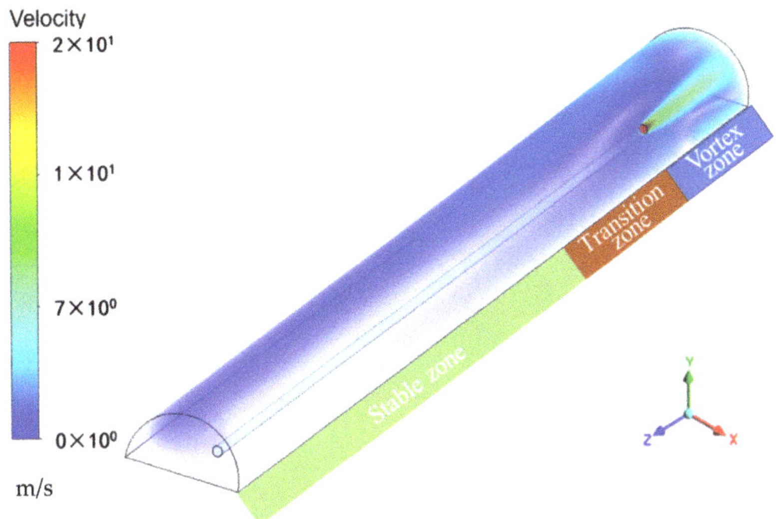

Figure 3. Wind speed body of the tunnel space.

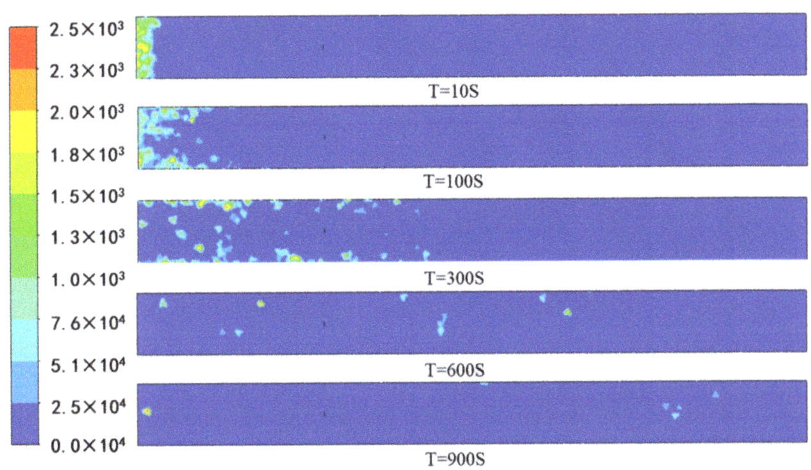

Figure 4. Distribution of blasting dust concentration at a 1.5 m height.

It can be seen from Figure 4 that:

(1) A large amount of dust generated by tunnel blasting was sprayed into the tunnel from the palm face at a high speed. At this time, the dust concentration was very high, up to 1400 mg/m^3. Within 2 s~10 s, the blasting dust was very dense, forming a "cloud" and gathering near the palm face. Under the action of the initial velocity and air flow, the blasting dust diffused to the tunnel outlet and spreads as far as about 20 m;

(2) Within 30 s~600 s of blasting, the dust "cloud" gradually diffused under the action of air flow and its own gravity. After that, the sedimentation occurred, and the dust was captured or discharged, leading to the dust concentration becoming sparse. The dust particles not captured and discharged were randomly distributed in the whole tunnel;

(3) With the passage of time, about 900 s later, basically all dust was captured and discharged, and the dust concentration was lower than 8 mg/m^3, which meet the

standard concentration value allowed by the regulations [31,32]. Workers can enter the working surface for the operation.

The dust concentration cloud maps at T = 10 s, T = 30 s, T = 90 s, and T = 120 s were observed for three horizontal surfaces at heights of Y = 0.5 m, Y = 1.5 m, and Y = 4 m, as shown in Figure 5. In addition, the peak dust concentrations of these three horizontal surfaces were monitored separately, and the monitoring results are shown in Figure 6.

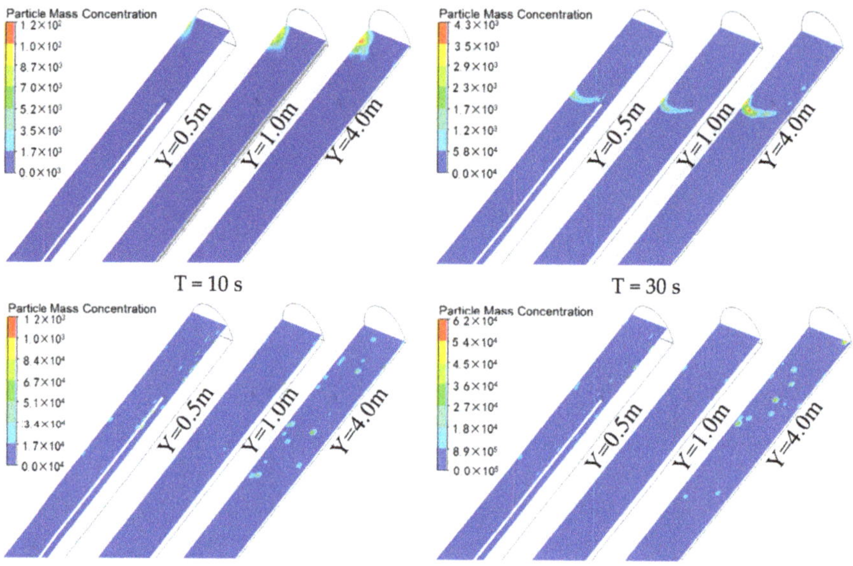

Figure 5. Dust concentration cloud map with time at three height levels.

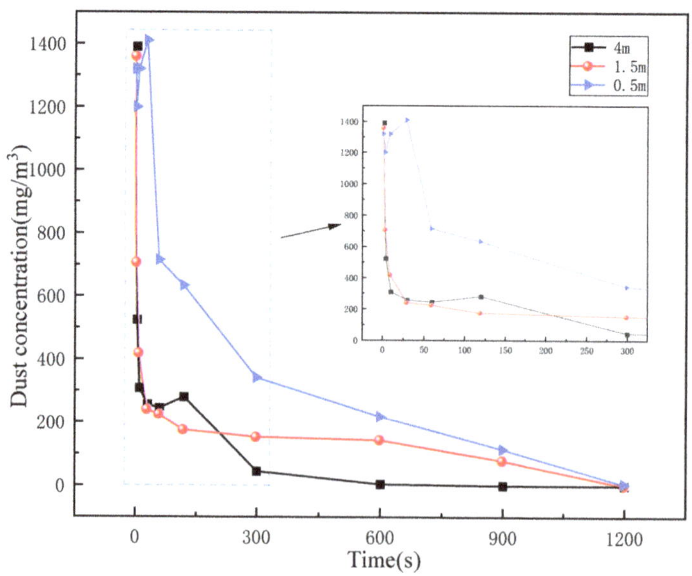

Figure 6. Curves of the blasting dust concentration with time at three height levels.

As seen in Figures 5 and 6:

Within 5 min after blasting, dust with a large particle size rapidly settled or was captured under the influence of gravity, resulting in a rapid decline in the dust concentration. From the cloud plot of the dust concentration in the horizontal plane at three heights, it can be concluded that the dust concentration decreased with increasing height. The peak value of dust concentration at a 4 m-high horizontal plane decreased fastest with time, and the dust concentration decreased slowly with the decrease of height. This was because the dust in the upper part of the tunnel was constantly sinking under the influence of gravity, and the air flow near the bottom plate was subject to large resistance and low wind speed, so the dust was not easily diffused, leading to the slow decline of the dust concentration at the 0.5 m height.

4.3. *Distribution Pattern of Dust of Different Particle Sizes along the Tunnel*

The Rosin–Rammler distribution function is the most authoritative model and method for describing the dust particle size distribution [33,34]. In this paper, the discrete phase model was used, and the particle size satisfied the R-R distribution, where the minimum particle size was 2 μm and the maximum particle size was 100 μm. Dust in the tunnel space within 300 s after blasting was selected for the particle size distribution analysis. At this time, the dust diffusion speed was fast and the particle size distribution range was wide, which is representative. The trajectories of particles with different particle sizes over time are shown in Figure 7. The following conclusions can be drawn from Figure 7:

(1) Large particles of dust settled near the palm face under their own gravity, while small particles continued to be suspended in the tunnel air and had random pulsation. During stochastic diffusion, the particles underwent capture and terminated their motion when they came in contact with the tunnel floor;
(2) The larger the particle size was, the faster the sedimentation velocity was, the shorter the horizontal movement distance was, and the shorter the sedimentation time was. Dust particles above 56 μm basically all settled within 25 m from the palm surface, a small portion of dust particles from 12–56 μm could settle, while most dust particles below 12 μm were suspended in the air due to their small particle size and randomly diffused outside the tunnel entrance;
(3) Over time, the large dust particles gradually settled, and the dust particles remaining in the tunnel space became smaller and smaller, while the overall quantity of dust particles gradually decreased. It took about 30 s, 120 s, and 300 s for the dust particles to diffuse to 50 m, 100 m, and 150 m, respectively.

In order to investigate the effect of different locations along the tunnel on the movement of particles of different sizes, the dust particle size distribution was statistically analyzed when the dust mass concentration had less variation. Six sections were taken for analysis at 10 m, 25 m, 50 m, 75 m, 100 m, and 150 m away from the palm surface. The distribution of the dust particle size in different sections during dust movement is shown in Figure 8.

It can be seen from Figure 8 that:

(1) As the distance from the palm surface increased, the large dust particles basically settled out, and the proportion of small dust particles gradually became larger, with the majority of the particles smaller than 12 μm. The overall particle size became smaller; the dust particle size distribution range was reduced; the particle size distribution was more concentrated, resulting in the increase of the R-R distribution index;
(2) The overall mass flow rate of the dust particles along the path decreased with increasing particle size. Within the first 25 m, the mass flow rate of dust from 2–12 μm accounted for 51.98% to 54.79%; the mass flow rate of dust less than 56 μm was as high as 90.45% to 96.25%; the mass flow rate of dust particle size above 75 μm was only 0.73% to 1.74%. The mass flow rate of large dust particles above 75 m from the palm face was almost 0.

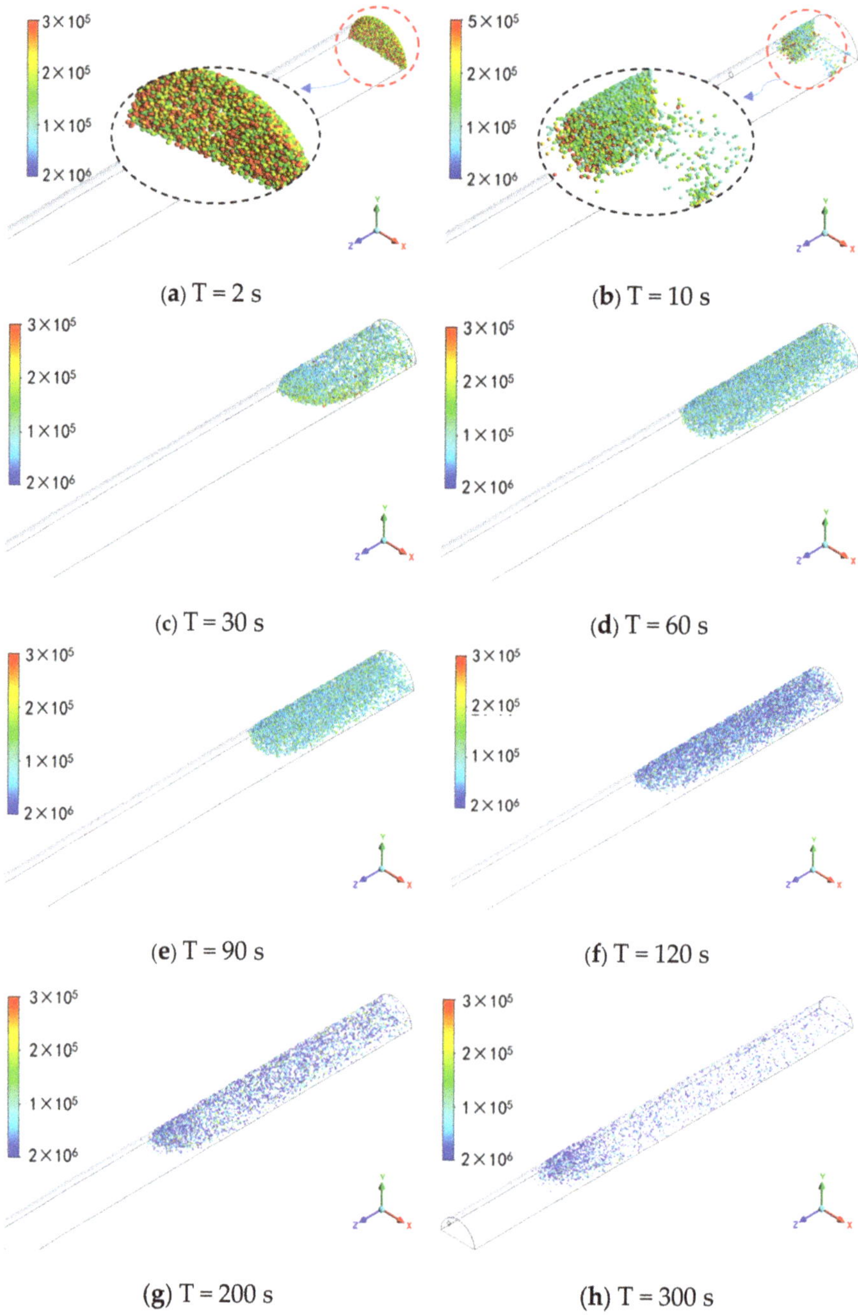

Figure 7. Trajectory of blasting dust over time.

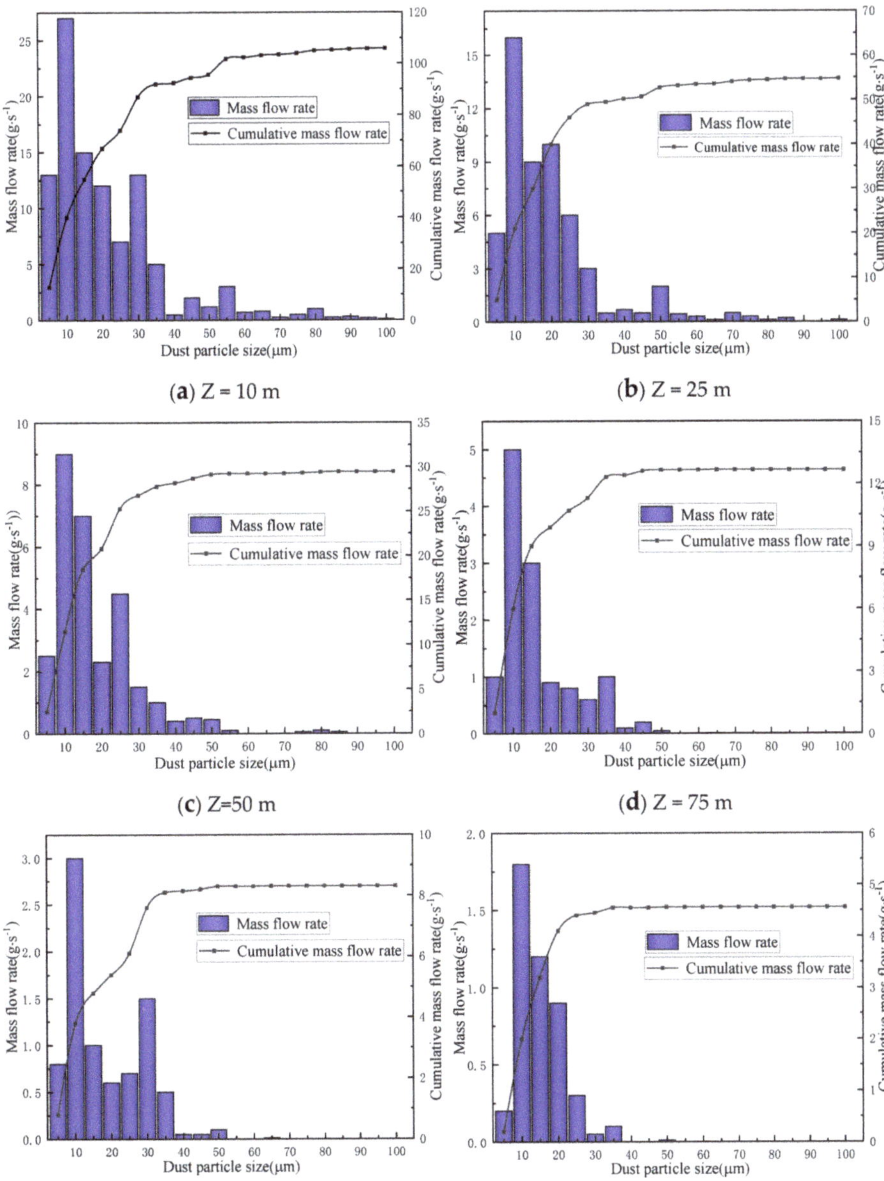

(**a**) Z = 10 m (**b**) Z = 25 m

(**c**) Z=50 m (**d**) Z = 75 m

(**e**) Z = 100 m (**f**) Z = 150 m

Figure 8. Distribution of the dust particle size at different locations along the tunnel.

4.4. Validation of the Numerical Simulation Results

In order to verify the accuracy of the numerical simulation results and optimize the setting of the simulation parameters, the simulation results of the dust concentration at the height of the tunnel breathing zone (Y = 1.5 m) were compared with the experimental data measured in [35], as shown in Figure 9.

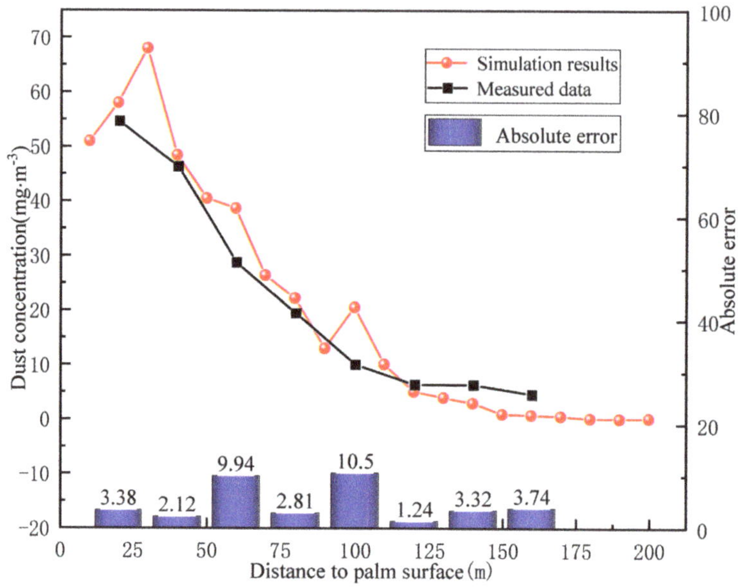

Figure 9. Comparison between the simulated and measured values of the dust concentration.

As can be seen from Figure 9, the numerical simulation results of the dust concentration were generally consistent with the measured data in terms of the trend, and the error was within the controllable range. Meanwhile, the average absolute error between the measured data and simulation results was 4.63, indicating that the error between the measured data and the simulated results was small and the simulated results were credible. However, there were still some errors in the values of the two, which were mainly due to the differences between the simulation parameter settings and the actual engineering background conditions, and the field interference factors affected the authenticity of the measured data. The comparative analysis showed that the numerical simulation method used in this paper was accurate and credible.

5. Conclusions

In this study, based on the theoretical research of gas–solid two-phase flow and the simulation of the air flow field, the FLUENT software was used to simulate the concentration change and particle size distribution of blasting dust in tunnel construction by the drilling and blasting method. On the basis of the previously used methods, the settling pattern of dust of different particle sizes along the tunnel was further observed at the microscopic level. Moreover, the simulation results were compared and analyzed with the measured data in [35]. The following conclusions can be derived:

(1) After the tunnel face blasting, the blasting dust was sprayed into the tunnel at a high speed under the action of shock wave and diffused with the wind to the portal. The dust concentration decreased gradually with the passage of time. The dust concentration decreased fastest within 2~120 s and then decreased slowly after 120 s. Then, 900 s after the blasting, the dust concentration in the tunnel decreased to less than 8 mg/m^3, which reached the specified standard dust concentration value;

(2) Dust particles will settle, be captured, or excluded along the tunnel under the wind flow. Large particles first settled rapidly under the action of gravity, while small particles pulse randomly in the tunnel and continued to diffuse forward with the wind flow. The size of suspended particles in the air gradually decreased; the particle size distribution range became concentrated; the R-R distribution index increased;

(3) At the microscopic level, dust particles above 56 μm mainly settled within 25 m from the palm face; dust particles from 12–56 μm were mainly distributed within 100 m; dust particles less than 12 μm were more evenly distributed along the tunnel. It can be concluded that the farther away from the tunnel face, the lower the mass flow rate, and the smaller the dust particle size is; the proportion of small particle dust increased, but the overall amount of dust decreased;

(4) The simulation results showed that the amount of blasting dust was large and difficult to reduce under natural conditions, and it took much time to reduce to the safety permission value. Therefore, reasonable and effective dust reduction measures need to be taken to improve construction efficiency and engineering benefits. At the same time, the simulation results were basically consistent with the measured data, which can be used to guide the layout of blasting water mist dust reduction measures in the tunnel. This conclusion can be used not only to explore a green and low-carbon mining technology, but also has important implications for improving the tunnel construction environment and enhancing the occupational safety of workers.

Author Contributions: Conceptualization, J.S.; methodology, H.A. and J.S.; writing—original draft preparation, W.Z., S.G. and H.A.; writing—review and editing, H.A.; supervision, J.S.; funding acquisition, H.A. All authors have read and agreed to the published version of the manuscript.

Funding: This research was partly funded by the Research Start-up Found for Talent of Kunming University of Science and Technology, Grant Number KKSY201867017, and the Fund from the Research Centre for Analysis and Measurement KUST (Analytic and Testing Research Centre of Yunnan), Grant Number 2020T20180040, and 2021M20202139010.

Institutional Review Board Statement: Not applicable.

Informed Consent Statement: Not applicable.

Data Availability Statement: The data used to support the findings of this study are available from the corresponding author upon request.

Conflicts of Interest: The authors declare no conflict of interest.

References

1. Zhu, S.G.; Tao, T.; Tian, F. Analysis and Prevention of fugitive dust from mine tailings. *J. Anhui Inst. Archit.* **2014**, *22*, 82–85.
2. Jiang, Z.A.; Li, T.; Zhang, Z. Numerical Simulation of the Airflow Field Rules in Tunnel Drill Blasting Construction. *Ind. Saf. Environ. Prot.* **2016**, *42*, 53–55.
3. National Health Commission of the People's Republic of China. National occupational disease report 2020. *China Occup. Med.* **2021**, *48*, 396.
4. Dong, J.; Wang, R.S.; Yu, G.S.; Li, X.H. Numerical Simulation of Dust Migration Law at Fully Mechanized Excavation Face of Rock Roadway with Forced Ventilation Based on DPM. *Saf. Coal Mines* **2016**, *47*, 190–193.
5. Lu, G.L.; Hu, H.; Xing, F.B. Numerical Simulation Study on the Movement Characteristics of Dust Flow in Mine Working Face. In Proceedings of the 2019 International Conference on Oil & Gas Engineering and Geological Sciences, Dalian, China, 28–29 September 2019; p. 012224.
6. Li, Y.C.; Liu, T.Q.; Zhi, L.I.; Zhao, M. Research on unsteady migration law of dust in blasting tunnelling space. *J. Saf. Sci. Technol.* **2014**, *10*, 33–38.
7. Zhang, G.L.; Jiang, Z.A.; Yang, B.; Yao, S.H.; Peng, Y.; Wang, Y.P. Numerical simulation of the minimum mine dust exhausting wind speed under high altitude environment. *J. China Coal Soc.* **2020**, *46*, 2294–2303.
8. Zhang, L.C.; Zhou, G.; Ma, Y.; Jing, B.; Sun, B.; Han, F.W.; He, M.; Chen, X. Numerical analysis on spatial distribution for concentration and particle size of particulate pollutants in dust environment at fully mechanized coal mining face. *Powder Technol.* **2021**, *383*, 143–158. [CrossRef]
9. Sa, Z.Y.; Li, F.; Qin, B.; Pan, X.H. Numerical simulation study of dust concentration distribution regularity in cavern stope. *Saf. Sci.* **2012**, *50*, 857–860. [CrossRef]
10. Kanaoka, C.; Furuuchi, M.; Inaba, J.; Myojo, T. Flow and dust concentration near working face of a tunnel under construction. *J. Aerosol. Sci.* **2000**, *31*, 31–32. [CrossRef]
11. Jiang, Z.A.; Wang, Y.P.; Men, L.G. Ventilation control of tunnel drilling dust based on numerical simulation. *J. Cent. South Univ.* **2021**, *28*, 1342–1356. [CrossRef]
12. Sun, Z.Q. Study on Dust Migration Regularity and Control Technology in Drilling and Blasting Method of Highway Tunnel Construction. Ph.D. Thesis, Beijing University of Science Technology, Beijing, China, 2015.

13. Quan, M.M. Study on Numerical Simulation of Blasting Fume Migration Law and Control in Tunnel with Drilling and Lasting Construction. Master's Thesis, Anhui University of Technology, Anhui, China, 2020.
14. Cao, Z.M.; Liu, X.; Niu, B.C. Migration Characteristics of Dust during Construction Stage in Highway Tunnels at High Altitude Areas. *China J. Undergr. Space Eng.* **2019**, *15*, 927–935.
15. Bian, J.Q.; Bi, J.Y.; Wang, H.D. Numerical simulation study on unsteady distribution law of dust in blasting roadway of inclined ditch mine. *China Min. Mag.* **2019**, *28*, 269–274.
16. Zhang, J.J. Gas-Solid Two-Phase Flow Simulation of Dust Migration and Deposition in Fully Mechanized Excavation Face. Master's Thesis, University of Mining and Technology, Xuzhou, China, 2015.
17. Hu, S.Y.; Feng, G.R.; Ren, X.Y.; Xu, G.; Chang, P.; Wang, Z.; Zhang, Y.T.; Li, Z.; Gao, Q. Numerical study of gas–solid two-phase flow in a coal roadway after blasting. *Adv. Powder Technol.* **2016**, *27*, 1607–1617. [CrossRef]
18. Wang, Y.P.; Jiang, Z.A.; Chen, J.S.; Chen, J.H.; Wang, M. Study of high-pressure air curtain and combined dedusting of gas water spray in multilevel ore pass based on CFD-DEM. *Adv. Powder Technol.* **2019**, *30*, 1789–1804. [CrossRef]
19. Zhou, G.; Zhang, Q.; Bai, R.; Xu, M.; Zhang, M. CFD simulation of air-respirable dust coupling migration law at fully mechanized mining face with large mining height. *J. China Univ. Min. Technol.* **2016**, *45*, 684–693.
20. Jiang, Z.A.; Deng, Q.L.; Shi, X.X.; Chen, J.S. Numerical Simulation of Dust Mass Concentration Distribution in Screening Workshop of Asbestos Concentrator. *J. Hunan Univ.* **2017**, *44*, 135–141.
21. Khaldi, N.; Marzouk, S.; Mhiri, H.; Bournot, P. Distribution characteristics of pollutant transport in a turbulent two-phase flow. *Environ. Sci. Pollut. Res.* **2015**, *22*, 6349–6358. [CrossRef]
22. Candra, K.J.; Pulung, S.A.; Sadashiv, M.A. Dust dispersion and management in underground mining faces. *Int. J. Min. Sci. Technol.* **2014**, *24*, 39–44.
23. Li, F.; Sa, Z.Y.; Wang, Y.; Zhang, H.N. Study on the dust concentration distribution and migration in flat-type chamber stope based on fluent software. *Min. Res. Dev.* **2010**, *30*, 77–80.
24. He, G.H. Dust Distribution Law of Super-long Roadway in the Process of Digging and Blasting. *Mod. Min.* **2016**, *32*, 21–27.
25. Jiang, Z.A.; Chen, M.L.; Chen, J.S. Numerical simulation of dust concentration distribution and changing regularities in roadway stope blasting. *J. Cent. South Univ.* **2013**, *44*, 1190–1196.
26. Niu, W.; Jiang, Z.A.; Liu, Y. Numerical simulation on dust movement regularities at fully-mechanized coal faces and its utilization. *J. Liaoning Tech. Univ.* **2010**, *29*, 357–360.
27. Dai, L.Y. Study on Function of Rosin-Rammler Particle size Distribution. *Nonferr. Min. Metall.* **2000**, *16*, 15–17.
28. Zheng, G.B.; Kang, T.H.; Chai, Z.Y.; Yin, Z.H. Applied the Rosin-Rammler distribution function to study on the law of coal dust particle-size distribution. *J. Taiyuan Univ. Technol.* **2006**, *37*, 317.
29. Pérez, K.; Toro, N.; Gálvez, E.; Robles, P.; Wilson, R.; Navarra, A. Environmental, economic and technological factors affecting Chilean copper smelters—A critical review. *J. Mater. Res. Technol.* **2021**, *15*, 213–225. [CrossRef]
30. Jiang, Z.A.; Chen, J.H.; Wang, M.; Chen, J.S. Numerical simulation on influencing factors of dust concentration in ore unloading station. *J. China Coal Soc.* **2018**, *43*, 185–191.
31. CCCC First Highway Engineering Group Co, Ltd. *Technical Specifications for Construction of Highway Tunnel*; JTG F60-2009; China Communications Press: Beijing, China, 2009.
32. Xu, S.Q.; Ren, H.Y.; Wang, M.S.; Li, J. New method of dust removal by negative ions in construction tunnel. *J. Transp. Eng.* **2018**, *18*, 84–93.
33. Hu, Z.W.; Bi, J.Y.; Wang, H.D. Study on the law of blasting dust migration in long distance tunneling face of Xiegou coal mine. *China Min. Mag.* **2019**, *28*, 340–345.
34. Gong, X.Y.; Mo, J.M.; Xue, X.Y.; Xue, H.E. Analysis on influence of fractal characteristics on distribution and settlement of dust flow in heading face ventilation. *Coal Technol.* **2017**, *36*, 112–114.
35. Ren, G.H. Research on Dust Migration and Dust Removal during Blasting of Long Inclined Tunnel under Complex Conditions. Master's Thesis, Shandong University of Science and Technology, Shandong, China, 2020.

metals

MDPI

Article

A Numerical Model for the Compressive Behavior of Granular Backfill Based on Experimental Data and Application in Surface Subsidence

Zhi-Hua Le, Qing-Lei Yu *, Jiang-Yong Pu, Yong-Sheng Cao and Kai Liu

Center for Rock Instability and Seismicity Research, Northeastern University, Shenyang 110819, China; 1710356@stu.neu.edu.cn (Z.-H.L.); 2110374@stu.neu.edu.cn (J.-Y.P.); 1810350@stu.neu.edu.cn (Y.-S.C.); 2000937@stu.neu.edu.cn (K.L.)

* Correspondence: yuqinglei@mail.neu.edu.cn

Abstract: Granular backfill is generally confined in stopes to bear underground pressure in metal mines. Its mechanical behavior under lateral confinement is vital for controlling stope wall behavior and estimating surface subsidence in backfill mining operations. In this paper, an experimental apparatus has been developed to explore the bearing process of granular material. Pebbles were selected to model granular backfill. A series of compression experiments of pebble aggregation were performed under lateral confinement condition using the experimental apparatus. The bearing characteristics of the pebble aggregation with seven gradations were analyzed. Based on the experimental data, a constitutive model that takes the real physical characteristics of granular material into account was proposed with variable deformation modulus. The constitutive model was implemented into the FLAC^{3D} software and verified basically by comparison with experimental results. The surface subsidence in backfilling mines was studied using the proposed model. The effects of the particle size of the granular backfill and the height and buried depth of mined-out stopes on surface subsidence have been clarified. The research results are of great significance for guiding backfill mining and evaluating surface subsidence and movement.

Keywords: granular backfill; bearing characteristics; numerical model; particle size; surface subsidence

Citation: Le, Z.-H.; Yu, Q.-L.; Pu, J.-Y.; Cao, Y.-S.; Liu, K. A Numerical Model for the Compressive Behavior of Granular Backfill Based on Experimental Data and Application in Surface Subsidence. *Metals* **2022**, *12*, 202. https://doi.org/10.3390/met12020202

Academic Editors: Lijie Guo, Rastko Vasilic, Petros E and Tsakirids

Received: 21 November 2021
Accepted: 18 January 2022
Published: 21 January 2022

1. Introduction

Granular backfill mining has been widely used because filling body performance plays a significant role in supporting the surrounding rock, preventing surface subsidence, and utilizing solid wastes. However, ground subsidence occurred at a nickel mine even with cemented backfilling in China after 18 years [1,2]. If the backfill material was granular (non-cemented filling) for this nickel mine, the surface subsidence might be worse due to the mainly particularly compressive characteristics of granular materials. Therefore, the deformation characteristics and mechanical behavior of granular backfill have attracted great attention. El-Sohby and Andrawes [3] divided the total deformation of granular material into elastic and sliding components by a conventional triaxial test, and the deformation was unrecoverable. On this basis, the compaction process was divided into a rapid compression deformation stage, slow compaction deformation stage, and stable compaction stage [4]. Then, the stress–strain curve of granular material was separated into the initial compaction stage with low stress and elastic compaction stage with high stress [5]. As a result, it has been concluded that the nonlinear behavior of granular materials was displayed [6–10]. Meanwhile, the deformation resistance of granular materials is affected by the particle shape, stiffness, dimension, and granular gradation [11–14].

The deformation modulus is an essential mechanical property of granular materials. In general, the deformation modulus has been analyzed and usually obtained by the existing theoretical methods and stress–strain curves. For example, Kiran et al. [15] proposed a method to calculate the deformation modulus of a granular pile by an in-situ load test, and Sharma and Gupta [16] found that the granular pile has non-uniform settlement according to the non-linear deformation modulus. In the laboratory, the deformation modulus of granular materials is usually obtained by compression experiments. It increases with increasing compaction deformation [17,18]. An exponential relationship between the deformation modulus and axial strain was revealed [19].

For numerical modelling, the deformation modulus is a key input parameter to simulate the mechanical behavior of granular backfill in mining engineering. Yu et al. [20], Wang et al. [21], and Pu et al. [22] adapted the Mohr–Coulomb criterion to gangue backfill and assigned a constant value to the deformation modulus of it in FLAC^{3D}. The simulation results showed that the filling body played a strong role in supporting roof strata, and the roof deformation decreased with increasing filling ratio. Li et al. [23] used the same model and reported that surface subsidence decreased by increasing the filling body's deformation modulus. The double-yield criterion [24] was also introduced to simulate the mechanical behavior of granular backfill. Li et al. [25] matched a constant value to the deformation modulus of granular backfill using the double-yield criterion and revealed that backfill materials composed of finer particles are beneficial to inhibit surface deformation and subsidence. Moreover, Zhu et al. [26] found that reducing the compression ratio of granular backfill is conducive to restraining surface subsidence.

However, experimental studies showed that the bearing process of granular backfill is dynamic and nonlinear under the loading of overlying strata. The stress–strain relationship was described by an exponential function [9]. The deformation modulus of granular backfill is not constant. In particular, in deep mining, the granular backfill is often subject to a high-stress state, and the evolution of the deformation modulus is crucial for the surface subsidence. The previous studies from numerical modelling that considered the deformation modulus of granular material as a constant did not reflect the actual mechanical behavior of the granular backfill. If engineering was involved in a large volume of granular backfill, the numerical results might differ from the physical ones.

Therefore, the purposes of this paper are to propose a constitutive model for granular material based on a series of confinement compression experiments. The proposed constitutive model that takes the actual physical characteristics of granular material into account was implemented in FLAC^{3D} and verified basically by comparison with experimental results. Then, the proposed model was applied to analyze surface subsidence in backfilling mines, and the effects of the particle size of the granular backfill and the height and depth of the mined-out stope on surface subsidence and movement were discussed.

2. Compression Experiment of Granular Backfill

2.1. Testing System

The test system used for determining the compaction characteristics of granular materials is shown in Figure 1, and it consists of a compactor, loading control unit, data acquisition unit, and acoustic emission (AE) unit. The device and method for testing the compression performance of granular materials were granted a Chinese patent (ZL202010685632.1).

The compactor is a self-designed steel cylinder with an inside diameter of 320 mm, a thickness of 25 mm, and a maximum filling height of granular materials of 500 mm. The axial loading comes from the YAW-5000 electro-hydraulic servo machine, and the loading process was carried out by a dowel bar and a loading platen. The steel cylinder, loading platen, and dowel bar are made of 45# carbon steel, of which the elastic modulus and Poisson's ratio are 207 GPa and 0.26, respectively. The strain gauges measured the lateral deformation transmitted to the cylinder wall, and the data acquisition automatically recorded the data. During the compression process of granular materials, the elastic wave caused by friction between particles and particle breakage is transmitted to the cylinder wall, and the AE sensor can convert the monitored surface displacement into an AE signal from which changes in the internal structure of granular materials are inferred. Therefore, the AE monitoring method was used to monitor the movement response and crushing of particles. Nine AE sensors were arranged in three layers and placed on the outer wall of the steel cylinder at equal intervals. The fixed threshold was set at 45 dB. The loading control, strain acquisition, and AE monitoring system work synchronously.

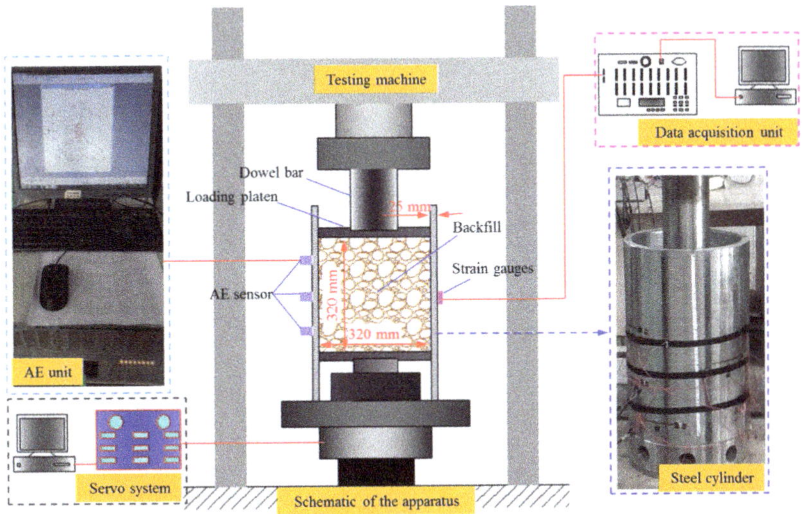

Figure 1. Testing system.

2.2. Sample Preparation

In this study, pebbles were selected to model the granular backfill, and they were sifted from the pebble formation of Yanshan Mine, Hebei Province, China. In the laboratory, the pebbles were further divided into seven gradations using standard test sieves. The gradations were 4.75–9.5 mm, 9.5–13.2 mm, 13.2–16 mm, 16–20 mm, 20–26.5 mm, 26.5–31.5 mm, and 31.5–37.5 mm, and a multi-function 3D scanner (EinScan-Pro) and image processing software were used to statistically analyze the shape of the pebbles of the seven gradations, as shown in Figure 2. According to statistical analysis, the pebble particles have the shapes of irregular convex polyhedrons with a relatively smooth surface, and the shapes of pebbles are also quite different in one gradation.

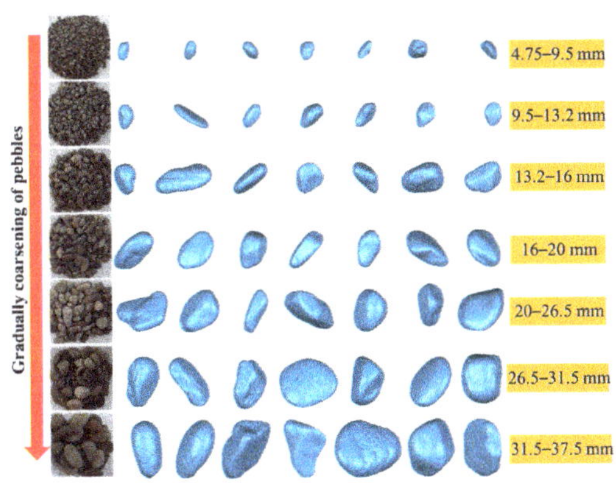

Figure 2. Seven gradations of pebbles and 3D scan image of particles.

2.3. Experimental Methods and Procedures

This study carried out the compression experiments of pebbles with seven gradations. In order to ensure that the pebble particle placement was relatively uniform, the pebbles were filled into the compactor layer by layer. The filling height in each test was about 320 mm. The initial porosity of pebble backfill with seven gradations and the test scheme are shown in Table 1. Relatively low strain rate compression favors homogeneous crushing of particles and observation of the whole compression process of granular materials. We restricted the strain rates of the compression experiments near and within the quasi-static regime, i.e., the strain rate condition was always well below 10^{-3}/s [27]. According to the filling height of granular backfill and in order to prolong the strain hardening time of granular materials as much as possible, the displacement-control loading model of the YAW-5000 electro-hydraulic servo machine was adopted, and the loading rate was 1 mm/min.

Table 1. Experimental design.

Pebble Gradations (mm)	Filling Heights (mm)	Initial Porosity	*D*/*d*
4.75–9.5	317	0.391	33.68–67.37
9.5–13.2	331	0.392	24.24–33.68
13.2–16	299	0.394	20–24.24
16–20	333	0.4	16–20
20–26.5	337	0.404	12.08–16
26.5–31.5	347	0.406	10.16–12.08
31.5–37.5	347	0.415	8.53–10.16

It should be noted that the mechanical behavior of granular materials is similar to the representative element volume (REV) of the rock mass. Investigations show that the ratio of compaction diameter to pebble particle size (D/d) should be at least 5 in order to represent granular material behavior [28–30]. In this experiment, the largest pebble grade was 31.5–37.5 mm, and the diameter of compaction was 320 mm. The minimum value of D/d ranged from 8.53 to 10.16 (Table 1) and therefore met the experimental requirements.

2.4. Experimental Results

2.4.1. Compaction Characteristics of Pebble Backfill

The essence of the compressive process of granular backfill under lateral confinement conditions is continuous reorganization of particles for granular material structure accompanied by friction and crushing between particles [31,32]. The stress–strain curves of pebble backfill with various gradations are shown in Figure 3, which exhibits certain volatility because of the movement and fragmentation of particles, showing a nonlinear growth trend similar to the coal gangue compression curve of different particle sizes [9]. Strain hardening occurs with increasing axial stress due to the compaction of pebble backfill. From the strain–stress curves of the pebble backfill with various gradations (Figure 3), it can be observed that a greater particle size indicated a greater strain value at the same stress level. The strain is caused by the shrinkage of gaps and the crushing between pebble particles during the compression of granular backfill. In the same volume, there are few contacts between larger particles. Intense stress concentration can be produced under the same loading, resulting in particle crushing, and then large deformation. Therefore, it can be concluded that the large particle size of granular material indicated large deformation and weak resistance to stress.

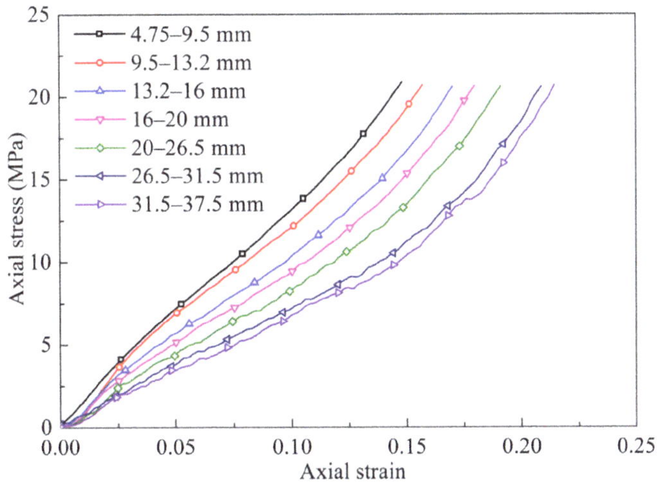

Figure 3. Stress–strain curves of pebbles with seven gradations.

2.4.2. Non-Constant Deformation Modulus

According to the stress–strain curves in Figure 3, it can be preliminarily determined that the deformation modulus (E_t) is a non-constant and changes with compression deformation. During the compression process of pebble backfill, the responses of particle crushing and moving were captured by the AE monitoring system and displayed in terms of released energy. The relationship between AE energy, deformation modulus (E_t), and the axial strain of the seven grades of pebble backfill were consistent, and Figure 4 shows this relationship for the pebble backfills with 4.75–9.5 mm, 16–20 mm, and 31.5–37.5 mm gradations. Based on the changes in AE energy, the deformation modulus (E_t) of the pebble backfills can be divided into the following three stages:

1. Rapid growth stage (Stage 1). The pebble backfill is initially in a loose and unstable state at lower axial stress, and AE energy is lower because particles have just rotated or moved. As axial stress increases, relatively stable contacts between particles within the pebble backfill are formed. The gaps between the particles shrink so rapidly that the deformation modulus (E_t) increases rapidly.
2. Diminishing stage (Stage 2). With axial stress gradually increasing, the crushing of many particles occurs inevitably due to stress concentration in this stage. A large amount of AE energy is released, and there are many high AE energy records that correspond to the strain level. As a result, the deformation modulus (E_t) decreases with increasing axial strain due to the crushing and softening of the pebbles [33].
3. Stable growth stage (Stage 3). With axial stress continuously increasing, the crushing and secondary crushing of particles are repeated and the gaps are filled by newly produced fine particles. Compared to the AE energy in stage 2, there is still a lot of AE energy released, but high AE energy records are reduced. This stage is primarily reflected by the strain hardening of the pebble backfill. Therefore, the deformation modulus (E_t) increases steadily with increasing axial strain.

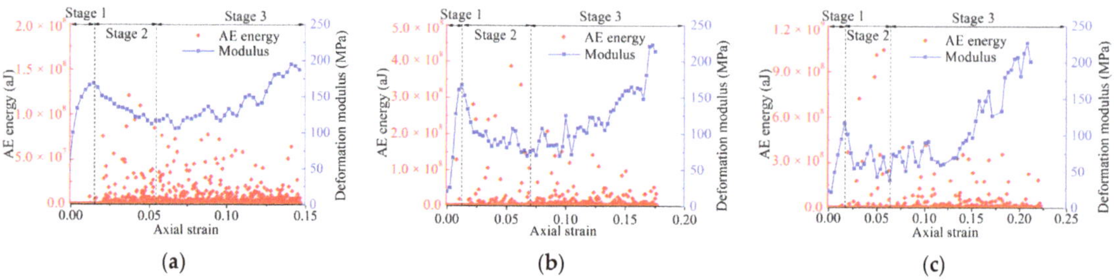

Figure 4. Deformation modulus (E_t) and AE energy of the pebbles in response to the axial strain: (**a**) 4.75–9.5 mm; (**b**) 16–20 mm; (**c**) 31.5–37.5 mm.

3. Numerical Model for Granular Material

3.1. Constitutive Model for Granular Backfill Based on Experimental Data

For the backfill mining, when granular backfill is filled into the mined-out area, the backfill is gradually compressed and shows strain hardening behavior due to the loading of overlying strata. The stress–strain curve of granular backfill is non-linear, and its deformation modulus is non-constant during the irreversible compaction (Figure 3). Therefore, this feature of the granular material should be considered in its constitutive model to describe the actual physical process of granular backfill. It is necessary for numerical simulations.

The deformation modulus (E_t) of the granular material is related to its composing structure as well as the elastic modulus of particles. During the compaction process, the change in the deformation modulus (E_t) of the granular material is strongly related to the density degree of the granular material. In this case, the volumetric strain (ε_v) of the granular material is an indicator of this. Therefore, the volumetric strain (ε_v) of the pebble backfill was selected as a variable to describe the deformation modulus (E_t). The relationship between the deformation modulus (E_t) and the volumetric strain (ε_v) of the pebble backfill can be established based on experimental data. The relationship of $E_t - \varepsilon_v$ of the pebble backfill with 4.75–9.5 mm, 16–20 mm, and 31.5–37.5 mm gradations is presented in Figure 5, and it can be fitted by the piecewise function (linear and exponential equations) of Equation (1).

$$E_t = \begin{cases} a + b\varepsilon_v & (\varepsilon_v < m_1) \\ c + d\varepsilon_v & (m_1 \leq \varepsilon_v \leq m_2) \\ e + k\exp(p\varepsilon_v) & (\varepsilon_v > m_2) \end{cases}, \quad (1)$$

where m_1 and m_2 represent the critical volumetric strain (ε_v) of stage 1, stage 2, and stage 3; a, b, c, d, e, k, and p are the experimental fitting coefficients. These coefficients can be obtained by fitting experimental data.

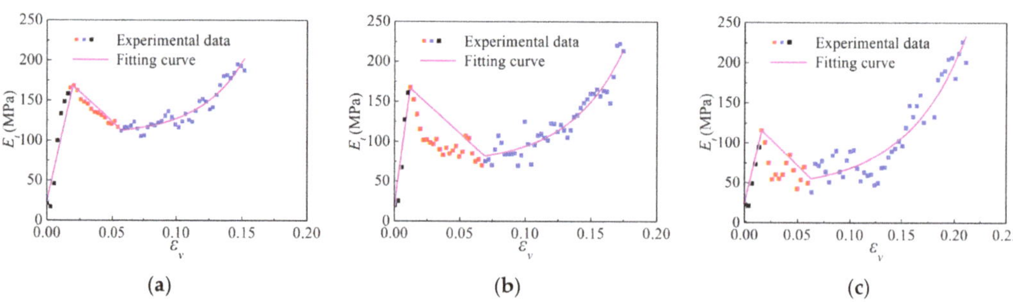

Figure 5. Relationship between E_t and ε_v. (**a**) 4.75–9.5 mm; (**b**) 16–20 mm; (**c**) 31.5–37.5 mm.

As the volumetric strain (ε_v) of the granular material increases when it is subject to external loading, the deformation modulus (E_t) of the granular material increases firstly, then decreases, and increases again. Equation (1) is used to describe the constitutive relationship for granular material to consider the dynamic mechanical property in numerical simulations.

3.2. Model Implementation

FLAC^{3D} is a three-dimensional explicit finite-difference program for engineering mechanics computation, and the explicit Lagrangian calculation scheme and the mixed-discretization zoning technique used in FLAC^{3D} ensure that plastic collapse and flow are modeled very accurately [34]. Therefore, the proposed constitutive model for granular material has been implemented into the FLAC^{3D} software using fish language to simulate the compressive behavior of the granular material. Its implementation in FLAC^{3D} is as follows:

1. During the numerical calculations, the volume strain (ε_v) of the granular material is monitored continuously, which usually increases gradually as granular material is compacted. According to the value of the volumetric strain (ε_v), an appropriate formula of piecewise function Equation (1) is selected to calculate the deformation modulus (E_t) of the granular material.
2. The deformation modulus (E_t) of the granular material is in a continuous iteration state due to the variation of the volume strain (ε_v). They are updated every ten time steps. The following calculation builds on the previous step in the numerical simulation of the granular material.

The numerical modeling procedure for granular material is demonstrated in Figure 6. Firstly, a numerical model of pebble backfill was built in FLAC^{3D}, and the same boundary conditions as the experiment were applied, i.e., the sides and bottom boundary of the model were constrained by displacement. Afterwards, based on the adopted Mohr–Coulomb criterion, the deformation modulus was assigned by the constitutive model for granular material. Thirdly, a velocity was applied to the top of the model to simulate the loading. Finally, the calculation continued until the axial strain and volume strain matched the experimental results by changing the time step.

3.3. Verification

To verify the model, compression tests of pebble backfill with seven gradations were simulated, corresponding to the experimental design in Table 1. The loading rate was 1 mm/min, and the x and z directions of the numerical model were fixed. It should be noted that the fitting coefficients and critical volume strain in the proposed constitutive model for pebble backfill (a, b, c, d, e, k, p, m_1, and m_2 in Equation (1)) were determined by fitting experimental data, as shown in Table 2.

Figure 7 shows the comparison of the axial stress–strain curves between experimental and numerical data. In Figure 7, the solid dotted lines denote the experimental stress–strain curves, while the hollow dotted lines denote numerical ones. It is observed that they were well matched, which indicated that numerical model for pebble backfill is capable of capturing the compressive behavior of the granular material under lateral confined conditions.

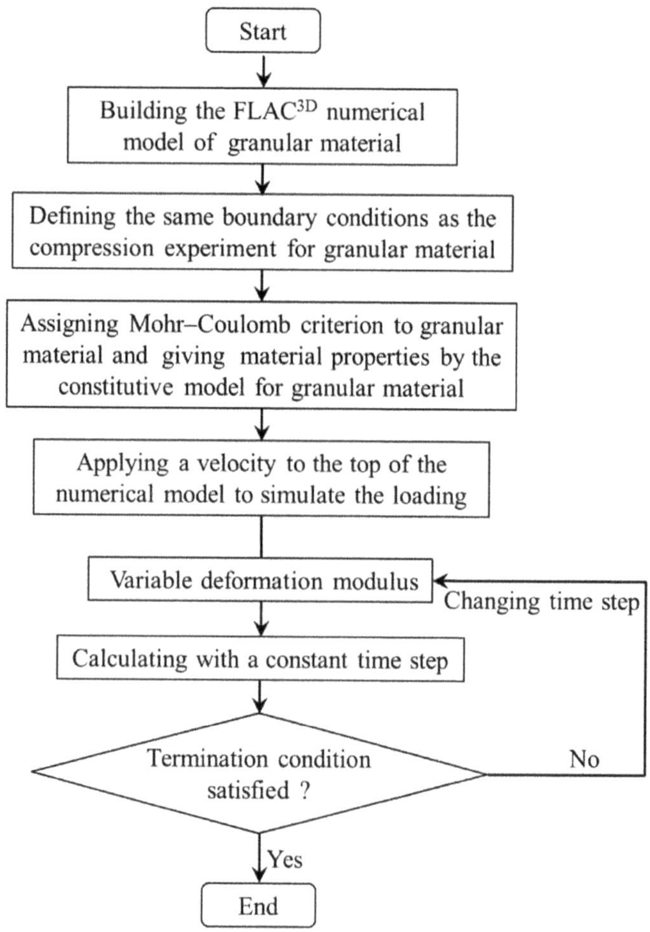

Figure 6. Numerical modeling procedure for granular material.

Table 2. Values of fitting coefficients and critical volume strain of seven gradations of granular backfill.

Pebble Gradations (mm)	Values of the Fitting Parameters and Critical Volume Strain								
	a	b	c	d	e	k	p	m_1	m_2
4.75–9.5	20.49	7836.44	191.13	−1352.77	107.18	1.01	29.83	0.01857	0.05787
9.5–13.2	14.88	6861.07	207.98	−1790.18	98.97	0.48	33.49	0.02232	0.05896
13.2–16	36.23	4554.79	167.02	−1239.95	84.43	1.11	28.45	0.02257	0.06146
16–20	18.85	3634.87	115.87	−601.65	66.71	2.72	22.77	0.0229	0.06281
20–26.5	24.26	2798.12	110.68	−525.93	69.58	1.43	24.96	0.026	0.06454
26.5–31.5	32.56	1796.91	102.38	−612.32	55.69	1.31	23.93	0.02898	0.06589
31.5–37.5	23.24	1232.09	62.64	−59.95	45.25	3.86	18.39	0.03049	0.06745

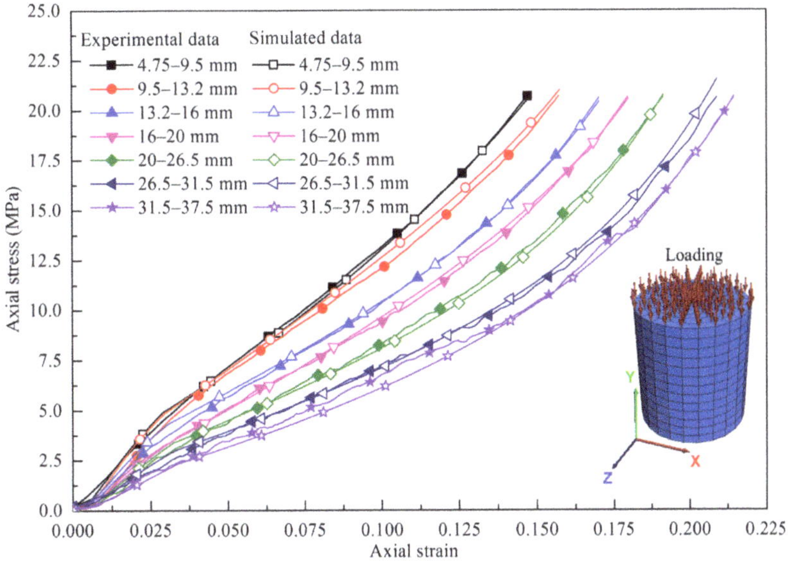

Figure 7. Comparison of the axial stress-axial strain curves obtained by experiment and simulation.

4. Application of Backfill Mining for Surface Subsidence

In this section, the proposed numerical model was used to analyze the surface subsidence for a mining operation where the mined-out stopes were filled with granular backfill. The effects of the particle size of the granular backfill and the height and depth of the mined-out stope on surface subsidence and movement were further discussed.

4.1. Numerical Simulation Model

The numerical model for surface subsidence analysis of backfill mining was set up as demonstrated in Figure 8. The dimensions are 1000 m × 450 m × 320 m (length × width × height). In light of the boundary effect, an orebody/stope with a 200 m × 50 m × 50 m (length × width × height) is located in the center of the model at a depth of 240 m. It consists of five groups of rock strata with different heights. There are 1,152,000 elements and 1,188,915 nodes for the numerical model, and the size of elements is about 5 m × 5 m × 5 m. The lateral and bottom boundaries of the model were fixed on the corresponding normal directions, and the top of the model was a free boundary. Mohr–Coulomb criterion was used for the rock strata, and the mechanical properties and thickness of the rock strata are listed in Table 3. They were assigned following the work of Cheng and Qiao [35].

Table 3. Properties of materials in the ideal numerical model.

Lithology	Bulk Modulus (GPa)	Shear Modulus (GPa)	Tensile Strength (MPa)	Cohesion (MPa)	Internal Friction angle (°)	Thickness (m)
Topsoil	0.08	0.02	0.0003	0.007	20	20
Peridotite	0.25	0.12	0.038	0.102	38	90
Slate	0.55	0.37	0.043	0.112	32	60
Migmatite	0.66	0.39	0.052	0.131	49	120
Marble	0.81	0.41	0.091	0.282	55	30
Granular backfill	Iterating	Iterating	0	0	18	-

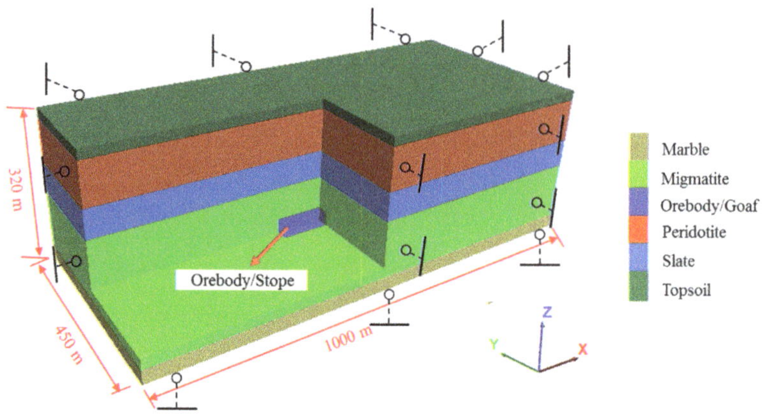

Figure 8. Numerical model in FLAC^{3D}.

The modeling procedure is as follows: the states of original rock stress in the model were first analyzed, and the initial displacement was cleared; then, the ore body was excavated, running 20 time steps. Afterwards, the mined-out area was filled with granular backfill, and the proposed constitutive model was applied for granular backfill. The values of the bulk modulus and shear modulus of the backfill were updated iteratively according to the volumetric strain of the granular backfill until the model reached equilibrium. For comparison, the double-yield model was also used for granular backfill to analyze surface subsidence.

Figure 9 shows the comparison of the surface subsidence and horizontal movement between the proposed model and double-yield model for granular backfill with particle sizes of 4.75–9.5 mm and 31.5–37.5 mm. The values of surface subsidence and horizontal movement obtained by the proposed model for granular material were greater than those obtained by the double-yield model after the mined-out stope was filled with granular backfill with particle sizes of 4.75–9.5 mm and 31.5–37.5 mm. This occurred because the constant deformation modulus of the granular backfill in the whole compaction process was assigned in the double-yield model and it cannot capture the large deformation of granular backfill in the initial compression stage [5]. However, the proposed model with variable deformation modulus overcomes this disadvantage and has the capability of describing the real physical characteristics of the granular material under compaction. Therefore, the results obtained by the proposed model for granular material should be more realistic.

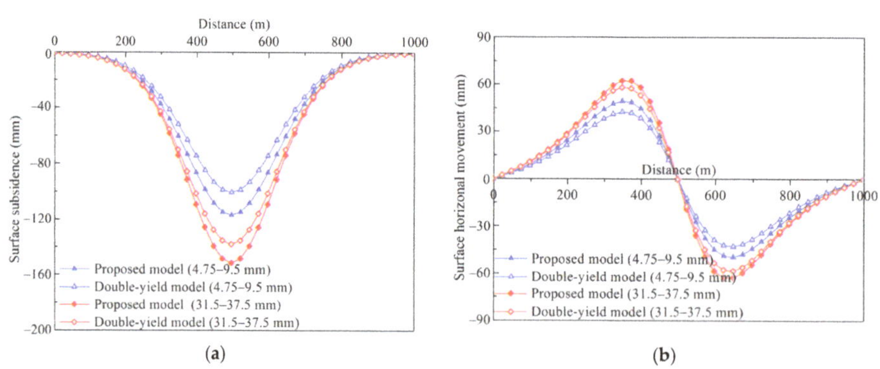

Figure 9. Comparison of the surface subsidence and horizontal movement obtained by the proposed model and the double-yield model: (**a**) Surface subsidence; (**b**) Surface horizontal movement.

4.2. Effect of Particle Size of the Granular Backfill on Surface Subsidence and Movement

Excavation causes tensile stress concentration in the surrounding rock of the stope (Figure 10a). After backfilling, the stress field changed slightly and most of the tensile zones disappeared (Figure 10b). Figure 11 presents the subsidence of overlying strata in the cases that the mined-out stope was unfilled and filled with pebbles of 4.75–9.5 mm. The subsidence values at the levels of 80 m (stope roof), 150 m, 210 m, 260 m, 300 m, and 320 m (surface) in the model (Figure 8) were monitored. It was found that the subsidence of the rock strata was distributed uniformly from the stope roof to the surface, and backfilling played an important role in restraining the subsidence of the overlying strata and surface.

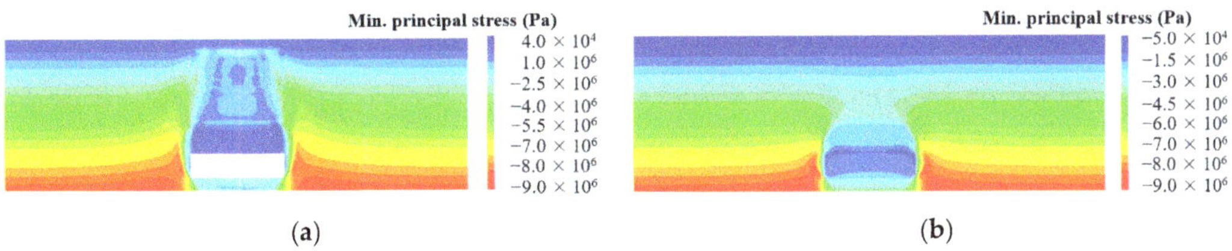

Figure 10. Stress nephogram along the length: (**a**) Mined-out stope unfilled; (**b**) Mined-out stope filled with pebbles of 4.75–9.5 mm.

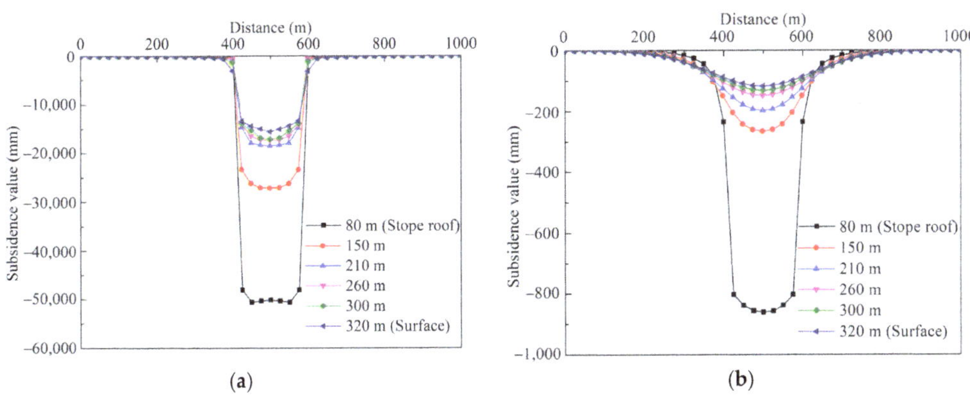

Figure 11. Subsidence of overlying strata: (**a**) Mined-out stope unfilled; (**b**) Mined-out stope filled with pebbles of 4.75–9.5 mm.

The curves of surface subsidence and horizontal movement were depicted after the mined-out stope was filled with the granular backfill with various particle sizes, as shown in Figure 12. As the particle size of the granular backfill increased, the values of surface subsidence and horizontal movement increased, which corresponds to the experimental result that the larger the particle size, the weaker the resistance to deformation (Figure 3).

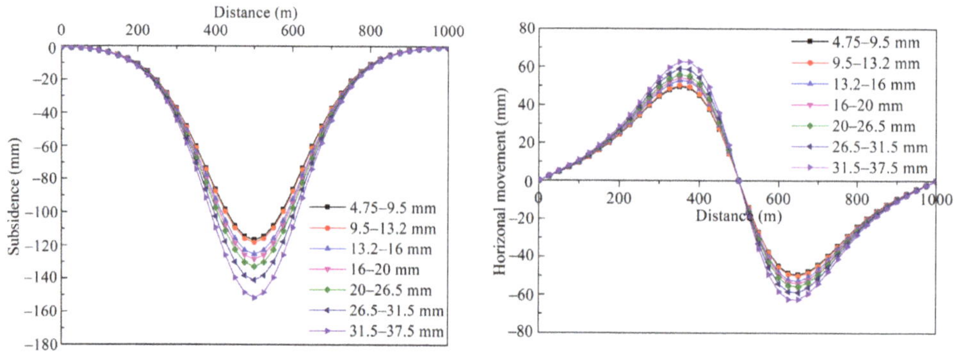

(a) (b)

Figure 12. Surface subsidence and horizontal movement with different particle sizes: (**a**) Subsidence; (**b**) Horizontal movement.

The indicators reflecting the characteristics of surface subsidence and movement include inclination, curvature, and horizontal deformation. Based on the values of surface subsidence and horizontal movement (Figure 12), they can be obtained from Equations (2)–(4), respectively [36].

Surface inclination deformation:

$$i_{n\sim n-1} = \frac{W_n - W_{n-1}}{L_{n\sim n-1}}, \tag{2}$$

where $i_{n\sim n-1}$ is the surface inclined deformation between points n and $n-1$; W_n and W_{n-1} are the surface subsidence of points n and $n-1$, respectively; $L_{n\sim n-1}$ is the horizontal distance between points n and $n-1$.

Surface curvature deformation:

$$K_{n+1\sim n\sim n-1} = \frac{i_{n+1\sim n} - i_{n\sim n-1}}{\frac{L_{n+1\sim n}+L_{n\sim n-1}}{2}}, \tag{3}$$

where $K_{n+1\sim n\sim n-1}$ is the surface curvature deformation between points n, $n-1$ and $n+1$.

Surface horizontal deformation:

$$\varepsilon_n = \frac{U_n - U_{n-1}}{L_{n\sim n-1}}, \tag{4}$$

where ε_n is the surface horizontal deformation between points n and $n-1$; U_n and U_{n-1} are the surface horizontal movement of points n and $n-1$, respectively.

Consequently, the surface inclination, curvature, and horizontal deformation in the cases that the mined-out stope was filled with seven gradations of granular backfill were calculated, and the corresponding curves are shown in Figure 13. The values of surface inclination, curvature, and horizontal deformation also increased with increasing particle size of the granular backfill. This was determined by the compressive performance of the granular backfill with various particle sizes under the loads imposed by the overlying strata.

In the numerical calculation, the value of the deformation modulus of the granular backfill is in a continuous iteration state and varies with Equation (1). The deformation modulus has an important influence on surface subsidence, especially for deeper workings. Figure 14 presents the correlation between the maximal surface subsidence and final deformation modulus of the granular backfill with seven gradations. It can be concluded that the surface subsidence decreases with increasing deformation modulus, and this can be described by linear equations; the correlation coefficient was 0.8885.

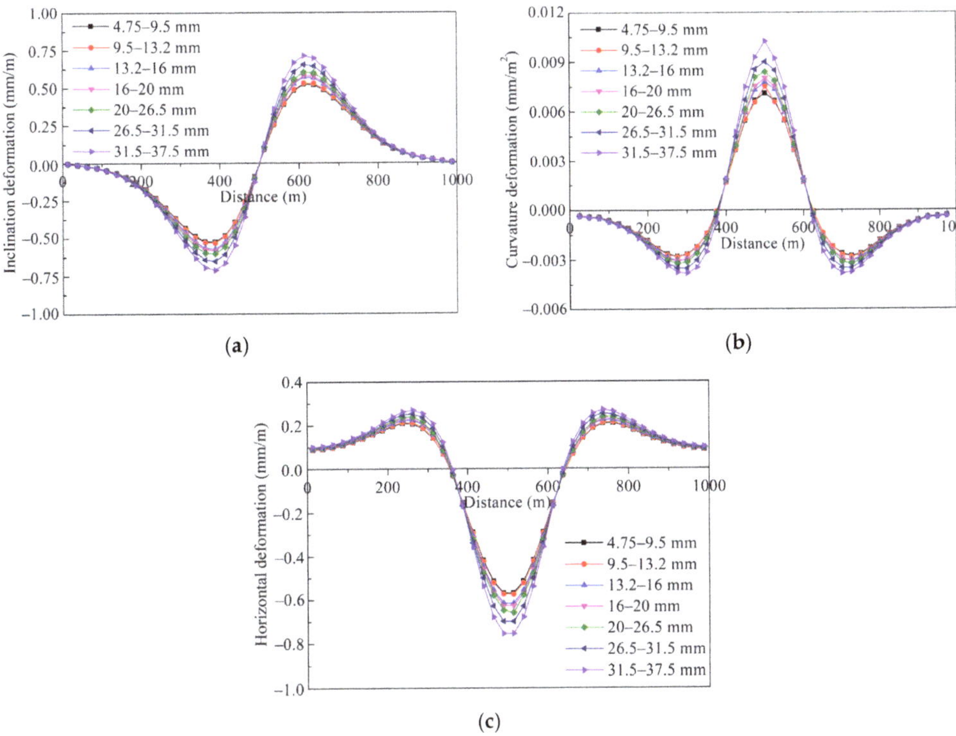

Figure 13. Characteristics of surface subsidence and movement with different particle sizes: (**a**) Inclination deformation; (**b**) Curvature deformation; (**c**) Horizontal deformation.

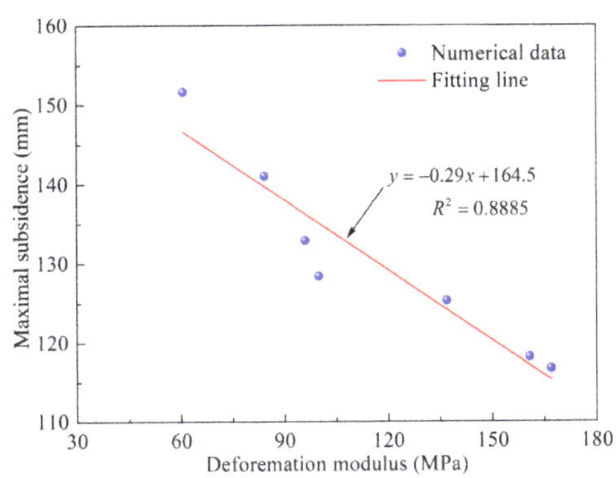

Figure 14. Relationship between maximal surface subsidence and deformation modulus of the granular backfill.

4.3. Effect of the Particle Size of the Granular Backfill on the Surface Subsidence Basin

In the mining operation, the surface subsidence basin has an important influence on the layout and stability of surface buildings. The surface subsidence basin was analyzed to determine the accurate influence range of surface subsidence after the mined-out stope was filled with granular backfill. Figure 15 shows the distribution of the surface subsidence for the granular backfill with various particle sizes. The influence area with various particle sizes was almost identical in the plane with a size of 1000 m × 450 m, while the extreme value of the surface subsidence basin increased with increasing particle size. Based on the criteria of the surface inclination, curvature, and horizontal deformation of surface buildings as shown in Table 4 [37], the maximum surface inclination, curvature, and horizontal deformation were less than the destruction extrema of damage level I of surface buildings. Therefore, the stability of surface buildings was not affected by filling mining in this case.

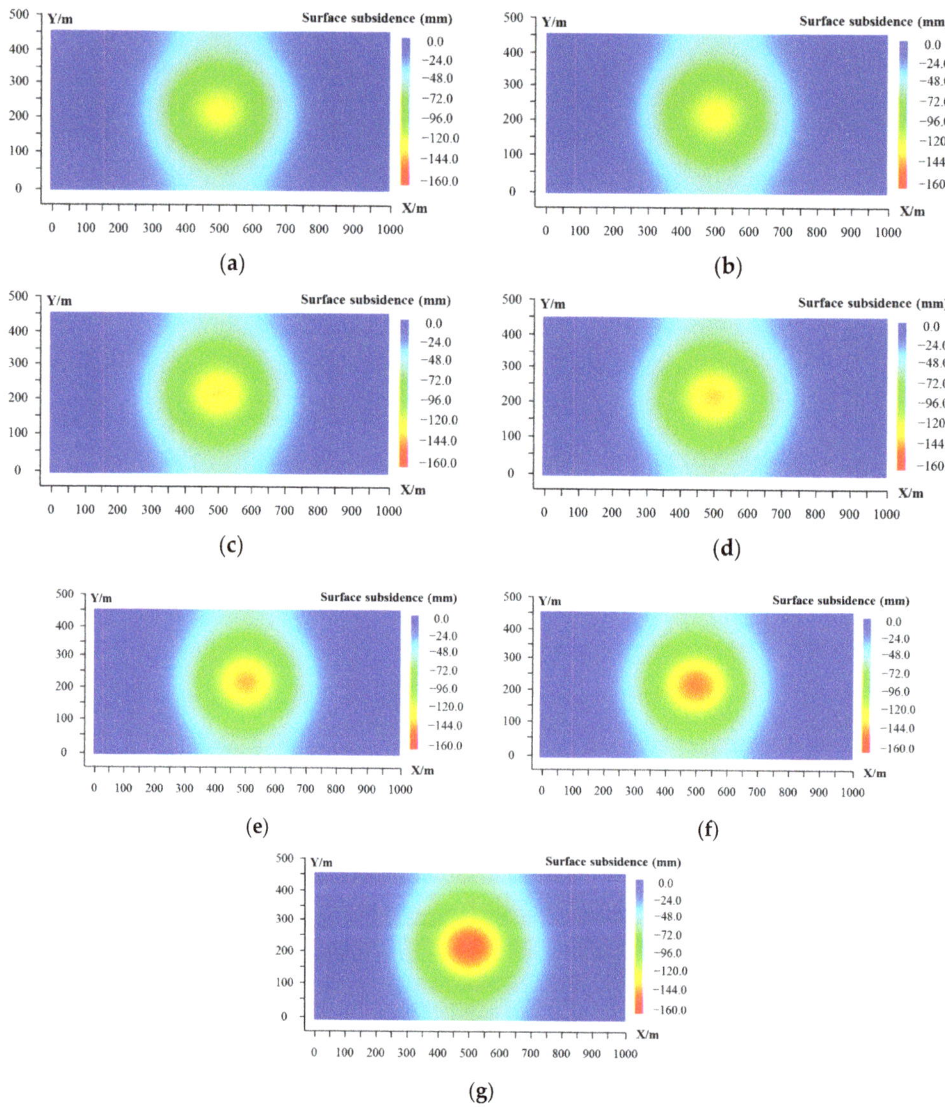

Figure 15. Surface subsidence basins for the granular backfill with various particle sizes: (**a**) 4.75–9.5 mm, (**b**) 9.5–13.2 mm, (**c**) 13.2–16 mm, (**d**) 16–20 mm, (**e**) 20–26.5 mm, (**f**) 26.5–31.5 mm, (**g**) 31.5–37.5 mm.

Table 4. Damage levels of surface buildings [37].

Damage Levels	Inclination Deformation (mm/m)	Curvature Deformation (mm/m^2)	Horizontal Deformation (mm/m)	Damage Classification
I	≤3.0	≤0.2	≤2.0	Very slight damage
II	≤6.0	≤0.4	≤4.0	Slight damage
III	≤10.0	≤0.6	≤6.0	Moderate damage
IV	>10.0	>0.6	>6.0	Heavy damage

4.4. Effect of the Height of the Mined-Out Stope on Surface Subsidence and Movement

In this section, the effect of the height of mined-out stope on surface subsidence and movement in the backfilling case is discussed using the proposed model. Based on the numerical model in Figure 8, the heights of the stope were set to be 40 m, 30 m, 20 m, and 10 m. The baseline for the heights was the stope roof, i.e., the buried depth of the stope being 240 m.

Figure 16 presents the maximal value of surface subsidence and movement in the backfilling cases with different particle sizes. With increasing height of the stope, the range of the fracture zone, crack zone, and bending zone of overlying strata increased, and the maximal values of the surface subsidence, horizontal movement, inclination, curvature, and horizontal deformation also increased. For the optimum gradation with 4.75–9.5 mm, as the height of the stope increased from 10 m to 50 m, the maximal inclination (Figure 16c), curvature (Figure 16d), and horizontal deformation (Figure 16e) increased by 177.9%, 180.2%, and 189.4%, respectively. For the case of the stope height of 50 m, as the particle size of the backfill increased from 4.75–9.5 mm to 31.5–37.5 mm, the maximal inclination, curvature, and horizontal deformation increased by 33.2%, 44.4%, and 39.3%, respectively, and the effect of the particle size of the granular backfill on surface subsidence increased with an increase in the stope height.

4.5. Effect of the Depth of the Mined-Out Stope on Surface Subsidence and Movement

Based on the numerical model in Figure 8, the depth of the stope was set to be 270 m, 300 m, 330 m, and 360 m. The effect of the depth of the mined-out stope on surface subsidence and movement in the backfilling case was discussed using the proposed model. The simulation results for the cases with various grades of granular backfill are shown in Figure 17. Consequently, the maximal values of surface subsidence, horizontal movement, inclination, curvature, and horizontal deformation decreased with increases in the depth of the stope. For the optimum gradation with 4.75–9.5 mm, as the depth of the stope increased from 240 m to 360 m, the maximal inclination (Figure 17c), curvature (Figure 17d), and horizontal deformation (Figure 17e) decreased by 41.0%, 52.2%, and 37.7%, respectively. For the stope depth of 360 m, as the particle size of the backfill decreased from 31.5–37.5 mm to 4.75–9.5 mm, the maximal inclination, curvature, and horizontal deformation decreased by 17.0%, 17.1%, and 18.1%, respectively, and the effect of the particle size of the granular backfill on surface subsidence was reduced by increasing the buried depth of the stope.

Admittedly, the above application is an ideal case, similar to coal mines with horizontal seams. Therefore, the analysis for surface subsidence may not be difficult. However, metal mines are often designed by multiple level mining, and the underground workings are complex. For example, a metal mine with a thick and steep orebody is designed by cut-and-fill mining. In this case, the proposed model that can capture the actual bearing process of the granular backfill may provide reliable results for the analysis for surface subsidence. The proposed model is likely to be applied widely for metal mines.

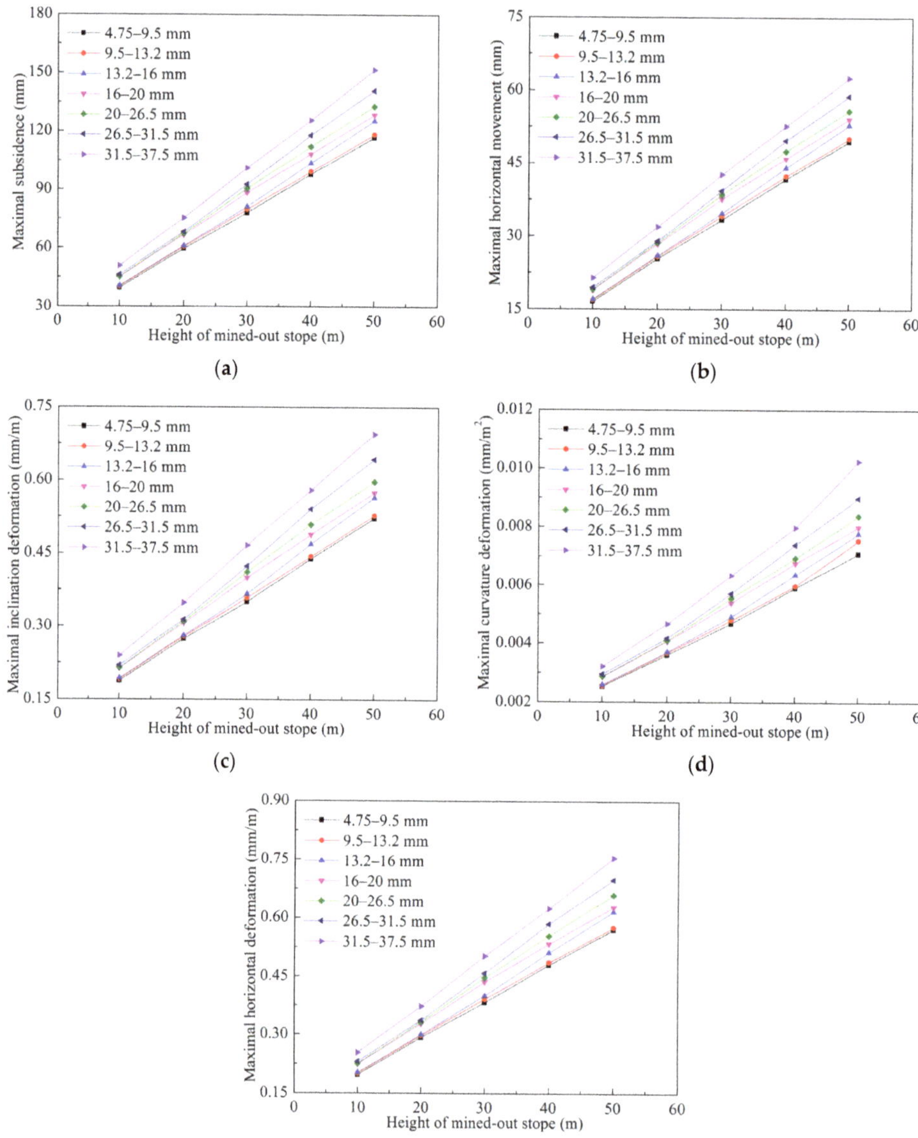

Figure 16. Maximal value of surface subsidence and movement in the backfilling cases (height of the stope) with different particle sizes: (**a**) Surface subsidence; (**b**) Horizontal movement; (**c**) Inclination deformation; (**d**) Curvature deformation; (**e**) Horizontal deformation.

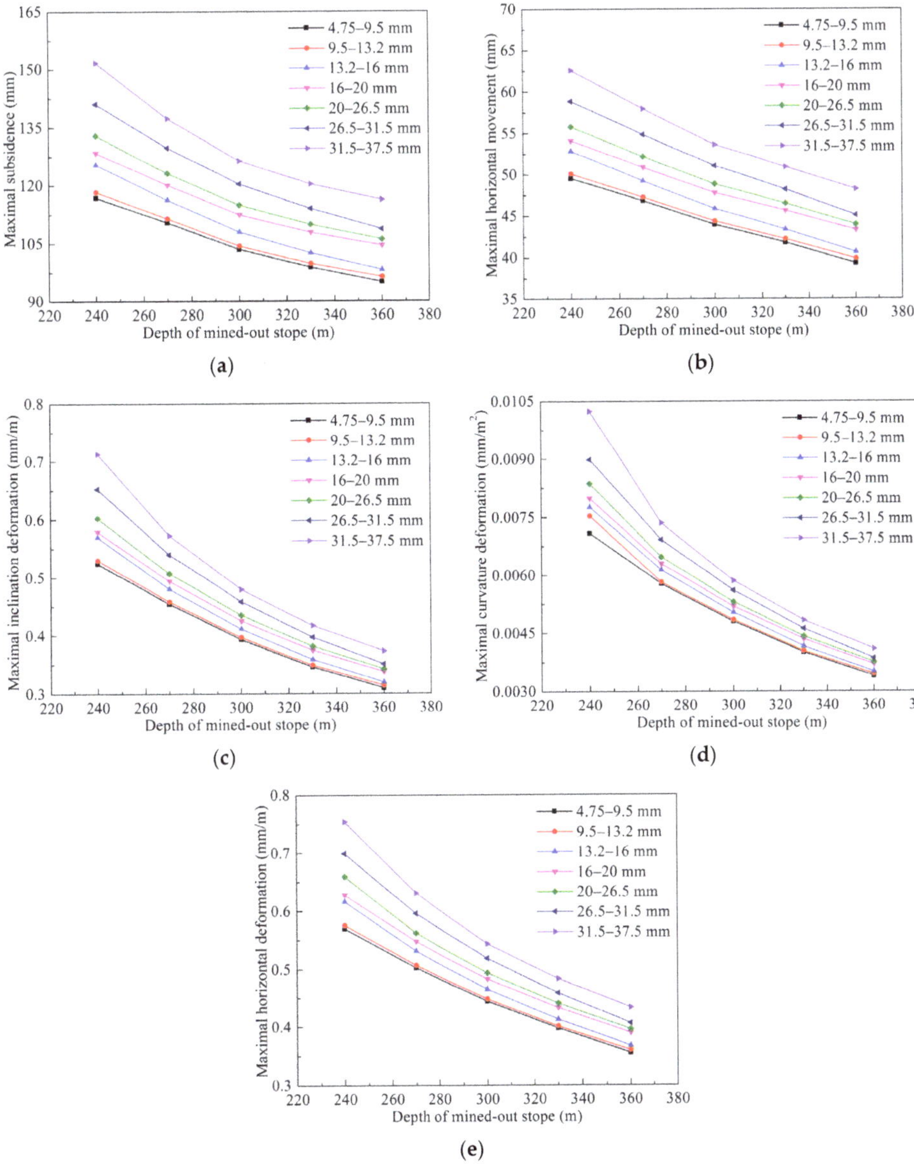

Figure 17. Maximal value of surface subsidence and movement in the backfilling cases (depth of the stope) with different particle sizes: (**a**) Surface subsidence; (**b**) Horizontal movement; (**c**) Inclination deformation; (**d**) Curvature deformation; (**e**) Horizontal deformation.

5. Conclusions

In this study, the bearing characteristics of pebbles with various gradations were investigated by a self-designed experimental apparatus. A constitutive model with variable deformation modulus was proposed for granular material and implemented into FLAC^{3D} software based on the experimental data. The proposed model was preliminarily applied to analyze the surface subsidence in backfilling mines. The influential factors of the surface subsidence and movement were discussed. The following conclusions can be drawn:

1. The change in the deformation modulus is the main feature of granular backfill during the compaction process. The deformation modulus of the granular backfill firstly increases with compaction, then decreases along with some large or sharp particles being fragmented due to contact stress concentration between them, and again increases along with the adequately composed backfill being compacted again.
2. The proposed constitutive model with variable deformation modulus can capture the physical characteristics of the granular material during the compaction process. It was implemented into the FLAC^{3D} software, which effectively explores the surface subsidence in backfilling mines.
3. The effect of the particle size of the granular backfill, stope height, and stope buried depth on surface subsidence was revealed using the proposed model for backfilling mines. The effect of the particle size of the granular backfill on surface subsidence can be enlarged by increasing the stope height and can be reduced by increasing the buried depth of the stope. This study provides references for the design of granular backfilling mines.

Author Contributions: Funding acquisition, Q.-L.Y.; Methodology, Q.-L.Y. and Z.-H.L.; Formal analysis, Z.-H.L.; Software, Z.-H.L. and J.-Y.P.; Writing—original draft preparation, Z.-H.L.; Investigation, Z.-H.L., J.-Y.P., Y.-S.C., K.L.; Writing—review and editing, Q.-L.Y. and J.-Y.P. All authors have read and agreed to the published version of the manuscript.

Funding: This research was funded by the Joint Funds of the National Natural Science Foundation of China (Grant No. U21A20106) and the Fundamental Research Funds for the Central Universities (Grant No. N2101036 and No. N2101045). The APC was funded by the National Science Foundation of China (Grant No. 52074060).

Institutional Review Board Statement: Not applicable.

Informed Consent Statement: Not applicable.

Data Availability Statement: All data and models of this study are available from the corresponding author upon reasonable request.

Conflicts of Interest: The authors declare no conflict of interest.

References

1. Li, X.; Wang, S.J.; Liu, T.Y.; Ma, F.S. Engineering geology, ground surface movement and fissures induced by underground mining in the Jinchuan Nickel Mine. *Eng. Geol.* **2004**, *76*, 93–107. [CrossRef]
2. Ma, F.S.; Zhao, H.J.; Yuan, R.M.; Guo, J. Ground movement resulting from underground backfill mining in a nickel mine (Gansu Province, China). *Nat. Hazards* **2015**, *77*, 1475–1490. [CrossRef]
3. El-Sohby, M.A.; Andrawes, K.Z. Deformation characteristics of granular materials under hydrostatic compression. *Can. Geotech. J.* **1972**, *9*, 338–350. [CrossRef]
4. Su, C.D.; Gu, M.; Tang, X.; Guo, W.B. Experiment study of compaction characteristics of crushed stones from coal seam roof. *Chin. J. Rock Mech. Eng.* **2012**, *31*, 18–26. (In Chinese)
5. Liu, A.; Liu, S.M.; Wang, G.; Elsworth, D. Continuous compaction and permeability evolution in longwall gob materials. *Rock Mech. Rock Eng.* **2020**, *53*, 5489–5510. [CrossRef]
6. Eslami, A.; Ahmadnezhad, M.; Amaneh, E.K.; Rezazadeh, S. Experimental study on stiffness properties of non-engineered clay and granular fills. *Arab. J. Geosci.* **2014**, *8*, 3065–3075. [CrossRef]
7. Zhang, B.Y.; Chen, T.; Peng, C.; Qian, X.X.; Jie, Y.X. Experimental study on loading-creep coupling effect in rockfill material. *Int. J. Geomech.* **2017**, *17*, 04017059. [CrossRef]
8. Li, M.; Zhang, J.X.; Wu, Z.Y.; Liu, Y.; Li, A.L. An experimental study of the influence of lithology on compaction behaviour of broken waste rock in coal mine backfill. *R. Soc. Open Sci.* **2019**, *6*, 182205. [CrossRef] [PubMed]
9. Zhou, N.; Han, X.L.; Zhang, J.X.; Li, M. Compressive deformation and energy dissipation of crushed coal gangue. *Powder Technol.* **2016**, *297*, 220–228. [CrossRef]
10. Meng, G.H.; Zhang, J.X.; Li, M.; Zhu, C.L.; Zhang, Q. Prediction of compression and deformation behaviours of gangue backfill materials under multi-factor coupling effects for strata control and pollution reduction. *Environ. Sci. Pollut. Res. Int.* **2020**, *27*, 36528–36540. [CrossRef]

11. Aminzadeh, A.; Hosseininia, E.S. A study on the effect of particle shape and fragmentation on the mechanical behavior of granular materials using discrete element method. *Powders Grains* **2013**, *1542*, 915–918.
12. Peng, H.H.; Lin, C.K.; Chung, Y.C. Effects of particle stiffness on mechanical response of granular solid under confined compression. *Procedia Eng.* **2014**, *79*, 143–152. [CrossRef]
13. Zhang, J.X.; Li, M.; Liu, Z.; Zhou, N. Fractal characteristics of crushed particles of coal gangue under compaction. *Powder Technol.* **2017**, *305*, 12–18. [CrossRef]
14. Li, M.; Li, A.L.; Zhang, J.X.; Huang, Y.L.; Li, J.M. Effects of particle sizes on compressive deformation and particle breakage of gangue used for coal mine goaf backfill. *Powder Technol.* **2020**, *360*, 493–502. [CrossRef]
15. Kiran, K.V.; Madhav, M.R.; Wissmann, K.J. Method for estimation of deformation modulus of granular pile from settlements at different depths. In Proceedings of the Sixth Indian Young Geotechnical Engineers Conference, NIT, Trichy, India, 10–11 March 2017; pp. 1–7.
16. Sharma, J.K.; Gupta, P. Analysis and settlement evaluation of an end-bearing granular pile with non-linear deformation modulus. *Studia Geotech. Et Mech.* **2018**, *40*, 188–201. [CrossRef]
17. Gui, Y.; Tao, H.; Zhu, P.N.; Yang, H.H.; Deng, T.F. Contrastive studies of the testing method for the bearing capacity and deformation modulus of dynamic consolidation backfill foundation. *Adv. Mat. Res.* **2012**, *446-449*, 1606–1614.
18. Li, M.; Zhang, J.X.; Gao, R. Compression characteristics of solid wastes as backfill materials. *Adv. Mater. Sci. Eng.* **2016**, *2016*, 2496194. [CrossRef]
19. Zhang, P.F.; Zhang, Y.B.; Zhao, T.B.; Tan, Y.L.; Yu, F.H. Experimental research on deformation characteristics of waste-rock material in ynderground backfill mining. *Minerals* **2019**, *9*, 102. [CrossRef]
20. Yu, W.J.; Feng, T.; Chen, X.Y. Law of ground subsidence and strata movement caused by backfill mining of "under three" coal. *Adv. Mater. Res.* **2013**, *807–809*, 2304–2308. [CrossRef]
21. Wang, H.W.; Wu, Y.P.; Liu, M.F.; Jiao, J.Q.; Luo, S.H. Roof-breaking mechanism and stress-evolution characteristics in partial backfill mining of steeply inclined seams. *Geomat. Nat. Haz. Risk.* **2020**, *11*, 2006–2035. [CrossRef]
22. Pu, H.; Zhang, J. Research on protecting the safety of buildings by using backfill mining with solid. *Procedia Environ. Sci.* **2012**, *12*, 191–198.
23. Li, J.; Zhang, J.X.; Huang, Y.L.; Zhang, Q.; Xu, J.M. An investigation of surface deformation after fully mechanized, solid back fill mining. *Int. J. Min. Sci. Techno.* **2012**, *22*, 453–457. [CrossRef]
24. Yavuz, H. An estimation method for cover pressure re-establishment distance and pressure distribution in the goaf of longwall coal mines. *Int. J. Rock Mech. Min. Sci.* **2004**, *41*, 193–205. [CrossRef]
25. Li, M.; Zhang, J.X.; Huang, Y.L.; Zhou, N. Effects of particle size of crushed gangue backfill materials on surface subsidence and its application under buildings. *Environ. Earth Sci.* **2017**, *76*, 603. [CrossRef]
26. Zhu, X.J.; Guo, G.L.; Wang, J.; Fang, Q.; Chen, T. Analysis of strata and ground subsidence in fully mechanized solid backfilling mining: A case study of Huayuan coal mine. *Min. Technol.* **2016**, *125*, 233–241. [CrossRef]
27. Das, S.K.; Das, A. Influence of quasi-static loading rates on crushable granular materials: A DEM analysis—ScienceDirect. *Powder Technol.* **2019**, *344*, 393–403. [CrossRef]
28. Hu, B.N.; Guo, A.G. Testing study on coal waste back filling material compression simulation. *J. China Coal Soc.* **2009**, *34*, 1076–1080. (In Chinese)
29. Wang, W.; Li, H.M.; Xiong, Z.Q.; Su, Y.S.; Zhang, P. Research on the influence of diameter gradation on compressive deformation characteristics of gangues. *China J. Undergr. Space Eng.* **2016**, *12*, 1553–1558. (In Chinese)
30. Yu, B.Y.; Chen, Z.Q.; Wu, J.Y. Experimental study on compaction and fractal characteristics of saturated broken rocks with different initial gradations. *J. Min. Saf. Eng.* **2016**, *33*, 342–347. (In Chinese)
31. Li, J.M.; Huang, Y.L.; Chen, Z.W.; Li, M.; Qiao, M.; Kizil, M. Particle-Crushing characteristics and acoustic-emission patterns of crushing gangue backfilling material under cyclic loading. *Minerals* **2018**, *8*, 244. [CrossRef]
32. Sitharam, T.G.; Nimbkar, M.S. Micromechanical modelling of granular materials: Effect of particle size and gradation. *Geotech. Geol. Eng.* **2000**, *18*, 91–117. [CrossRef]
33. Soroush, A.; Jannatiaghdam, R. Behavior of rockfill materials in triaxial compression testing. *Int. J. Civ. Eng.* **2012**, *10*, 153–161.
34. Itasca Consulting Group, Inc. *Fast Language Analysis of Continua in 3 Dimensions, Version 3.0, User Manual*; Itasca Consulting Group, Inc.: Minneapolis, MN, USA, 2005.
35. Cheng, H.Y.; Qiao, D.P. Discussion on filling mining of Longshou Mine in Jinchuan and the laws of surface subsidence. *Metal Mine* **2012**, *11*, 32–35. (In Chinese)
36. Li, H.Z.; Guo, G.L.; Zhai, S.C. Mining scheme design for super-high water backfill strip mining under buildings: A Chinese case study. *Environ. Earth Sci.* **2016**, *75*, 1017. [CrossRef]
37. Yao, W.H.; Song, X.M. Study on impact of mining subsidence on ground buildings. *China Coal.* **2013**, *39*, 57–59. (In Chinese)

metals

MDPI

Article

Copper and Zinc Recovery from Sulfide Concentrate by Novel Artificial Microbial Community

Xinglan Cui [1,2,3,4,†], Xuetao Yuan [1,2,†], Hongxia Li [1,2,3,4], Xiaokui Che [1,2], Juan Zhong [1,2,3,4], Lei Wang [1,2,3,4], Ying Liu [1,2,*], Xuewu Hu [1,2,5], Qidong Zhang [1,2], Rongzhen Jin [1,2,5] and Qi Zheng [1,2]

1 The National Engineering Research Center for Environment-Friendly Metallurgy in Producing Premium Non-Ferrous Metals, GRINM Resources and Environmental Technology Corporation Limited, Beijing 101407, China; cuixinglan@grinm.com (X.C.); yuanxt@grinm.com (X.Y.); lihongxia@grinm.com (H.L.); xkche@grinm.com (X.C.); zhongjuan@grinm.com (J.Z.); wanglei@grinm.com (L.W.); b20170142@xs.ustb.edu.cn (X.H.); zhangqidong@grinm.com (Q.Z.); kimyeongjin@hotmail.com (R.J.); Mr1311@grinm.com (Q.Z.)

2 GRINM Group Corporation Limited, Beijing 100088, China

3 GRIMAT Engineering Institute Corporation Limited, Beijing 100088, China

4 State Key Laboratory of Vanadium and Titanium Resources Comprehensive Utilization, Panzhihua 617000, China

5 Beijing Key Laboratory of Resource-Oriented Treatment of Industrial Pollutants, University of Science and Technology Beijing, Beijing 100083, China

* Correspondence: rddly@grinm.com

† These authors contributed equally to this work.

Abstract: Exploring efficient methods to enhance leaching efficiency is critical for bioleaching technology to deal with sulfide concentrate. In our study, a novel artificial microbial community was established to augment the bioleaching efficiency and recovery of copper (Cu) and zinc (Zn). The optimum parameters in bioleaching experiments were explored according to compare a series of conditions from gradient experiments: the pH value was 1.2, temperature was 45 °C, and rotation speed was 160 r/min, which were different with pure microorganism growth conditions. Under optimal conditions, the result of recovery for Cu and Zn indicated that the average leaching rate reached to 80% and 100% respectively, which almost increased 1.8 times and 1.2 times more than control (aseptic condition) group. Therefore, this method of Cu and Zn recovery using a new-type artificial microbial community is expected to be an environmentally-friendly and efficient bioleaching technology solution, which has the potential of large-field engineering application in the future.

Keywords: copper; zinc; recovery; sulfide concentrate; artificial microbial community

Citation: Cui, X.; Yuan, X.; Li, H.; Che, X.; Zhong, J.; Wang, L.; Liu, Y.; Hu, X.; Zhang, Q.; Jin, R.; et al. Copper and Zinc Recovery from Sulfide Concentrate by Novel Artificial Microbial Community. *Metals* **2022**, *12*, 45. https://doi.org/10.3390/met12010045

Academic Editor: Antoni Roca

Received: 7 October 2021
Accepted: 22 December 2021
Published: 25 December 2021

1. Introduction

Mineral resources are the economic and material basis of human society. In recent years, with the proposal of green mine development concept, efficient exploitation of mineral resources and reduction of ecological environmental impact have been important principles for future exploitation and utilization of mineral resources [1]. However, most mines in China have backward mining technology, enormous destruction and waste, too low recycling proportion, and low management efficiency [2]. As is known to us, with continuous decline in copper rich ore reserves, reprocessing of low-grade copper-bearing sulfide ores, accounting for >70% of the global copper reserves, has become inevitable [3–5]. Unfortunately, metal extraction from these low-grade ores using traditional smelting techniques is uneconomical. Therefore, most low-grade ores have been discarded [6–8].

In the last two decades, extensive efforts have been made to apply microbes to sulfides leaching due to demand of ores and low costs [9,10]. Biological processes are effective methods and the interaction between microorganisms and minerals has been a hot topic to reveal the transformation of key elements affecting bioleaching efficiency [11,12]. Moreover, bioleaching attracts increasingly more attention for its simplified operation, low cost, and environmentally friendly benefits [13,14]. In the bioleaching process, structural metals in solid materials are transformed into soluble and extractable ones by microorganisms through biological oxidation or complexation processes for easy recovery [15]. Bioleaching has been actively studied for years to deal with metal-contaminated sludge and environments, such as sediment and soil, and the removal efficiency of most heavy metals is usually higher than 85% [16–19]. Bioleaching facilitates metal mobilization from solid sources via different biologically catalyzed reactions mediated by different microbial leaching agents such as (in)organic acids. A wide range of microorganisms such as chemolithoautotrophic bacteria, archaea, and fungi have been applied for bioleaching of metals from different solid materials [20–23].

In the bioleaching technology, microorganisms, such as *Acidithiobacillus ferrooxidans, Acidithiobacillus thiooxidans, Acidithiobacillus caldus, Leptospirillum ferrooxidans,* and *Ferroplasma* spp., are often employed to enhance the dissolution process of copper sulfide ores [24]. The detailed bioleaching efficiency for metals of several microbes is listed in Table 1. Obviously, the community which consists of *Acidithiobacillus thiooxidans*, *Acidithiobacillus ferrooxidans,* and *Leptospirillum ferriphilum* had lower leaching efficiency for Cu and Zn and pH value was 2.0–2.2 in 25 days, which only reached 69–83% and 4.1–14% respectively [25]. Further research used two bacteria including *Acidithiobacillus thiooxidans* and *Acidithiobacillus ferrooxidans*. The leaching efficiency of Cu and Zn reached 38% and Zn 67% [26]. Furthermore, another previous study showed that Bacillus megaterium QM B1551 can extract valuable metals such as Co, Ni, and Cu from a complex sulfide flotation concentrate, but the Cu only leached 39.8% [27]. Meanwhile, the bioleaching potential of indigenous *Aspergillus fumigatus* to remove metals was evidenced and the leaching efficiency reached 100%, 62%, 58%, 56%, and 54% for Mn, As, Fe, Pb, and Zn, respectively [28]. Moreover, *Aspergillus niger* was used to treated waste LCD panels, and the indium bioleaching efficiency could be improved from 12.3% to 100% by fermentation method optimization [29], but the method process speed time was more than 15 h. Additionally, the microbial leaching of heavy metal from municipal sludge was studied in a continuously stirred tank reactor, and about 62% of Cu and about 77% of Zn were dissolved [30]. Above all, the majority of bioleaching experiments employed single wild microorganisms, which means that the leaching efficiency was relatively limited. Some researchers improved the experimental system with microbes used as the target and adopted the mixed community system, which greatly improved the leaching efficiency [26,29,31,32]. However, to our knowledge, there is little research focus on exploring the simultaneous improvement of Cu and Zn leaching efficiency in polymetallic ore remains.

Table 1. The advance in leaching metals technological scheme.

Microbe or Method	Result (Leaching Efficiency)	Reference
Aspergillus fumigatus	As (62%), Fe (58%), Mn (100%), Zn (54%)	Bahi Jalili Seh-Bardan et al., 2012 [28]
Acidithiobacillus thiooxidans; Acidithiobacillus ferrooxidans; Leptospirillum ferriphilum	Cu (69~83%); Zn (4.1~14%)	Olli H. Tuovinen et al., 2015 [25]
Acidithiobacillus thiooxidans; Acidithiobacillus ferrooxidans	Cu (38%); Zn (67%)	Van Khanh Nguyen and Jong-Un Lee., 2015 [26]
Bacillus megaterium QM B1551	Co (60.7%); Ni (76.3%); Cu (39.8%)	Xinlan Cui et al., 2016 [27]
Acidithiobacillus sp., *Leptopirillum* sp., *Ferroplasma* sp.	Pyrite (69.29%); Chalcocite (65.02%); Covellite (84.97%)	Shang He et al., 2021 [24]
Aspergillus niger	Indium (100%)	Jiaying Cui et al., 2021 [29]
Chemical leaching	Zn (85%); Cu and Fe (10%)	Jia Li et al., 2021 [33]
Acidithiobacillus ferrooxidans	Cu (54%); Ni (75%), Fe (55%)	Mahdokht Arshadi et al., 2021 [34]

In our study, we aim to adopt a new microbial culture method to cultivate a new-type artificial microbial community to leach Cu and Zn from sulfide concentrate, and explore the influence of different conditions (temperature, pH value, and rotation speed) on metal recovery, to further research and development of environmentally friendly and efficient bioleaching technology. An artificial microbial community consisting of *Sulfobacillus thermotolerans 6Y-1, Leptosirillum ferriphilum MJ-CL,* and *Acidithiobacillus caldus OY* (the concentration ratio is 1:1:1) was established for targeted and efficient recovering of copper and zinc from sulfide concentrate. Furthermore, different pH value, temperature, and rotation speed gradient were determined for the optimal conditions by the artificial microbial community. Then, the preferred conditions were chosen for bioleaching efficiency of Cu and Zn, using this novel artificial microbial community, which provided a new-type and high-efficiency method for recovering of copper and zinc from low-grade ores.

2. Materials and Methods

2.1. Sample Source and Mineral Component Analysis

The sulfide concentrate used in our study was from Qinghai Derni flotation sulfide concentrate. The mineral component was characterized by X-ray diffraction (XRD) on a Philips Diffractometer (model: X'Pert-Pro MPD; Philips, Eindhoven, The Netherland) using CuKα radiation (λ = 0.15418 nm, 45 kV, 200 mA). Diffraction patterns covering a 2θ range of 10°–90° were measured at a step size of 0.02°. The scanning speed was 8°/min and all the tests were carried out at room temperature.

The chemical and elemental composition of the sulfide concentrate sample was characterized by X-ray fluorescence spectrometer (XRF) (Model: ARL Advant XP, Thermo Electron Corporation, Berne, Switzerland). All measurements have been done with the excitation power of 4.2 kW, 50 kV excitation voltage and 55 mA excitation current.

The chemical analyses of the screened samples and mineralogical data were obtained from mineral liberation analyses (MLA, 650 F) equipped with a Bruker EDX (energy dispersive X-ray) system and MLA suite 3.1 for data acquisition.

2.2. Culture for the Artificial Microbial Community

Bacteria of the genus *Sulfobacillus* are successfully used in biotechnologies of treatment of sulfide ore materials [32]. Cultures deposited in China Center for Type Culture Collection (CCTCC) were used in the study. The artificial microbial community was constituted by *Sulfobacillus thermotolerans 6Y-1* (CCTCC No. M2010279), *Leptosirillum ferriphilum MJ-CL* (CCTCC No. M2011019), and *Acidithiobacillus caldus OY* (CCTCC No. M2010356) (with a concentration ratio for 1:1:1). The artificial microbial community was cultured with a medium containing: 44.7 g/L of $FeSO_4{\cdot}7H_2O$, 3 g/L of $(NH_4)_2SO_4$, 0.1 g/L of KCl, 0.5 g/L of K_2HPO_4, 0.5 g/L of $MgSO_4{\cdot}7H_2O$, 0.01 g/L of $Ca(NO_3)_2$. The value of the medium was adjusted by using 1 mol/L sulfuric acid. The artificial microbial community was aerobically inoculated in 100 mL medium at 45 °C condition with shaking speed of 160 r/min for 3 days. Then it was inoculated with 5 mL microbial suspension into 100 mL medium and reactivated two times until the optical density at 600 nm (OD_{600}) reached about 1.5. The number of bacteria in bacterial liquid was directly counted by blood counting plate under light microscope (Nikon ECL IPSE 50I). The pH value of the solution was measured by using a pH meter (Orion 868A, ThermoFisher, Waltham, MA, USA).

2.3. Chemical Analysis Procedure of Leached Elements

As for leaching process experiment, 20 g sulfide concentrate sample, 90 mL medium, and 10 mL logarithmic phase microbial community solution were placed into 250 mL conical flask. The initial pH and different temperature conditions were controlled. During leaching process, 5 mL samples of suspension were aseptically removed at 24 h intervals. The distilled water was added to replenish the evaporated solution before sampling. After the leaching finished, the residual ore concentrate was separated into solids. The supernatant immediately by centrifugation (Eppendorf Centrifuge 5804R, Hamburg, Germany) with 11,000 r/min for 10 min to obtain the supernatant and sediment. The quantitative analysis of the main elements was carried out by ICP-OES (Inductively coupled plasma emission Spectrometer, Agilent 725-ES, Santa Clara, CA, USA).

3. Results and Discussion

3.1. Mineral Composition for Sulfide Concentrate

With the aim of understanding the chemical constituents of sulfide concentrate, XRF was employed in the study. As shown in Table 2, the sulfide concentrate contained large amounts of SO_3 (47.6%), Fe_2O_3 (37.6%), SiO_2 (5.94%), MgO (4.24%), and CaO (2.27%). In addition, there were several limited content chemical compounds, which including ZnO (0.730%), CuO (0.540%), Al_2O_3 (0.380%), and other oxides content was less than 0.100%. Therefore, there were minute amount of oxides content of Cu and Zn in the sulfide concentrate.

Table 2. XRF Results for sulfide concentrate (wt%).

Oxides	**Fe_2O_3**	**SiO_2**	**MgO**	**CaO**	**Na_2O**	**ZnO**	**CuO**
Content	37.6	5.94	4.24	2.27	0.540	0.730	0.540
Oxides	**Cr_2O_3**	**TiO_2**	**MnO**	**K_2O**	**NiO**	**Al_2O_3**	**SO_3**
Content	0.0100	0.0100	0.0500	0.0100	0.0400	0.380	47.6

Furthermore, chemical elements in the sulfide concentrate were determined by ICP-OES. As shown in Table 3, the main elements in the sulfide concentrate were Fe (48.5%), S (32.3%), Ca (1.41%), Mg (1.32%), Zn (0.890%), Cu (0.600%), and other lower content (less than 0.1%) elements, such as Al, Mn, Ni, Na, Ti, Cr, Pb, and K.

Table 3. Main chemical constituents of sulfide concentrate (wt%).

Elements	**Fe**	**S**	**Si**	**Ca**	**Mg**	**Zn**	**Cu**	**Al**
Content	48.5	32.3	2.61	1.41	1.32	0.890	0.600	0.0600
Elements	**Mn**	**Ni**	**Ti**	**Cr**	**Pb**	**K**	**Na**	
Content	0.0400	0.0300	0.0100	0.0100	0.0100	0.0100	0.0100	

The powder X-ray diffraction (XRD) patterns for sulfide concentrate are shown in Figure 1. As can be seen, the main minerals of the sample are pyrite, pyrrhotite, chalcopyrite, sphalerite, and magnetite, similar to previous research [33]. In detail, the diffraction peaks were observed at 28.5°, 33.0°, 37.0°, 40.7°, 47.4°, 56.3°, 61.7°, 64.3°, 83.6°, and 88.3° corresponding to the lattice planes of pyrite (111), (200), (210), (211), (220), (311), (023), (321), (332), and (422), respectively (PCPDF: 99-0087). Meanwhile, diffraction peaks appeared at 29.9°, 33.8°, 43.8°, 53.1°, 62.2°, and 71.2° corresponding to the lattice planes of pyrrhotite (200), (205), (2010), (220), (400), and (1120), respectively (PCPDF: 29-0724). In addition, 29.4°, 48.6°, and 81.6° were assigned to three characteristic diffraction planes of chalcopyrite (112), (220), and (404) (PCPDF: 26-1116), and 47.5°, 59.1°, 76.8° corresponding to planes of sphalerite (220), (222), and (331) (PCPDF: 05-0566). The XRD data revealed that pyrite and pyrrhotite have perfect crystallinity, but chalcopyrite and sphalerite are imperfect according to the sharpness differentiation of the peak shape.

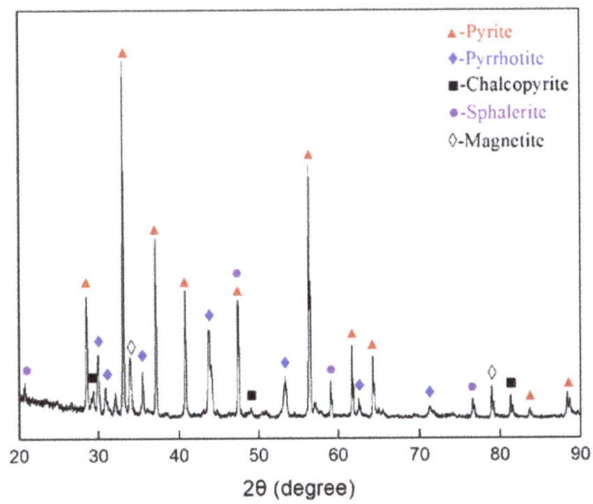

Figure 1. Powder XRD pattern for sulfide concentrate.

3.2. MLA and Liberation Classes for Sulfide Concentrate

Mineral Liberation Analysis (MLA) is the effective modern method to analyze mineral properties, especially the liberation of minerals which is very important to the separation of different minerals. The chemical analyses and main mineral constituent data obtained from MLA are presented in Table 4 and the minerals in sulfide concentrate could be divided into five categories as sulfide, oxide, carbonate minerals, sulfate minerals, and silicate minerals. Analyzing their particle size distribution in Figure 2a, the sulfide minerals including pyrite, pyrrhotite, chalcopyrite, sphalerite, and galena as a group and their particle size distribution is mainly between 10 μm and 100 μm (13–93%). The particle size distribution of oxide minerals such as magnetite, hematite, chromite, and ilmenite with the percent of 26.5% lower than 10 μm and all of them were lower than 125 μm. Moreover, the size distribution property of carbonate minerals (dolomite and calcite) and silicate minerals (enstatite, quartz, biotite, hornblende, and pyroxene) was similar to the oxide minerals. As far as sulfate minerals (anhydrite and $FeCaMgSO_4$) are concerned, their size was lower than 99 μm and most of them were lower than 10 μm (60%). The size distribution results indicated that the sulfide concentrate was suitable for bioleaching [24].

Table 4. MLA results for main mineral constituents in sulfide concentrate (wt%).

Mineral	Content	Mineral	Content
Pyrite	40.0	Quartz	0.720
Pyrrhotite	31.0	Albite	0.0600
Chalcopyrite	1.77	Hornblende	0.360
Sphalerite	1.31	Pyroxene	0.0300
Galena	0.0100	Muscovite	0.0300
Magnetite, Hematite	13.5	Biotite	0.610
Chromite	0.0100	Enstatite	3.86
Ilmenite	0.0100	Calcite	0.0400
Anhydrite	0.0800	Dolomite	4.07
$FeCaMgSO_4$	2.50	-	-

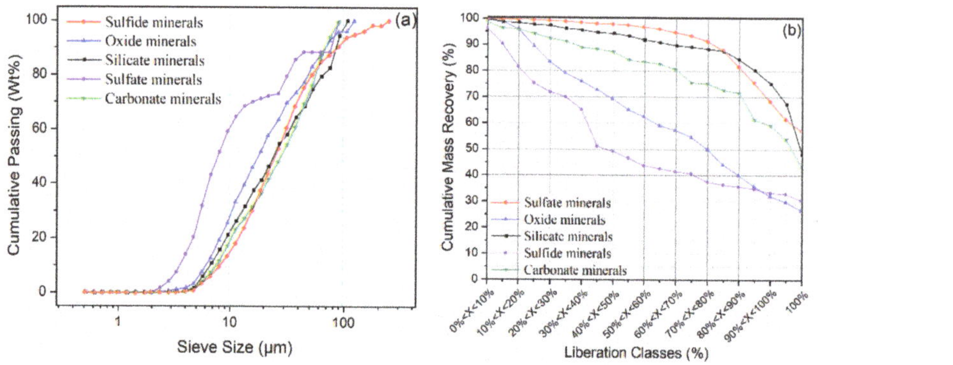

Figure 2. (**a**) Sieve size distribution for different minerals in sulfide concentrate; (**b**) liberation classes results for sulfide concentrate.

Analyzing the mineral liberation classes results in Figure 2b, the liberation characteristic of five categories was ordered: sulfide minerals > silicate minerals > carbonate minerals > oxide minerals > sulfate minerals. Most importantly, the sulfide minerals present the best liberation characteristic and 57.3% of them had 100% liberation degree, which was higher than the rest of the other four type minerals. Because the Cu and Zn were mainly abundant in chalcopyrite and sphalerite, the suitable size and great liberation provide possibilities for recovery form sulfide concentrate by the artificial microbial community.

3.3. Artificial Microbial Community Growth Characteristic Analysis

It is widely known that microbial community composition mainly depended on environmental factors, such as pH, temperature, and oxygen content. In order to better understand growth condition of the artificial microbial community, different pH, temperature, and cultural revolving speed were determined, respectively. As shown in Figure 3a, the growth rate of microbial significant changed with pH ranging from 1.0 to 1.8. After 5 days, the artificial microbial community was in logarithmic phase. Moreover, better growth rates were observed with pH vales of 1.8 and 1.5, the average logarithmic cell concentration was 8.3×10^7 cell/mL and 7.7×10^7 cell/mL respectively. A lower growth rate was observed when pH value was 1.2 with average logarithmic cell concentration was 6.5×10^7 cell/mL. The lowest growth rate was observed at pH = 1.0 and the cell concentration was only 2.4×10^7 cell/mL. We could conclude that the most suitable pH value should be 1.8, which was similar to previous study (optimal pH value was 1.5) [35].

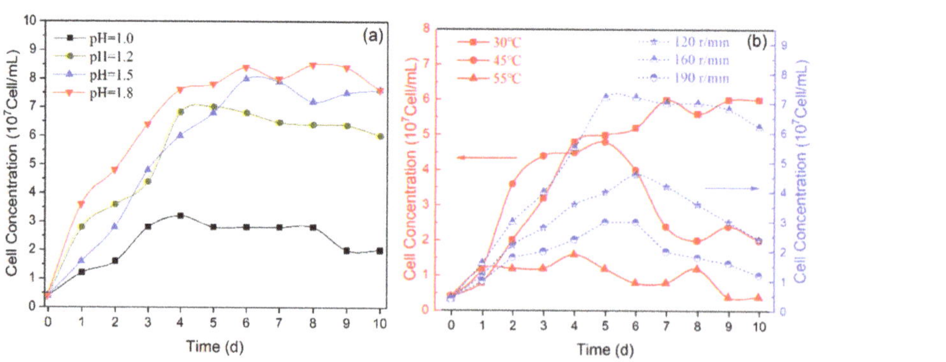

Figure 3. Culture condition results of (**a**) pH, (**b**) temperature, and revolving speed for artificial microbial community.

Furthermore, another two environmental factors of temperature and revolving speed were studied as well [28,29]. As shown in Figure 3b (Red Lines), the microorganism is significantly affected by culture temperature. The best growth rate presented under 30 °C, and the number of microbials continued increasing within 1–10 days, which reached the maximum value about 6.1×10^7 cell/mL. When the temperature was 45 °C, the growth rate was the fastest in the first three days, and the number of microorganisms attaining the maximum was about 4.7×10^7 cell/mL on the fifth day. However, the number of microorganisms decreased rapidly five days later, which reached to 2.2×10^7 cell/mL. When the temperature was 55 °C, an unsatisfactory growth curve was performed and the cell concentration did not exceed 1.5×10^7 cell/mL in the whole culture process.

As far as rotation speed is concerned, the growth of microorganisms should be influenced notably. As shown in Figure 3b (Blue Lines), the fastest growth rate was observed with 160 r/min condition, and the number of microorganisms which reached the maximum value about 6.3×10^7 cell/mL on the fifth day. However, when the rotation speed was adjusted to 120 r/min or 190 r/min, the average cell concentration was 2.8×10^7 cell/mL and 1.2×10^7 cell/mL respectively. Based on the above results, we conclude that the artificial microbial community preferred values of pH, temperature, and cultural revolving speed of 1.8, 30 °C, and 160 r/min respectively.

3.4. Reaction Condition Exploration for Leaching Process

Based on the artificial microbial community growth characteristic analysis results, the leaching efficiency of Cu and Zn was studied under different pH (pH = 1.0, 1.2, 1.5, 1.8) and different temperature (35 °C, 45 °C, 55 °C) conditions. As show in Figure 4 (left part), the highest leaching efficiency of Cu and Zn were observed when pH value was 1.2, and the value attained was 80% and 100% respectively. In addition, the leaching efficiency of Cu is similar when pH value was 1.0, 1.5, and 1.8, only reached to 65%, 62%, 61%, respectively. Meanwhile, the leaching efficiency of Zn is similar when pH value was 1.5 and 1.8, which reached to 95% and 94.5% respectively. However, the lowest leaching efficiency of Zn only came up to 85% when pH value was 1.0. Therefore, the optimal pH value was 1.2, which was different with the cell-culture value (1.8).

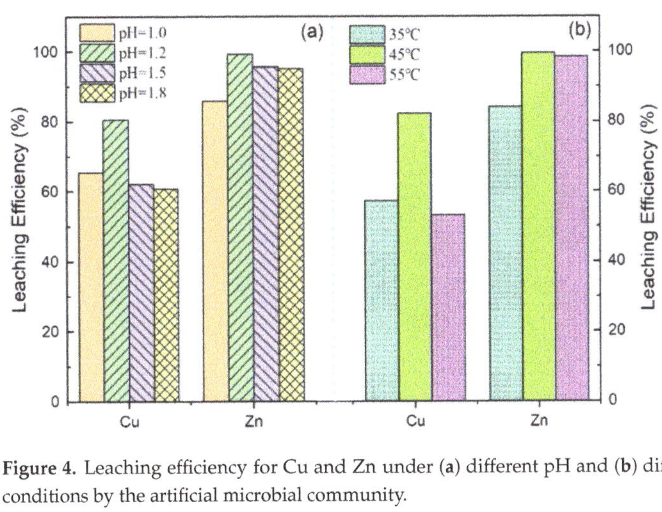

Figure 4. Leaching efficiency for Cu and Zn under (**a**) different pH and (**b**) different temperature conditions by the artificial microbial community.

Furthermore, the Figure 4 (right part) presented the highest leaching efficiency of Cu and Zn under 45 °C, which reached to 84% and 98% respectively. However, when temperature value reduced to 35 °C, the leaching efficiency of Cu was 58%, and the value of Zn was 81%. Meanwhile, when temperature raised to 55 °C, the leaching efficiency of Cu

Zn reached to 51% and 97%, respectively. Therefore, the preferred value of temperature was also changed from 35 °C to 45 °C in the experiments, compared with cell-culture process. We could conclude that the artificial microbial community has positive effect on the leaching efficiency of Cu and Zn. However, the conditions of microbial growth and degradation need to be further corrected in practical engineering applications. Compared with the previous microorganism growth conditions, the preferred values of pH and temperature were changed in the experiments, the highest leaching efficiency of Cu and Zn which occurred in the value of pH and temperature was 1.2 and 45 °C respectively, and the conditions have a slight difference with regard to previous related study [29].

In order to further improve the leaching efficiency of Cu and Zn, the study of essential factors for ore pulp density and inoculation quantity was necessary [29]. The leaching efficiency of Cu and Zn in various of ore pulp density (5%, 10%, 15%, 20%, 25%) were explored, while the pH value was 1.2, temperature was 45 °C, and rotation speed was 160 r/min. For the ore pulp density result presented in Figure 5 (red lines), the leaching efficiency of Cu and Zn increased at first and then decreased rapidly with the increasing of ore pulp density. The highest leaching efficiency of Cu and Zn reached to 100% and 79% on 20% ore pulp density. Meanwhile, when the ore pulp density was lower than 20%, the leaching efficiency of Cu and Zn only attained to 74% and 84% respectively. Moreover, with 25% ore pulp density, the leaching efficiency of Cu and Zn decreased rapidly as well, which reached to 53% and 55% respectively.

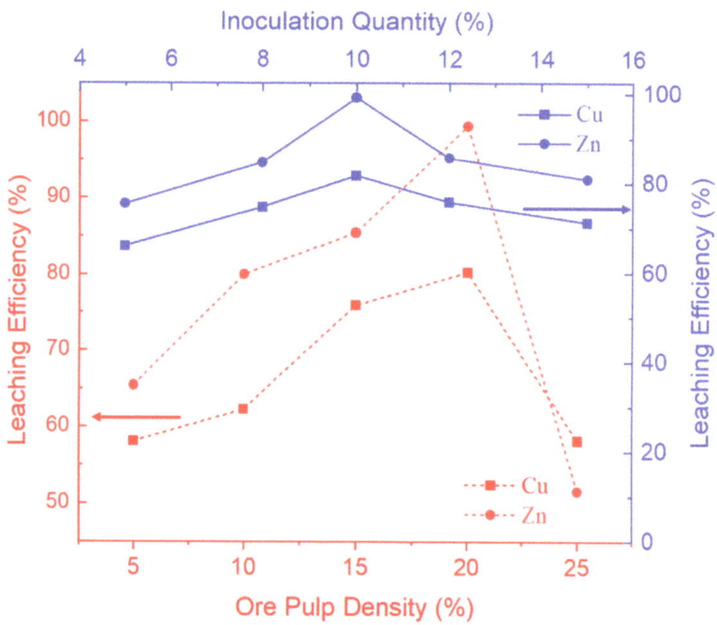

Figure 5. Leaching efficiency curves determined by ore pulp density and inoculation quantity.

In addition, the inoculation quantity of microbial community is another essential factor for the leaching efficiency of Cu and Zn. In this part, different inoculation quantity (5%, 8%, 10%, 12%, and 15%) were explored in the same conditions, with the value of ore pulp density is 20%. Figure 5 (blue lines) shows the inoculation quantity results: with increasing inoculation quantity, the leaching efficiency of Cu and Zn increased at first and then decreased. The highest value occurred on 10% inoculation quantity, which reached to 82% and 100%, present a non-linear relationship. At the same time, if the inoculation quantity (8%) was smaller than 10%, the leaching efficiency of Cu and Zn decreased to 77%

and 85% respectively, and further reduced to 68% and 72% with the inoculation quantity of 5%, when the inoculation quantity (12%) was higher than 10%, the leaching efficiency of Cu and Zn reduced to 78% and 87% respectively, and further reduced to 72% and 82% with the inoculation quantity of 15%.

The trend of the leaching efficiency is inseparable with ore pulp density and inoculation quantity. We speculate that there are two reasons. Firstly, lower concentration of ore pulp density will not be enough to satisfy the adherent area of microorganisms, but higher concentration will hinder the exchange of microorganisms and ore pulp. As far as inoculation quantity is concerned, limited microorganisms result in lower leaching efficiency of Cu and Zn, but the dissolved oxygen could not support superfluous microorganisms to maintain optimal activity. Based on the above results, we could conclude that ore pulp density and inoculation quantity were key factors affecting the leaching efficiency of Cu and Zn from sulfide concentrate. Therefore, 20% ore pulp density and the inoculation quantity of 10% are optimal conditions for engineering application in system.

3.5. Efficient Performance for Novel Artificial Microbial Community

Based on above experiment results, when the leaching efficiency of Cu and Zn have reached maximum, the optimal conditions were: pH was 1.2, temperature was 45 °C, artificial microbial community inoculation quantity was 10%, and ore pulp density was 20%. In order to explore the leaching efficiency performance by the novel artificial microbial community, control groups (leaching under aseptic conditions) without microbial were set up in the same time, which was used to compare with the experimental group.

As shown in Figure 6, the artificial microbial community had a significant effect on the leaching of Cu and Zn. The points were as follows. For Cu, the leaching rate of the control group increased steadily in the first eight days, and reached the maximum on the eighth day, with the average leaching rate of 45%. Compared to control group, in the experimental group, the leaching rate increased steadily in the first ten days, and the growth rate was greater than the control group. The leaching rate of Cu reached the maximum on the tenth day, with an average leaching rate of 80%. The increase of artificial microbial community lengthened the leaching cycle of Cu, but the leaching rate of Cu was significantly increased, and the leaching rate of Cu in the experimental group was 1.8 times that of the control group. For Zn, the leaching rate of control group and experimental group was higher than the Cu. In the control group of Zn, the leaching rate reached the maximum on the eighth day, with an average leaching rate of 80%. However, after the addition of artificial microbial community, all Zn leaching on the sixth day and leaching rate was 100%. The addition of artificial microbial community not only shortened the leaching cycle of Zn, but also increased the leaching rate by 1.2 times. In a single microbial reaction system, the leaching rates of Cu and Zn by microorganisms were up to 62% and 77% [30]. Compared with the single microbial system, the novel artificial microorganism improves the leaching rate of Cu and Zn effectively, and the leaching efficiency is increased by 30% and 29%.

All in all, comprehensive analysis shows that the artificial microbial community can significantly enhance the leaching rate of Cu and Zn from mine tailings, and the leaching rate of Zn was significantly higher than of Cu. In addition, an artificial microbial community has the advantage of high efficiency and no pollution in leaching Cu and Zn, which has important research significance in improving mineral recovery and degrading heavy metal pollution in the future.

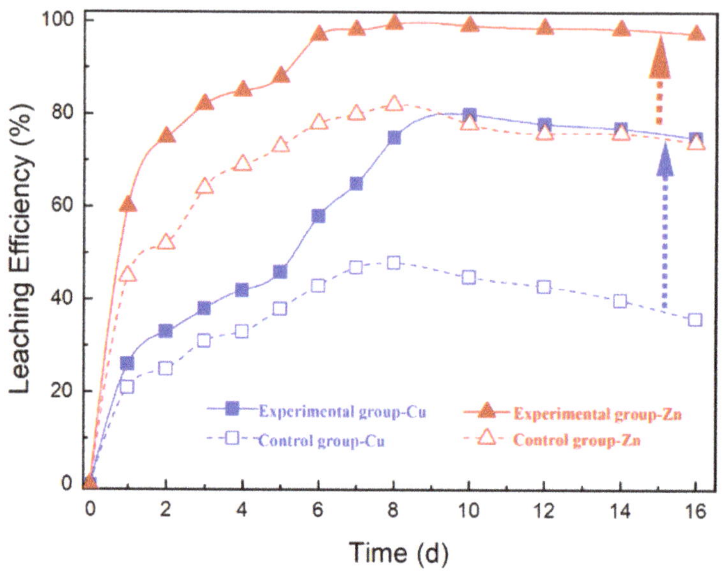

Figure 6. Copper and zinc recovery by the novel artificial microbial community.

4. Conclusions

As an environmentally friendly and high-efficiency bioleaching technologies scheme, the artificial microbial community's enhancement of the recovery of Cu and Zn was studied for the first time. The optimal experimental parameters (pH value, temperature, and rotation speed) are unlike those between microorganism growth conditions and bioleaching experiments. Under the optimal conditions, the average recovery rate of Cu and Zn reached to 80% and 100% respectively, which almost increased 180% and 120% more than the control group. Further studies plan to optimize the novel artificial microbial community to leach more kinds of metal ions and provide different solutions for various kinds of low-grade ore.

Author Contributions: Conceptualization, X.C. (Xinglan Cui) and X.Y.; methodology, X.C. (Xiaokui Cui); software, H.L. and Q.Z. (Qidong Zhang); validation, J.Z., L.W. and X.H.; formal analysis, Q.Z. (Qi Zheng); investigation, R.J.; resources, Y.L.; data curation, X.C. (Xinglan Cui); writing—original draft preparation, X.C. (Xinglan Cui) and X.Y.; writing—review and editing, Y.L. All authors have read and agreed to the published version of the manuscript.

Funding: This research was funded by the National Key Research and Development Project (No. 2019YFC1805900), the National Key Research and Development Project (No. 2020YFC1807700), the Open Foundation of State Key Laboratory of Vanadium and Titanium Resources Comprehensive Utilization (No. 2021P4FZG13A), the Youth Fund Project of GRINM (No. 12008, No12119 and No12120), National Natural Science Foundation of China (No. 51704028), and the Open Foundation of State Key Laboratory of Mineral Processing (No. BGRIMM-KJSKL-2020-07).

Institutional Review Board Statement: Not applicable.

Informed Consent Statement: Not applicable.

Data Availability Statement: The data presented in this study are available on request from the corresponding author.

Conflicts of Interest: The authors declare no conflict of interest.

References

1. Chen, J.; Jiskani, I.M.; Jinliang, C.; Yan, H. Evaluation and future framework of green mine construction in China based on the DPSIR model. *Sustain. Environ. Res.* **2020**, *30*, 448–457. [CrossRef]
2. Wei, X.; Liu, H.; Li, G. Influencing Factors of Eco-Environmental Safety of Mines and Their Green Development: A Case Study of Taoshan Coal Mine in Heilongjiang Province, China. *Nat. Environ. Polution Technol.* **2020**, *19*, 831–838.
3. Martinez, P.; Vera, M.; Bobadilla-Fazzini, R. Omics on bioleaching: Current and future impacts. *Appl. Microbiol. Biotechnol.* **2015**, *99*, 8337–8350. [CrossRef]
4. Bobadilla-Fazzini, R.A.; Pérez, A.; Gautier, V.; Jordan, H.; Parada, P. Primary copper sulfides bioleaching vs. chloride leaching: Advantages and drawbacks. *Hydrometallurgy* **2017**, *168*, 26–31. [CrossRef]
5. Hernandez, S. *The World of Copper Factbook*; International Copper Study Group: Lisbon, Portugal, 2017; pp. 1–58.
6. Hallberg, K.B.; Gonza´lez-Toril, E.; Johnson, D.B. *Acidithiobacillus ferrivorans*, sp. nov.; facultatively anaerobic, psychrotolerant iron-, and sulfur-oxidizing acidophiles isolated from metal mine-impacted environments. *Extremophiles* **2010**, *14*, 9–19. [CrossRef]
7. Ilyas, S.; Chi, R.-A.; Lee, J.-C. Fungal Bioleaching of Metals from Mine Tailing. *Miner. Process. Extr. Met. Rev.* **2013**, *34*, 185–194. [CrossRef]
8. Panda, S.; Akcil, A.; Pradhan, N.; Deveci, H. Current scenario of chalcopyrite bioleaching: A review on the recent advances to its heap-leach technology. *Bioresour. Technol.* **2015**, *196*, 694–706. [CrossRef]
9. Feng, S.; Yang, H.; Wang, W. Insights to the effects of free cells on community structure of attached cells and chalcopyrite bioleaching during different stages. *Bioresour. Technol.* **2016**, *200*, 186–193. [CrossRef]
10. Africa, C.; Van Hille, R. Investigation and in situ visualisation of interfacial interactions of thermophilic microorganisms with metal-sulfides in a simulated heap environment. *Miner. Eng.* **2013**, *48*, 100–107. [CrossRef]
11. Wang, X.; Ma, L.; Wu, J.; Xiao, Y.; Tao, J.; Liu, X. Effective bioleaching of low-grade copper ores: Insights from microbial cross experiments. *Bioresour. Technol.* **2020**, *308*, 123273. [CrossRef]
12. Ke, Z.; Zhao, X.; Feng, Y.; Wang, Q.; Wang, Y.; Zhang, C.; Zhu, B.; Ren, G. Semiconducting Minerals Participated Extracellular Electron Transfer Process in High-Altitude Red Soil from Gansu, China. *Geomicrobiol. J.* **2021**, *38*, 905–913. [CrossRef]
13. Zeng, J.; Gou, M.; Tang, Y.-Q.; Li, G.-Y.; Sun, Z.-Y.; Kida, K. Effective bioleaching of chromium in tannery sludge with an enriched sulfur-oxidizing bacterial community. *Bioresour. Technol.* **2016**, *218*, 859–866. [CrossRef] [PubMed]
14. Zheng, C.; Huang, Y.; Guo, J.; Cai, R.; Zheng, H.; Lin, C.; Chen, Q. Investigation of cleaner sulfide mineral oxidation technology: Simulation and evaluation of stirred bioreactors for gold-bioleaching process. *J. Clean. Prod.* **2018**, *192*, 364–375. [CrossRef]
15. Holmes, D. Review of International Biohydrometallurgy Symposium, Frankfurt, 2007. *Hydrometallurgy* **2008**, *92*, 69–72. [CrossRef]
16. Chen, S.-Y.; Pan, S.-H. Simultaneous metal leaching and sludge digestion by thermophilic microorganisms: Effect of solids content. *J. Hazard. Mater.* **2010**, *179*, 340–347. [CrossRef]
17. Chen, S.-Y.; Lin, J.-G. Bioleaching of heavy metals from contaminated sediment by indigenous sulfur-oxidizing bacteria in an air-lift bioreactor: Effects of sulfur concentration. *Water Res.* **2004**, *38*, 3205–3214. [CrossRef]
18. Deng, X.; Chai, L.; Yang, Z. Bioleaching of heavy metals from a contaminated soil using indigenous *Penicillium chrysogenum strain* F1. *J. Hazard. Mater.* **2012**, *233–234*, 25–32. [CrossRef]
19. Blais, J.; Tyagi, R.; Auclair, J. Bioleaching of metals from sewage sludge: Microorganisms and growth kinetics. *Water Res.* **1993**, *27*, 101–110. [CrossRef]
20. Hopfe, S.; Flemming, K.; Lehmann, F.; Möckel, R.; Kutschke, S.; Pollmann, K. Leaching of rare earth elements from fluorescent powder using the tea fungus Kombucha. *Waste Manag.* **2017**, *62*, 211–221. [CrossRef]
21. Kaksonen, A.; Boxall, N.; Gumulya, Y. Recent progress in biohydrometallurgy and microbial characterization. *Hydrometallurgy* **2018**, *180*, 7–25. [CrossRef]
22. Pathak, A.; Morrison, L.; Healy, M.G. Catalytic potential of selected metal ions for bioleaching, and potential techno-economic and environmental issues: A critical review. *Bioresour. Technol.* **2017**, *229*, 211–221. [CrossRef]
23. Asghari, I.; Mousavi, S.M.; Amiri, F.; Tavassoli, S. Bioleaching of spent refinery catalysts: A review. *J. Ind. Eng. Chem.* **2013**, *19*, 1069–1081. [CrossRef]
24. Shang, H.; Gao, W.-C.; Wu, B.; Wen, J.-K. Bioleaching and dissolution kinetics of pyrite, chalcocite and covellite. *J. Cent. South Univ.* **2021**, *28*, 2037–2051. [CrossRef]
25. Tuovinen, O.; Särkijärvi, S.; Peuraniemi, E. Acid Leaching of Cu and Zn from a Smelter Slag with a Bacterial Consortium. *Adv. Mater. Res.* **2015**, *1130*, 660–663. [CrossRef]
26. Nguyen, V.K.; Lee, J.U. A comparison of microbial leaching and chemical leaching of arsenic and heavy metals from mine tailings. *Biotechnol. Bioprocess Eng.* **2015**, *20*, 91–99. [CrossRef]
27. Cui, X.; Wang, X. Bioleaching of a complex Co-Ni-Cu sulfide flotation concentrate by *Bacillus megaterium* QM B1551 at neutral pH. *Geomicrobiol. J.* **2016**, *3*, 734–741. [CrossRef]
28. Seh-Bardan, B.J.; Othman, R.; Ab Wahid, S.; Husin, A.; Sadegh-Zadeh, F. Bioleaching of Heavy Metals from Mine Tailings by *Aspergillus fumigatus*. *Bioremediat. J.* **2012**, *16*, 57–65. [CrossRef]
29. Cui, J.; Zhu, N.; Mao, F.; Wu, P.; Dang, Z. Bioleaching of indium from waste LCD panels by *Aspergillus niger*: Method optimization and mechanism analysis. *Sci. Total. Environ.* **2021**, *790*, 148151. [CrossRef]
30. Couillard, D.; Mercier, G. Bacterial leaching of heavy metals from sewage sludge-bioreactors comparison. *Environ. Pollut.* **1990**, *66*, 237–252. [CrossRef]

31. Liang, Y.L.; Yin, H.Q.; Xiao, Y.H.; Tang, M.; Feng, X.; Xie, Z.Y.; Qiu, G.Z.; Liu, X.D. Microbial Leaching of Copper from Tailings of Low Grade Sulphide Ores in Zambia. *Adv. Mater. Res.* **2015**, *1130*, 473–476. [CrossRef]
32. Srichandan, H.; Mohapatra, R.K.; Parhi, P.K.; Mishra, S. Bioleaching approach for extraction of metal values from secondary solid wastes: A critical review. *Hydrometallurgy* **2019**, *189*, 105122. [CrossRef]
33. Li, J.; Meng, X.; Zhao, H.; Zhang, Y.; Liu, R.; Gu, G. Efficient separation of zinc from zinc containing copper sulfide concentrate by chemical leaching. *Chem. Erde-Geochem.* **2021**, *107*, 125773. [CrossRef]
34. Arshadi, M.; Pourhossein, F.; Mousavi, S.M.; Yaghmaei, S. Green recovery of Cu-Ni-Fe from a mixture of spent PCBs using adapted A. ferrooxidans in a bubble column bioreactor. *Sep. Purif. Technol.* **2021**, *272*, 118701. [CrossRef]
35. Mäkinen, J.; Heikola, T.; Salo, M.; Kinnunen, P. The Effects of Milling and pH on Co, Ni, Zn and Cu Bioleaching from Polymetallic Sulfide Concentrate. *Minerals* **2021**, *11*, 317. [CrossRef]

metals

MDPI

Article

Experimental Study on Factors Influencing the Strength Distribution of In Situ Cemented Tailings Backfill

Xiaopeng Peng [1,2], Lijie Guo [1,2,*], Guangsheng Liu [1,2], Xiaocong Yang [1,2] and Xinzheng Chen [1,2]

1 BGRIMM Technology Group, Institute of Mining Engineering, Beijing 100160, China; pengxiaopeng@bgrimm.com (X.P.); liuguangsheng@bgrimm.com (G.L.); yxcong@bgrimm.com (X.Y.); chenxinzheng@bgrimm.com (X.C.)
2 National Center for International Joint Research on Green Metal Mining, Beijing 100160, China
* Correspondence: guolijie@bgrimm.com

Abstract: Previous studies have found that the strength of in situ cemented tailings backfill usually presents an S-shaped distribution, which decreases first, then increases, and decreases thereafter along the direction of slurry flow. In this study, to explore the factors determining the distribution, a similar model test of cemented tailings backfill was carried out. The distribution law of grain size composition and the cement content of backfill materials along the flow direction were experimentally studied, and the comprehensive factor influencing the strength distribution was analyzed. The results show that, firstly, near the feeding point, there are more coarse particles, whereas the content of fine particles is higher farther away. The measured maximum median particle size can be more than three times the minimum value. Secondly, the cement content increases gradually along the flow direction and reaches the peak at the end of the model, which can be more than twice the minimum value, indicating that the degree of segregation is significant. Thirdly, the strength distribution of cemented backfills is comprehensively determined by both the particle size distribution (PSD) and the cement content. The maximum value appears neither at the point with peak median particle size, nor at the point with the highest cement content. Lastly, there is a strong linear correlation between the strength of cemented backfills and the strength factor (SF), which is defined as the product of the uniformity coefficient and cement content of filling materials, indicating that the SF can be used to quantitatively reflect the comprehensive effects of PSD and cement content on the strength. As SF is a comprehensive quantitative index reflecting the distribution of strength, it will be further studied in later research to acquire more experimental results of the relationship between sample strength and SF, which will be meaningful for the quality evaluation of in situ cemented backfills, and the optimization of backfill system.

Keywords: backfill slurry; strength of cemented backfill; inhomogeneity of cemented backfill; cemented tailings backfill

Citation: Peng, X.; Guo, L.; Liu, G.; Yang, X.; Chen, X. Experimental Study on Factors Influencing the Strength Distribution of In Situ Cemented Tailings Backfill. *Metals* **2021**, *11*, 2059. https://doi.org/10.3390/met11122059

Academic Editor: Jean François Blais

Received: 16 November 2021
Accepted: 16 December 2021
Published: 20 December 2021

1. Introduction

Mining with backfill can effectively utilize the solid wastes of mines to prepare backfill slurries/pastes and backfill them into mined out voids, such that the purposes of waste management, stope backfilling, and surface subsidence control can be achieved [1–3]. Cemented tailings backfill (CTB) is a widely applied backfill method in China. The backfill slurry of CTB mainly consists of tailings, cementitious materials (commonly cements), water, and other additives (such as flocculants or plasticizers) [4,5]. Once all the ingredients are well mixed, the prepared backfill slurries are transported and backfilled into stopes through pipelines. Therefore, to guarantee the flowability and transportability of slurries, the solid content of slurries should be suitable, and all backfill materials should be suspended in and move with the slurries [4,6,7]. When backfill slurries are placed into a stope, because of the gravity acting on solid materials and the influence of the environment, the materials will flow along the stope and form a slop shape, whereas, along the settling direction, the suspended solid materials will segregate and be rearranged as a function of the material properties. The flowing and segregation of suspended materials will absolutely lead to inhomogeneities of the backfill body after cement hydration, thus affecting the strength distribution of CTB in the stope [8–10]. This inhomogeneous strength distribution of CTB will have adverse impacts on the further analysis of the stability of cemented backfills, as it is commonly treated as a homogeneous material [11,12]; it may even result in hidden hazards, thereby endangering safe production in the mine. Thus, to adequately evaluate the inhomogeneities and control the related hazards, it is important to study the principles and factors affecting the flowing and segregation mediating the inhomogeneous strength distribution of CTB.

There have been some studies focusing on the inhomogeneous strength distribution of CTB. By drilling coring holes to obtain in situ samples, Le [13] and Thompson [14] directly studied the inhomogeneities of CTB including the strength distribution, permeability, porosity, and pore water pressure. Similarly, Gan [15] and Wei [16] obtained in situ CTB cores from a mine and systematically studied the strength distribution of cores along the flowing direction. They all experimentally proved that the strength of in situ CTB is not a fixed value but varies along with the flowing and settling direction. In addition to drilling in situ CTB cores, similar model tests are accepted as a valuable study method. Ghirian [17,18] built a wooden cube model with the same size ratio as a real mine and used the prepared tailings paste to backfill the model to simulate and analyze the backfilling process. The flume model [8,19–21] has also been widely utilized to simulate and test the flowing process of backfill slurries. By launching flume tests, Xu [22] studied the strength distribution of CTB along the flowing direction and found that the strength did not vary linearly; instead, the distribution commonly has three phases, namely it decreased first, then increased to a certain value, and decreased thereafter to the end, presenting an S-shaped distribution. Lu [23] and Li [24] carried out tailings flume tests and identified a similar S-shaped distribution of strength. Shi [25] tested and compared the uniaxial compressive strength (UCS) of a series of backfill cores obtained from a copper mine and concluded that, for real backfilled stopes, along the flowing direction, there is also an S-shaped fluctuation of the strength distribution.

The aforementioned studies uncovered a non-linear distribution of CTB strength, but few researchers have explored the reasons for and mechanism of this distribution. Li [26,27] discussed and attributed this distribution mainly to the inhomogeneities of cement content. However, as a result of settling and segregation, the particle size distribution (PSD) of tailings will also vary dramatically, as revealed in the study by Yilmaz [28], which found that, near the feeding point, the content of coarse tailings is much higher than that of fine tailings, and vice versa. Wu [29] pointed out that the inhomogeneous distribution of tailings PSD can affect the microstructure of cemented paste backfill (CPB) and, thus, its strength; when the PSD matches the Talbol theoretic curve, the strength reaches its peak. Fall [30] and Ke [31] studied the effects of fine tailings on the hydration of cement and found that the increase in fine tailings content may have a negative impact on strength development. Furthermore, Yilmaz [32,33] described how the pore structures and porosities of cemented backfills can affect the structural response in UCS tests and, therefore, the strength. Therefore, it can be concluded that the inhomogeneous strength distribution is a result of various factors, including the distribution of cement content, tailings PSD, and porosity.

Taking into account the above research results, in this study, to quantitatively analyze the S-shaped distribution of CTB strength, a flume model test was carried out, and both the cement content and the tailings PSD of samples obtained along the slurry flowing direction were determined. By comparing and analyzing the experimental results, the reasons for the S-shaped distribution of CTB are discussed, and a relatively effective parameter is fitted to describe the comprehensive influences of backfill materials on the strength distribution.

2. Experimental Program

2.1. Design of Flume Model Test System

The flume model, with reference to the measured size of a real stope (50 m × 15 m, L × W), was designed as 3.3 m × 1 m × 0.8 m (L × W × H), using a 15-fold reduction. A picture of the flume model is shown in Figure 1a.

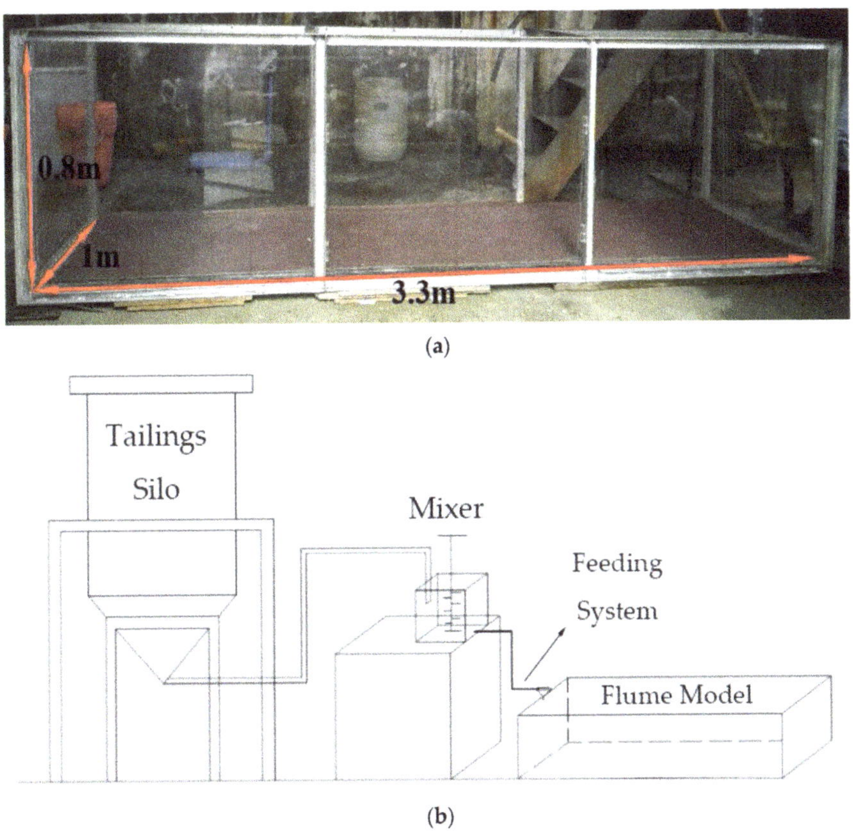

Figure 1. Diagram of the flume model test system. (**a**) Picture of the flume model; (**b**) the components of the test system.

In addition to the flume model, the test system consisted of a small silo and a tailings mixer. The tailings slurries were supplied by a mineral processing plant and transported through pipelines into the small silo. Then, the settled tailings in the silo were pumped into the mixer along with cement to prepare backfill slurries. After mixing, the backfill slurries were fed into the flume model to simulate the backfilling process. The test system is shown in Figure 1b.

To simulate the real backfilling and slurry flowing process, the feeding rate was set to match the similarity criterion [34–36]. The backfill rate of the reference mine is 200 m^3/h (similar to a real mine backfilling system), and the transport pipeline is a steel pipe DN100 (diameter 100 mm). During the backfilling process, the slurries are mainly affected by gravity. Therefore, the feeding rate of the flume model was calculated accordingly, yielding a value of 3.82 L/min.

2.2. Materials Used

This study mainly focuses on the simulation of the real backfilling process, while testing and analyzing the influence factors of the non-linear CTB strength distribution.

Therefore, the ingredients were similar to the reference mine, including tailings, binders, and mixing water, with no other additives or aggregates.

The tailings utilized were unclassified and obtained from the reference mine. The tailings samples were dried at 105 °C first for 24 h, and then grained to match the sample requirements of components analysis. The chemical properties of the tailings were tested using an ICP-OES (inductively coupled plasma emission spectrometer). The brand of this apparatus was Agilent 725-ES from Agilent company Santa Clara, California, America, and the detection limit of each element was under 0.01 ppm. This apparatus is widely utilized in the chemical component analysis of solid materials (including tailings) [37,38].

The test results are presented in Table 1.

Table 1. Chemical components of tailings.

Mineral	CaO	SiO_2	Al_2O_3	MgO	FeO	MnO	SO_3	Other
Content (wt.%)	4.2	41.4	10.4	11.5	21.3	0.2	2.1	8.9

As shown in Table 1, the main component of the tailings was SiO_2 (41.4 wt.%), whereas the content of SO_3 was less than 4 wt.%, indicating that the sulfate effects could be neglected [39]. Therefore, the tailings were suitable for backfilling.

The initial PSD of the tailings was monitored using a laser particle size analyzer (Mastersizer 2000 series from Malvern company, Worcestershire, United Kingdom), and the results are presented in Figure 2 and Table 2.

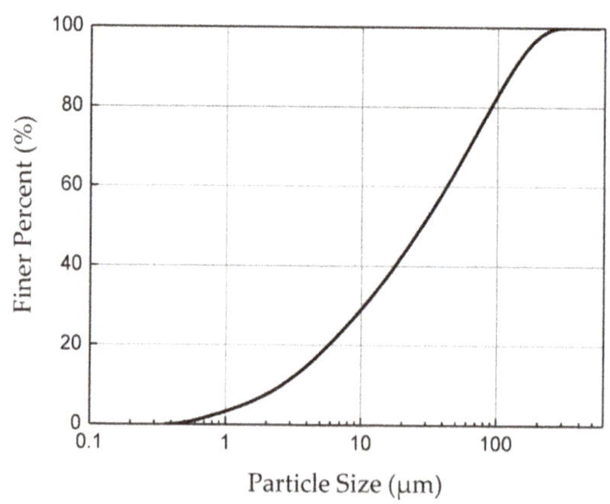

Figure 2. Particle size distribution (PSD) of the tailings used in the study.

Table 2. PSD data of tailings.

D_{10} (μm)	D_{30} (μm)	D_{50} (μm)	D_{60} (μm)	D_{90} (μm)
2.47	10.14	25.43	37.61	117.18

From Figure 2, it can be found that the diameters of more than 40% of the tailings were smaller than 20 μm, indicating several fine particles. Moreover, as shown in Table 2, the D_{10} and D_{60} values of tailings were 2.47 and 37.61 μm, respectively; thus, the coefficient of uniformity (C_u), i.e., the quotient of D_{60} and D_{10}, was approximately 15.22 (much larger than 5), indicating a large disparity between coarse and fine tailings.

The binder used in this study was PO 42.5 Portland cement, which is the binder typically used in the reference mine. As the chemical effect of mixing water was not the topic of this study and the consumption of water was large in this flume test, tap water was used to prepare the slurries.

2.3. Mix Recipes and Feeding Method

The mix recipes of slurries were in line with the bottom layer backfill recipe of the reference mine. The solid content of slurry was 70 wt.%, and the cement content was 20% of the total solid mass. This cement content is much higher than ordinary, because it is only used for the bottom layer to guarantee the safety of bulkheads. In this study, a higher cement content would allow for a larger difference in CTB strength distribution to be derived, which is beneficial for testing and analysis.

The supply rate was calculated as 3.82 L/min. As shown in Figure 1b, the mixer was placed at a higher position than the flume model, and the prepared backfill slurries flowed via pipes to the feeder due to gravity. To ensure relatively stable feeding, there was a large funnel at the end of the pipes to balance the supply and feeding rate. After passing through the funnel, the backfill slurries flowed into the flume model to simulate the real backfilling process. The test continued until the tailings in the silo were used up.

2.4. Testing and Monitoring

2.4.1. Sampling Scheme

To evaluate the changes in properties along the flowing direction, three rows of samples, parallel to the slurry flow direction, were evenly arranged and obtained. Each sample was labeled with the row number and sampling hole number; for example, 2-1 denotes that the sample was derived from the first sampling hole of row 2. The details of the sampling hole positions and intervals are illustrated in Figure 3.

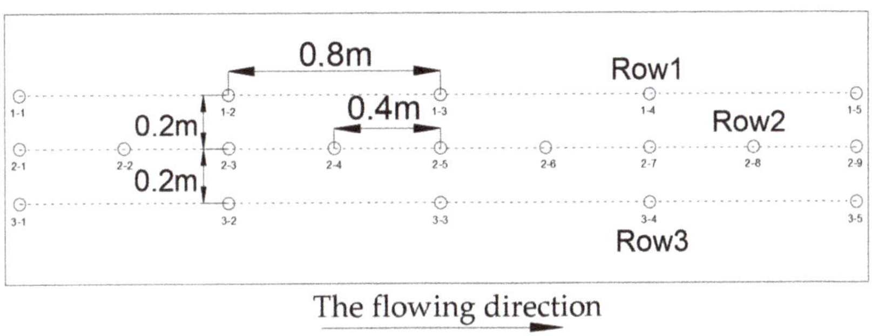

Figure 3. Illustration of sampling scheme of model test from top view.

As plotted in Figure 3, row 2 was set at the center of model, whereas rows 1 and 3 were placed on either side of row 2 with the identical interval of 0.2 m. Nine sampling holes were drilled evenly in row 2 with an interval of 0.4 m. The cement content, PSD, and UCS of these samples were tested and recorded. By contrast, five sampling holes were drilled evenly in row 1 and row 3 with an interval of 0.8 m. The cement content of these samples was tested and recorded. By comparing the measured results of samples at the same distance from the feeding point but in different rows, the reliability and stability of the testing method and results could be rechecked and ensured.

2.4.2. Particle Size Distribution (PSD) Monitoring

Because of the addition of cement, the sampling process needed to be quick, and some pretreatments were applied to prevent cement hydration. Once the flowing of slurries was completed, samples were quickly taken from the nine positions in row 2 (see Figure 3). To prevent cement hydration, the nine samples were soaked in absolute ethanol for 12 h. Subsequently, the samples were placed in a drying oven at 105 °C for 24 h. Thereafter, the PSD of each sample was monitored using a laser particle size analyzer.

2.4.3. Cement Content Monitoring

The cement content of each sample was monitored via titration with EDTA-2Na, which is a standard cement content test method for cement-based materials [40]. It is a chemical measurement method, using 10 wt.% NH_4Cl solution to react with $Ca(OH)_2$, the main hydration product of cement, thus generating $CaCl_2$. Then, using an EDTA-2Na standard solution, the amount of Ca^{2+} produced from the dissolution of $CaCl_2$ could be titrated and measured, thereby reflecting the cement content of the sample. Chen [41] experimentally proved the applicability of this method for the cement content measurement of CTB.

Once the flowing of slurries was completed, nine samples were taken from row 2 and five samples were taken from rows 1 and 3 at the designated positions. Each sample was more than 300 g. The samples derived from row 2 were tested to evaluate the effects of cement distribution on the UCS, whereas the measurement results of samples from rows 1 and 3 were used to validate the reliability of this test method.

2.4.4. Uniaxial Compressive Strength (UCS) Testing

One day (24 h) after the flume test was completed, nine sampling cores from row 2 were drilled, and each core was machined into at least two specimens (diameter 5 × length 10 cm). Thereafter, the derived specimens were placed in a curing room at 25 ± 5 °C and a relative humidity of 95% ± 5% for 60 days. After curing, the UCS of each sample was measured using an HM-5030 uniaxial loader. The load cell of this apparatus has a measurement range of 0–50 kN and an accuracy of ±0.01 kN, thus meeting the UCS testing accuracy requirements.

3. Results and Discussion

3.1. Strength Distribution of Cemented Tailings Backfill (CTB)

The flume test was carried out following the proposed scheme, and the final profile of backfill slurries is presented in Figure 4. Backfill materials gathered in greater abundance near the feeding point in contrast to the far end, resulting in an inclined shape.

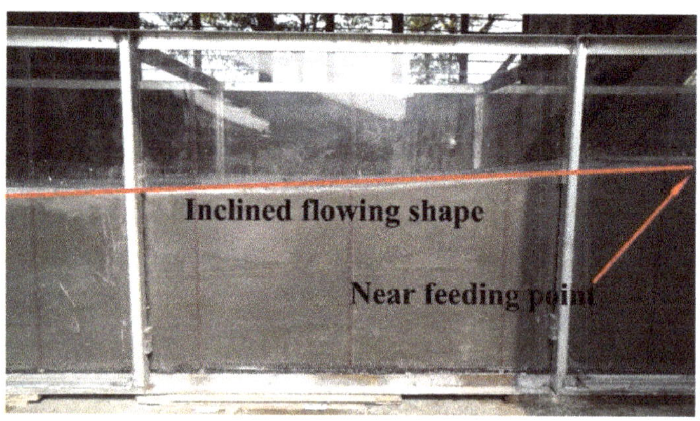

Figure 4. The final profile of backfill slurries after flowing in the test model.

As a result of the inclined shape, the sampling cores near the feeding point were longer than those obtained at the far end, whereby even the shortest core could be machined into two standard specimens, thus fulfilling the requirements of UCS testing. The specimens obtained and the UCS testing process are presented in Figure 5.

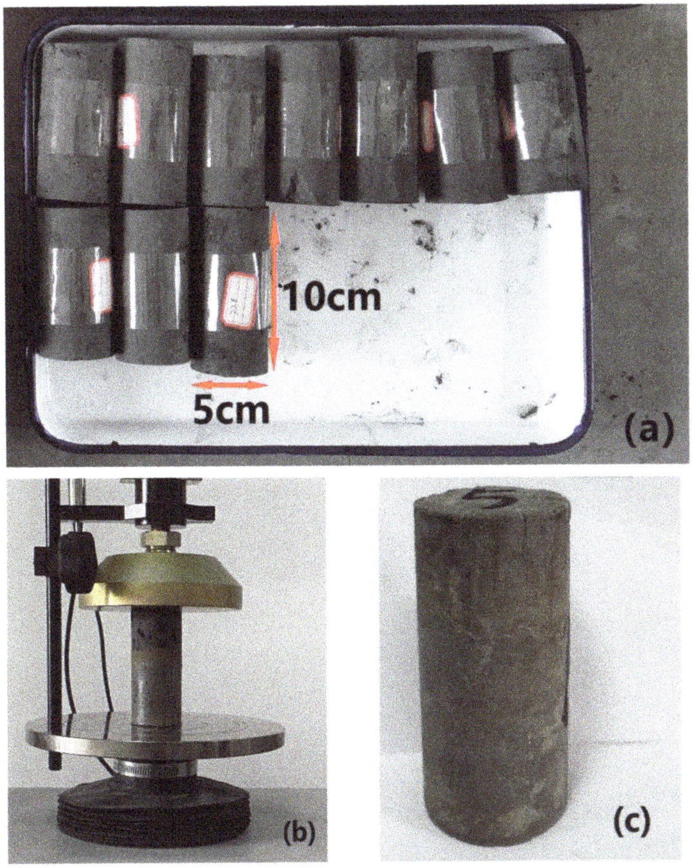

Figure 5. The cemented tailings backfill (CTB) specimens drilled from different areas after the flume test. (**a**) The specimens obtained; (**b**) the uniaxial compressive strength (UCS) testing process; (**c**) the specimen after failure.

The UCS of each specimen was tested and collected, which can be seen in Table 3. Using the position of the feeding point as the origin, the distance from the sampling point to the feeding point as the abscissa value, and the derived UCS as the ordinate value, the distribution of strength along the flowing direction is plotted in Figure 6.

Table 3. The obtained mean UCS values of samples in row 2.

No.	2-1	2-2	2-3	2-4	2-5	2-6	2-7	2-8	2-9
Mean UCS (MPa)	3.58	2.59	2.40	3.04	2.96	3.31	4.13	3.54	3.94

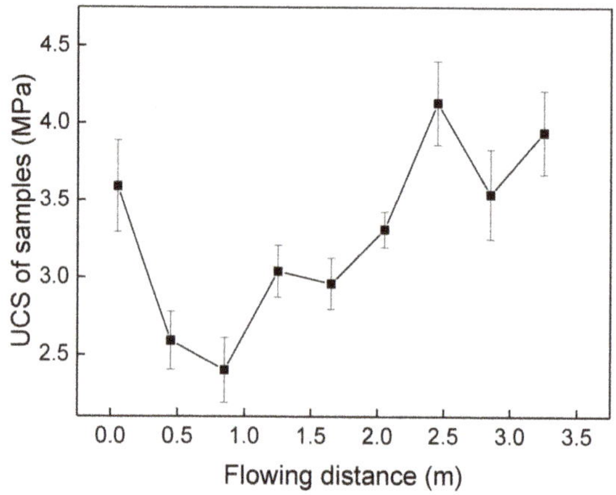

Figure 6. The strength distribution of backfill samples along the flowing direction.

Figure 6 shows that, along the flowing direction, the minimum and maximum values appeared at 0.8 m and 2.4 m away from the feeding points, i.e., 2.40 MPa and 4.13 MPa, respectively. Furthermore, it can also be seen that the values fluctuated along the flowing direction, decreasing first, then increasing for a while, and decreasing thereafter, resulting in an S-shaped UCS distribution, consistent with the aforementioned results in the literature.

3.2. Cement Content Distribution of CTB

The cement contents of sampling cores from all three rows were tested. Firstly, to check the consistency of the recorded results, the values of the corresponding positions in different rows are plotted and compared in Figure 7.

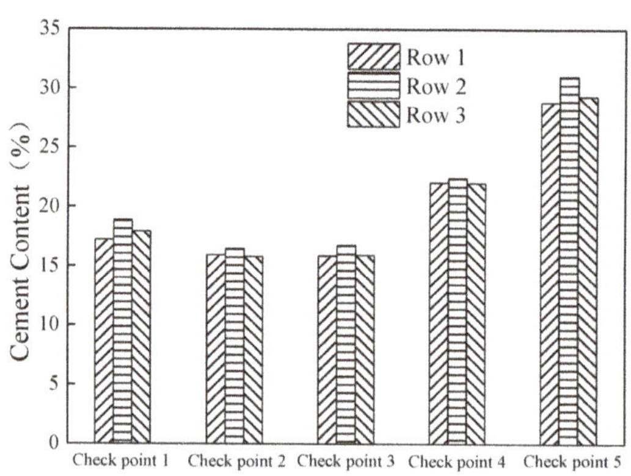

Figure 7. Verification chart of the cement contents of sampling cores in different rows.

It can be obviously observed from Figure 7 that there were no significant differences among the values in different rows at any point. Moreover, the maximum discrepancy rate was less than 2%, demonstrating the consistency and reliability of the recorded cement contents.

Therefore, using the recorded results in row 2, see the values in Table 4, Figure 8 shows the UCS and cement contents of the nine samples as a function of the distance from the feeding point.

Table 4. The cement content values obtained for samples in row 2.

No.	2-1	2-2	2-3	2-4	2-5	2-6	2-7	2-8	2-9
Mean cement content (wt%)	18.87	17.67	16.48	18.71	16.76	20.72	22.44	24.09	31.04

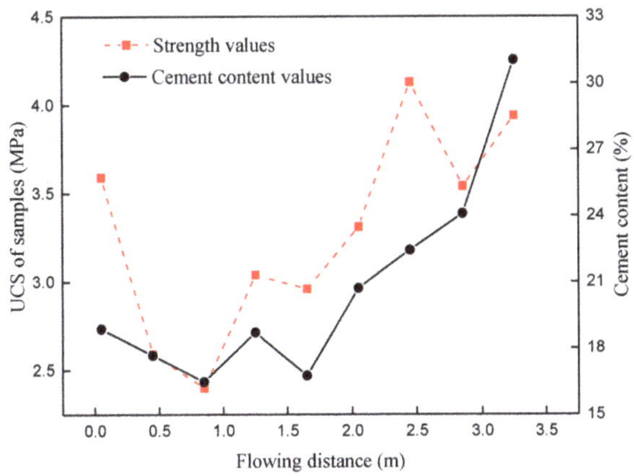

Figure 8. Comparison chart of the cement contents and UCS of samples in row 2 along the flowing direction.

Figure 8 indicates that, along the flowing direction, except for the fourth sample, the cement contents basically decreased first and then increased. The abnormal value of the fourth sample may be attributed to the turbulence of slurry flowing. In addition to the distribution, the figure shows that the peak value, located at the far end, was over 31%, which is much higher than the design value (20%). In contrast, the cement contents within 2 m of the origin were all lower than 20%, and the minimum value was 16.48% (almost half of the peak value). The main reason for the above phenomenon is that the cement particle sizes were much smaller compared to the tailings, thus being transported with the slurry flow, resulting in accumulation to the far end of the test model.

In terms of the relationship between UCS and cement content, it is well accepted that, for CTB with a fixed material content, a higher cement content results in a greater UCS. However, Figure 8 shows that the distribution of UCS differed from that of the cement content. The peak value of UCS appeared 2.4 m from the origin, with a cement content of ~22%, i.e., only 2% higher than the designed content. Moreover, to the far end with the peak cement content, the UCS was also not the highest. This demonstrates that the UCS of CTB in a stope is not only determined by the cement content, as discussed in previous studies [29–31]; some other properties, such as the PSD of tailings or pore structures can also affect the strength.

3.3. Particle Size Distribution of Tailings

As explained above, cement has more fine particles than tailings and is transported farther during flowing, resulting in more cement accumulating to the far end of the testing model. Obviously, for tailings particles, there would also be an inhomogeneous distribution of particle size, as described in previous studies [29,30]; this inhomogeneous PSD of tailings affects the pore structure of CTB and, thus, influences the UCS.

To evaluate these effects, using the proposed sampling scheme, the PSD of the samples in row 2 was monitored; then, to quantitatively describe and compare the features of the PSD, the median particle size (d_{50}) of each sample was selected as the representative parameter for further study, which has been presented in Table 5. Figure 9 shows the change in median sample size along the flowing direction, along with the UCS of corresponding samples.

Table 5. The median particle size of samples in row 2.

No.	2-1	2-2	2-3	2-4	2-5	2-6	2-7	2-8	2-9
Median particle size (μm)	59.85	94.26	104.35	106.10	129.18	77.60	63.92	67.20	39.83

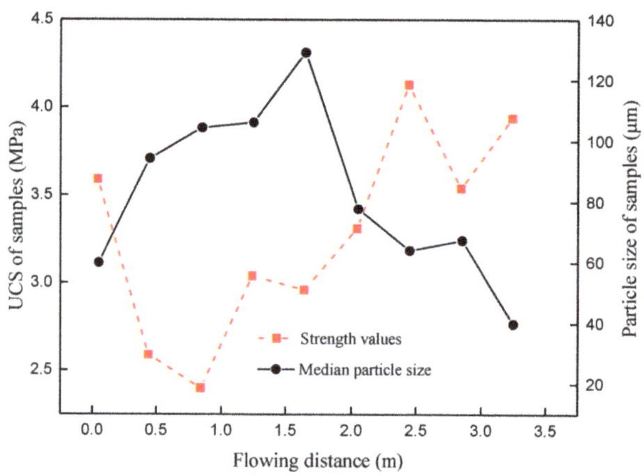

Figure 9. Comparison chart of the median particle sizes and UCS of samples in row 2 along the flowing direction.

From the figure, it can be found that, along the flowing direction, the median particle sizes increased first and then decreased gradually to the minimum value. This trend is mainly attributed to the impediment of particles, particularly coarse particles, due to their collisions and settling, leading to quicker sedimentation. Accordingly, more coarse particles will accumulate during slurry flow, resulting in an increase in median size. Upon the coarse particles settling, the content of fine particles will increase, leading to a decrease in median size to the far end.

However, as shown in the figure, the distribution of UCS did not match that of median particle size. For instance, at a distance of 0–0.8 m, the median sizes increased from ~60 to 100 μm, whereas the values of UCS decreased from ~3.6 MPa to 2.3 MPa. At a distance of 1.6–2.4 m, the median sizes decreased gradually from the peak value to around 62 μm. On the contrary, the UCS values increased from around 2.9 MPa to the peak value. This discrepancy is somewhat inconsistent with previous research conclusions [29,30], which argued that coarse particles can help to increase the strength. Moreover, these findings also indicate that, unlike lab samples, for CTB in stopes, due to material segregation caused by slurry flowing, the inhomogeneous distribution of cement content and the PSD of tailings both comprehensively determine the UCS of CTB.

3.4. Analysis of the Comprehensive Influence Factor of Strength Distribution

As discussed above, the strength distribution of CTB in stopes is comprehensively affected by both the distribution of cement content and the particle size of tailings. In fact, according to the relevant strength theories of concrete [42], some researchers [43–45] have pointed out that tailings particles in cemented backfill mainly constitute the framework of inner structures, whereas hydration products are introduced into the framework to strengthen the structure of the cemented body. Accordingly, when comparing the values in Figures 8 and 9, it can be found that, at the distance with the peak median particle size, the corresponding cement content was only 18%, which is less than the designed cement content, thus leading to a decrease in hydration products and weak connections of the tailings particles in the structure, resulting in a decrease in UCS. Similarly, at the far end, the cement content reached its peak, whereas the median particle size was the minimum, thus affecting the framework structure and leading to a decrease in UCS.

In studies of the soil mechanism [46] and mine backfill (e.g., [1]), the uniformity coefficient (C_u) of tailings was commonly utilized to comprehensively describe the distribution of fine and coarse particles. It is well accepted that the C_u can affect the compactness of the framework of a cemented body, whereby, for higher C_u, fine particles can more easily enter the pores resulting from the accumulation of coarse particles, leading to a more compact framework. Moreover, when the C_u matches the Talbot curve [42,46,47], the most compact structure is formed. Therefore, this coefficient can be used as a quantitative index to evaluate the particle distribution.

Taking into account the above analysis, for a given CTB sample, the cement content can directly reflect the hydration effect, and C_u can be used to evaluate its framework. Therefore, the product of the two parameters can be defined as the comprehensive influencing factor of strength distribution, expressed as:

$$SF = C_c \times C_u \quad (1)$$

where SF is the strength factor, and C_c is the cement content.

Equation (1) and the testing results from samples in row 2 were used to derive the SF, and the results are presented in Table 6.

Table 6. The obtained SF values of samples in row 2.

No.	2-1	2-2	2-3	2-4	2-5	2-6	2-7	2-8	2-9
SF (%)	2.561	1.965	1.870	2.453	2.542	2.845	3.086	2.968	3.452

A comparison chart is plotted in Figure 10 to further demonstrate the relationship between SF and CTB strength as a function of the UCS.

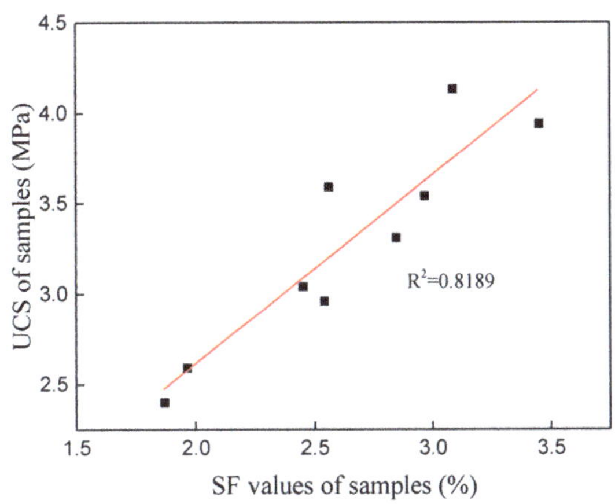

Figure 10. Comparison chart of the strength factor (SF) values and UCS of samples in row 2.

As shown in the figure, the linear correlation coefficient (R^2) of the two parameters was >0.8, indicating a strong linear relationship between SF and UCS. This also experimentally proves that the strength of CTB is comprehensively determined by the distribution of hydration products and the structural framework. Moreover, as the SF can linearly reflect the UCS of CTB to a certain extent, combined with the flume test and in situ coring, analyses of the variation and distribution of SF values can be used in the future as a more effective method to evaluate the quality of the backfill body. In addition to reflecting the strength fluctuation, the SF value can indicate the cement content and tailings PSD. Thus, any decrease in SF can be investigated to determine the cause, allowing the development of a targeted strategy. This is very meaningful for the optimization of backfill processes in a mine.

4. Conclusions

(1) During the flow of backfill slurries in stopes, solid materials settle and become redistributed. With respect to tailings, coarse particles have a larger flowing resistance and are deposited near the feeding point, leading to the median sizes of particles increasing first and then decreasing thereafter. The peak value of median size can be more than three times that of the minimum value.
(2) In contrast to tailings, cement commonly consists of fine particles, facilitating flow to the far end of the stope with backfill water. Accordingly, the experimental results showed that the cement content of CTB first decreased and then increased to the peak at the farthest point. The maximum cement content can be around two times that of the minimum value.
(3) The strength of CTB is comprehensively determined by the distribution of cement content and tailings particles. At the distance with the peak median particle size, the cement content is relatively low, thus reducing the strength. At the distance with maximum cement content, the median size is minimum, thus also reducing the strength. Due to the comprehensive influence of both factors, CTB presents an S-shaped strength distribution.
(4) It was experimentally proven that the SF (strength factor) defined in this study has a strong linear relationship with the UCS; thus, it can be used as a quantitative parameter to reflect the comprehensive effect of cement content and tailings distribution on strength. Thus, any decrease in SF can be investigated to determine the cause, allowing the development of a targeted strategy.

As discussed above, the SF defined in this study can reflect the comprehensive influences of CTB properties on the UCS. It is still an initial analysis of this index, and the data obtained in this study still remain limited. Thus, to further enrich the research content and form a more practical method of quality evaluation of CTB, more experimental studies focusing on the relationship between SF and UCS of CTB will be launched in future research.

Author Contributions: Conceptualization and methodology, X.P.; investigation, X.C.; writing—original draft preparation, X.P.; writing—review and editing, L.G. and G.L.; visualization, X.C.; supervision, X.Y.; project administration, L.G. All authors have read and agreed to the published version of the manuscript.

Funding: This research was funded by the National Key Research and Development Program of China, grant number 2018YFE0123000. The APC was funded by the Key Research Fund of BGRIMM, grant number 02-1911.

Data Availability Statement: All data, models, or code supporting the findings of this study are available from the corresponding author upon reasonable request.

Acknowledgments: The authors would like to acknowledge the Anqing Copper Mine of China for providing tailings, binders, and process water. We would also like to thank Ming Wei from Tongling Nonferrous Metals Corporation, China for his technical support.

Conflicts of Interest: The authors declare no conflict of interest.

References

1. Belem, T.; Benzaazoua, M. An overview on the use of paste backfill technology as a ground support method in cut-and-fill mines. In Proceedings of the 5th International Symposium on Ground Support in Mining and Underground Construction, Perth, Australia, 28–30 September 2004; Villaescusa, Potvin, Ed.; Tayler & Francis Group: London, UK, 2004; pp. 637–650.
2. Sheshpari, M. A review of underground mine backfilling methods with emphasis on cemented paste backfill. *Electron. J. Geotech. Eng.* **2015**, *20*, 5183–5208.
3. Yang, X.C.; Guo, L.J.; Xu, W.Y.; Yang, C.; Hou, G.Q.; Li, Z.N.; Shi, C.X.; Li, W.C.; Liu, G.S.; Peng, X.P.; et al. Introduction. In *Comprehensive Utilization Technology of Tailings and Waste Rocks*, 1st ed.; Ceng, K.F., Liu, M.H., Eds.; Chemical Industry Press: Beijing, China, 2018; pp. 1–3.
4. Huang, Y.C. Introduction. In *Theory and Technology of Mine Backfill*, 1st ed.; Qin, X.Y., Ed.; Metallurgical Industry Press: Beijing, China, 2014; pp. 7–9.
5. Yang, L.; Yilmaz, E.; Li, J.; Liu, H.; Jiang, H. Effect of superplasticizer type and dosage on fluidity and strength behavior of cemented tailings backfill with different solid contents. *Constr. Build. Mater.* **2018**, *187*, 290–298. [CrossRef]
6. Liu, L.; Fang, Z.Y.; Wu, Y.P.; Lai, X.P.; Wang, P.; Song, K.I. Experimental investigation of solid-liquid two-phase flow in cemented rock-tailings backfill using Electrical Resistance Tomography. *Constr. Build. Mater.* **2018**, *175*, 267–276. [CrossRef]
7. Cao, S.; Yilmaz, E.; Song, W. Evaluation of viscosity, strength and microstructural properties of cemented tailings backfill. *Minerals* **2018**, *8*, 352. [CrossRef]
8. Fitton, T.G.; Chryss, A.G.; Bhattacharya, S.N. Tailings beach slope prediction: A new rheological method. *Int. J. Min. Reclam. Environ.* **2006**, *20*, 181–202. [CrossRef]
9. Cao, S.; Song, W.D.; Xue, G.L.; Wang, Y.; Zhu, P.-R. Tests of strength reduction of cemented tailings filling considering layering character. *Rock Soil Mech.* **2015**, *36*, 2869–2876.
10. Béket Dalcé, J.; Li, L.; Yang, P. Experimental study of uniaxial compressive strength (UCS) distribution of hydraulic backfill associated with segregation. *Minerals* **2019**, *9*, 147. [CrossRef]
11. Cui, L.; Fall, M. Multiphysics modeling and simulation of strength development and distribution in cemented tailings backfill structures. *Int. J. Concr. Struct. Mater.* **2018**, *12*, 25. [CrossRef]
12. Lu, G.D.; Fall, M. A coupled chemo-viscoplastic cap model for simulating the behavior of hydrating cemented tailings backfill under blast loading. *Int. J. Numer. Anal. Methods Geomech.* **2016**, *40*, 1123–1149. [CrossRef]
13. Le, R.K.; Bawden, W.F.; Grabinsky, M.F. Field properties of cemented paste backfill at the Golden Giant mine. *Min. Technol.* **2005**, *114*, 65–80.
14. Thompson, B.D.; Bawden, W.F.; Grabinsky, M.F. In-situ measurements of cemented paste backfill at the Cayeli Mine. *Can. Geotech. J.* **2012**, *49*, 755–772. [CrossRef]
15. Gan, D.Q.; Shen, M.F.; Sun, G.H.; Liu, Z.; He, S. Experimental Study on Strength Distribution of Filling Body in Metal Mine. *Met. Mine* **2016**, *45*, 32–35.
16. Wei, X.M.; Guo, L.J.; Li, Z.N. Study on the strength difference of cemented backfill based on macro-micro coupling analysis. *China Min. Mag.* **2018**, *6*, 95–99.
17. Ghirian, A.; Fall, M. Coupled thermo-hydro-mechanical-chemical behaviour of cemented paste backfill in column experiments. Part I: Physical, hydraulic and thermal processes and characteristics. *Eng. Geol.* **2013**, *164*, 195–207. [CrossRef]
18. Ghirian, A.; Fall, M. Coupled thermo-hydro-mechanical-chemical behaviour of cemented paste backfill in column experiments. Part II: Mechanical, chemical and microstructural processes and characteristics. *Eng. Geol.* **2014**, *170*, 11–23. [CrossRef]
19. Kwak, M.; James, D.F.; Klein, K.A. Flow behavior of tailings paste for surface disposal. *Int. J. Miner. Process.* **2005**, *77*, 139–153. [CrossRef]
20. Deschamps, T.; Benzaazoua, M.; Bruno, B.; Aubertin, M. Laboratory study of surface paste disposal for sulfidic tailings: Physical model testing. *Miner. Eng.* **2011**, *24*, 794–806. [CrossRef]
21. Azimi, A.H.; Zhu, D.Z.; Rajaratnam, N. Experimental study of subaqueous sand deposition from slurry wall jets. *J. Eng. Mech.* **2014**, *140*, 296–314. [CrossRef]
22. Xu, W.Y.; Yang, X.C.; Guo, L.J. Experimental Research on Inhomogeneous Simulation of Fill body. *Met. Mine* **2011**, *5*, 18–22.
23. Lu, H.J.; Liang, P.; Gan, D.Q.; Zhang, S.L. Research on flow sedimentation law of filling slurry and mechanical characteristics of backfill body. *Rock Soil Mech.* **2017**, *38*, 263–270.
24. Li, W.C.; Guo, L.J.; Chen, X.Z.; Li, Z.N.; Li, X. Similarity Simulation Experiment on Strength Distribution in Three-Dimensional Spatial of Cemented Tailing Backfill. *Gold Sci. Technol.* **2018**, *26*, 528–534.
25. Shi, C.X.; Guo, L.J.; Chen, X.Z. Experimental Study on the Law of Flow and Segregation of Filing Slurry in Stope. *Gold Sci. Technol.* **2018**, *26*, 520–527.
26. Li, W.C.; Guo, L.J.; Xu, W.Y. Exploratory study on early strength in-situ monitoring of cemented tailing backfill sample. *China Min. Mag.* **1991**, *27*, 139–143.
27. Li, W.C.; Fall, M. Strength and self-desiccation of slag-cemented paste backfill at early ages: Link to initial sulphate concentration. *Cem. Concr. Compos.* **2018**, *89*, 160–168. [CrossRef]
28. Yilmaz, E.; Benzaazoua, M.; Bussière, B.; Pouliot, S. Influence of disposal configurations on hydrogeological behavior of sulphidic paste tailings: A field experimental study. *Int. J. Miner. Process.* **2014**, *131*, 12–25. [CrossRef]

29. Wu, J.; Feng, M.; Chen, Z.; Mao, X.; Han, G.; Wang, Y. Particle size distribution effects on the strength characteristic of cemented paste backfill. *Minerals* **2018**, *8*, 322. [CrossRef]
30. Fall, M.; Benzaazoua, M.; Ouellet, S. Experimental characterization of the influence of tailings fineness and density on the quality of cemented paste backfill. *Miner. Eng.* **2005**, *18*, 41–44. [CrossRef]
31. Ke, X.; Zhou, X.; Wang, X.; Wang, T.; Hou, H.; Zhou, M. Effect of tailings fineness on the pore structure development of cemented paste backfill. *Constr. Build. Mater.* **2016**, *126*, 345–350. [CrossRef]
32. Yilmaz, E.; Benzaazoua, M.; Belem, T.; Bussière, B. Effect of curing under pressure on compressive strength development of cemented paste backfill. *Miner. Eng.* **2009**, *22*, 772–785. [CrossRef]
33. Yilmaz, E.; Belem, T.; Benzaazoua, M. Effects of curing and stress conditions on hydromechanical, geotechnical and geochemical properties of cemented paste backfill. *Eng. Geol.* **2014**, *168*, 23–37. [CrossRef]
34. Yin, G.Z.; Jing, X.F.; Wei, Z.A.; Li, X.S. Study of model test of seepage characteristics and field measurement of coarse and fine tailings dam. *Chin. J. Rock Mech. Eng.* **2010**, *29*, 3710–3718.
35. Yin, G.Z.; Li, G.; Wei, Z.A.; Wan, L.; Shui, G.; Jing, X. Stability analysis of a copper tailings dam via laboratory model tests: A Chinese case study. *Miner. Eng.* **2011**, *24*, 122–130. [CrossRef]
36. Ma, W.; Iaconangelo, C.; de la Torre, J. Model similarity, model selection, and attribute classification. *Appl. Psychol. Meas.* **2016**, *40*, 200–217. [CrossRef]
37. Bech, J.; Roca, N.; Tume, P.; Ramos-Miras, J.; Gil, C.; Boluda, R. Screening for new accumulator plants in potential hazards elements polluted soil surrounding Peruvian mine tailings. *Catena* **2016**, *136*, 66–73. [CrossRef]
38. Demir, F.; Derun, E.M. Modelling and optimization of gold mine tailings based geopolymer by using response surface method and its application in Pb2+ removal. *J. Clean. Prod.* **2019**, *237*, 117766. [CrossRef]
39. Li, W.C.; Fall, M. Sulphate effect on the early age strength and self-desiccation of cemented paste backfill. *Constr. Build. Mater.* **2016**, *106*, 296–304. [CrossRef]
40. Chen, X.Z.; Yang, X.C.; Guo, L.J.; Peng, X. Experimental Study on the Determination of Cement Content in Cemented Tailings Backfill. *Nonferrous Met. Eng.* **2018**, *8*, 69–74.
41. Chen, X.Z.; Yang, X.C.; Guo, L.J. Effect of Curing Age on the Determination of Cement Content in Cemented Tailings Backfill by EDTA Titration Method. *Nonferrous Met. Eng.* **2019**, *9*, 93–99.
42. Mehta, P.K.; Monteiro, P.J. The microstructures of concrete. In *Concrete: Microstructure, Properties, and Materials*, 4th ed.; Ouyang, D., Ed.; China Building Materials Industry Press: Beijing, China, 2016; pp. 21–35.
43. Kesimal, A.; Yilmaz, E.; Ercikdi, B.; Alp, I.; Deveci, H. Effect of properties of tailings and binder on the short and long-term strength and stability of cemented paste backfill. *Mater. Lett.* **2005**, *59*, 3703–3709. [CrossRef]
44. Xu, W.B.; Pang, W.D.; Ding, M. Experiment on evolution of microstructures and long-term strength model of cemented backfill mass. *J. Cent. South Univ. (Sci. Technol.)* **2015**, *46*, 2333–2341.
45. Liu, L.; Fang, Z.Y.; Qi, C.C.; Zhang, B.; Guo, L.; Song, K.I. Experimental investigation on the relationship between pore characteristics and unconfined compressive strength of cemented paste backfill. *Constr. Build. Mater.* **2018**, *179*, 254–264. [CrossRef]
46. Li, G.X.; Zhang, B.Y.; Yu, Y.Z. The Physical properties and engineering classification of soil. In *Soil Mechanics*, 2nd ed.; Qin, N., Ed.; Tsinghua University Press: Beijing, China, 2013; pp. 8–10.
47. Talbot, A.N.; Brown, H.A.; Richart, F.E. *The Strength of Concrete: Its Relation to the Cement, Aggregates and Water*; University of Illinois at Urbana Champaign: Urbana Champaign, IL, USA, 1923.

metals MDPI

Article

Response of Floc Networks in Cemented Paste Backfill to a Pumping Agent

Jiaqi Zhu [1], Shunchuan Wu [1,2], Haiyong Cheng [1,2,*], Xiaojie Geng [1] and Jin Liu [1]

1 Faculty of Land Resources Engineering, Kunming University of Science and Technology, Kunming 650093, China; 20192101073@kust.edu.cn (J.Z.); wushunchuan@ustb.edu.cn (S.W.); 20212002@kust.edu.cn (X.G.); 20192201121@kust.edu.cn (J.L.)

2 Key Laboratory of Ministry of Education of China for Efficient Mining and Safety of Metal Mines, Beijing 100083, China

* Correspondence: chy@kust.edu.cn; Tel.: +86-13811828651

Abstract: Cemented paste backfill is critical for the development of green mines, the safe exploitation of mineral resources deep underground, and the efficient disposal of solid wastes produced by mining. In this paper, the mechanism underpinning how the pumping agent works was studied. The number, area, and fractal dimension of pores in the microstructure of fresh paste were quantitatively analyzed using scanning electron microscopy (SEM), image processing, and fractal theory, and the response of flocs was investigated. The results show that floc networks disintegrated and the liquid network became the dominant structure under the action of the pumping agent, which enhanced the lubrication and promotion of multi-scale particles. In addition, the force chains became fragile and scattered, diminishing the yield stress of the paste. The pores had a more homogenized dimension and the porosity was 15.52% higher. The increase in the fractal dimension of the pores indicated that there was a higher self-similarity, in terms of microstructure, with a strengthened liquid network. The migration of floc structures contributed to the enhancement of the fluidity and rheology of the paste. This study provides insights into the effects of floc and liquid networks on the performance of paste, and it is of engineering significance in terms of realizing safe and efficient CPB operations.

Keywords: cemented paste backfill; ESEM; picture processing; floc networks; pumping agent; pipeline transportation; fractal dimension

Citation: Zhu, J.; Wu, S.; Cheng, H.; Geng, X.; Liu, J. Response of Floc Networks in Cemented Paste Backfill to a Pumping Agent. *Metals* **2021**, *11*, 1906. https://doi.org/10.3390/met11121906

Academic Editors: Lijie Guo and Jean François Blais

Received: 29 October 2021
Accepted: 23 November 2021
Published: 26 November 2021

1. Introduction

Currently, mineral exploitation activities are moving from shallow zones to deep underground in China, and mining enterprises are dedicated to finding new ideas, technologies, and theories for the coordinated development of resources, the economy and the environment under significant administrative and social responsibility-related pressures. Green mining and deep mining have become the two major themes that will be focused on for a long time [1]. However, there are still many challenges in green mining and deep mining. Firstly, considerable safety hazards are hidden in deep mining. Disasters including hard rock bursts and soft rock collapses generally result from high stress in surrounding original rocks and the severe disturbance of tectonic stress, which appear to a severe degree in deep mining activities. In these cases, the backfilling method becomes an indispensable alternative to support the safe mining of deep resources [2]. Secondly, mining operations cause serious environmental pollution. Presently, the tailings produced by mining and mineral processing total 14.6 billion tons and occupy more than 8.7 billion m^2 of land; the amount of mine sewage discharged exceeds 10 billion t/a, which causes serious pollution to rivers and leads to significant excess contents of heavy metals in grains [3]. The backfilling method will provide significant support to the construction of green mines [4]. In order to eliminate potential safety hazards in deep underground mining and protect the environment [5], the backfilling method is promising to become the optimum solution for deep mining and green mining in the future [6].

Cemented paste backfill is a process in which multi-scale solids are mixed with binders, admixtures and water to prepare a paste with high concentration and a low segregation rate, which is transported to underground mined-out areas to achieve a predetermined function [7]. Pipeline transportation plays a key role in cemented paste backfill (CPB) operations [8]. As a material with high mass concentration and large yield stress [9], the paste is likely to cause problems such as pipeline wear, high energy consumption and pipe blockage. To solve these problems, it is necessary to improve the fluidity of the paste, which is an important index in paste quality evaluation due to its pipeline transportation [10,11]. Adding a pumping agent can effectively improve the fluidity of the paste and improve the pipeline's transportation capacity. Therefore, it is of great significance to study the mechanism of pumping agents on paste.

The pumping agent, generally a compound composed of a water-reducing agent, a setting retarder, an air-entraining agent and lubricant, etc., can improve the slurry performance in transportation, so that it can smoothly pass through pipelines without segregation and blockage. It has the functions of fluidization and lubrication, which are appropriate for improving the transportation of slurries with a high concentrations of solids [12]. In recent years, many studies have focused on the effect and mechanism of pumping agents on paste in order to achieve better application effects [13]. Liu, Wang [14] tested the effects of various dosages of a pumping agent on the rheological performance of paste and concluded that it can effectively improve rheological performance and is suitable for CPB in the mine of interest. Ercikdi, Cihangir [15] studied the rheological and mechanical properties of CPB with the addition of three different water-reducing agents, and proposed that the use of a pumping agent can not only ensure the good fluidity of paste but also greatly increase its concentration, thus enhancing the mechanical strength. Yang, Wang [16] discussed the reduction in the resistance of CPB with the use of pumping agent and the mechanism of paste slump loss under long-time action, and it was suggested that the pumping agent can be effectively applied to reduce resistance and alleviate pipe wear in the pipeline transportation of paste. Wu, Ruan [17] found through experiments and numerical simulation that the loss of resistance of pipeline transportation is much higher for paste with high mud content but without pumping agent compared to that with pumping agent, and to that with low mud content but without pumping agent. It is particularly important to add pumping agent to paste with a high mud content.

The manner in which the pumping agent improves the fluidity of pastes and reduces pipeline resistance lies in its impact on the CPB microstructure. Wu, Liu [18] found that there is a three-dimensional floc network in the paste, which shows a structural flow. The rheological properties of paste and the pipeline resistance are closely related to its internal structures. Li, Wang [19] believe that the pumping agent affects the setting of paste by changing the number and size of flocs, as well as the porosity and hydration reactions. Zheng and Gu [20] measured the total porosity and pore size distribution of hardened CPB by using a mercury intrusion method and concluded that the porosity and average pore size were reduced by adding water-reducing agent. Zuo, Wang [21] proposed that the water-reducing agent in the pumping agent aims to destroy the flocs formed by the aggregation of powder particles, and that the particles will be dispersed and uniformly distributed. Thereafter, the water enclosed in flocs can be released, which will improve the paste's fluidity. Zheng, Zhao [22] found that cement and fine tailings net slurry have similar floc structures without water-reducing agent under microscope, and that the water-reducing agent can destroy the flocs with a strong bonding, leading to more free water being released and better paste fluidity being obtained, and thus, the agent will be a prominent admixture for the preparation of CPB. Yang, Wu [23] constructed a microstructure model of fresh CPB and proposed that the pumping agent works by breaking flocs to release enclosed water. Wu, Ai [24] confirmed that, according to their observations of microscopic images of CPB, the flocs were destroyed by pumping agent.

Zhang, Wu [25] also found that the pumping agent destroyed the floc and the yield stress was reduced by using the environmental scanning electron microscope (ESEM).

Pumping agents have a significant influence on the fluidity and rheological properties of pastes; however, at the micro-level, the mechanism underpinning this influence has not been fully figured out and it also needs to be further studied from a mechanical point of view. Quantitative analysis and mechanical analysis of the mechanism of the pumping agent at the micro-level can provide a deeper understanding of its action on the microstructure of the paste and allow better preparation of CPB, based on its fluidity, for the application and development of the CPB technology.

In this study, the mechanism of the pumping agent and the rheological properties of the paste were analyzed by combining the macroscopic rheological properties and microstructure of paste on the theoretical basis of paste flocculation mesh. By adding different dosages of pumping agent, the paste's microstructure will gradually change. Combined with an environmental electron microscope scanning microscopic experiment, image processing technology and fractal theory, the change in state of the paste's microstructure will be quantitatively analyzed. Meanwhile, the gradual change of the microstructure will lead to corresponding changes in the paste's rheological properties. The mechanism of action of the pumping agent and the mechanism underpinning the changes in the paste's rheological properties were studied from the micro scale by associating the paste's microstructural change with the corresponding macro rheological property state change.

2. Materials and Methods

2.1. Materials

Unclassified tailings were taken from a mineral processing plant in Gansu, China. Gansu is located in the hinterland of Northwest China, between 92 degrees (20 min)–108 degrees (42 min) east longitude and 32 degrees (31 min)–42 degrees (50 min) north latitude. The mining area is located in the northwest of Gansu, and the location information is shown in Figure 1. The whole mining area is characterized by monoclinic structures and developed fault poles. The particle size distribution of unclassified tailings was analyzed by using a TopSizer laser particle analyzer (OMEC, Zhuhai, China), as shown in Figure 2. It is established that fine particles will affect the rheological properties of the paste [26]. The contents of −5 μm, −10 μm, and −20 μm were 6.29%, 12.21% and 21.67%, respectively. D_{10}, D_{30} and D_{60} were the diameters of sieve holes with accumulative contents of 10%, 30% and 60%, respectively. The values were D_{10} = 8.1 μm, D_{30} = 30.7 μm, d_{60} = 72.4 μm. The tailing was composed of particles of different sizes. C_U and the curvature coefficient C_C were used to characterize the uniformity of particle size composition of the material. Generally, when $C_U \geq 5$, the size distribution of sand particles was large and the grading was good. C_C represents the continuous gradation of sand. Generally, when C_C = 1~3, the gradation was good and the density of sand was good. Therefore, C_U and C_C were used to evaluate the composition of the tailings. The non-uniformity coefficient C_U was 8.94 and the curvature coefficient C_C was 1.61, indicating that the particles were distributed over a wide range of size with good compactness, continuity and gradation. The mean density of the tailings was 2.852 t/m^3. The loose and compact bulk density were measured to be 1.229 t/m^3 and 1.545 t/m^3, and the loose and compact porosity were 56.9% and 45.8%, respectively. The chemical components of the tailings are shown in Table 1. The tailings could be classified as weakly alkaline inert materials with an alkali coefficient of 0.699 and an activity coefficient of 0.213.

Crushed mine waste rocks, with diameters below 12 mm, were utilized as coarse aggregates. The average particle size was 4.12 mm. The mean density of the rocks was 2.809 t/m^3. The loose and compact bulk density were measured to be 1.615 t/m^3 and 1.844 t/m^3, and the loose and compact porosity were 42.5% and 34.4%, respectively.

The composite Portland cement, P.C32.5R, was utilized as the binder, of which the chemical compositions are shown in Table 1.

Figure 1. Location map of mining area.

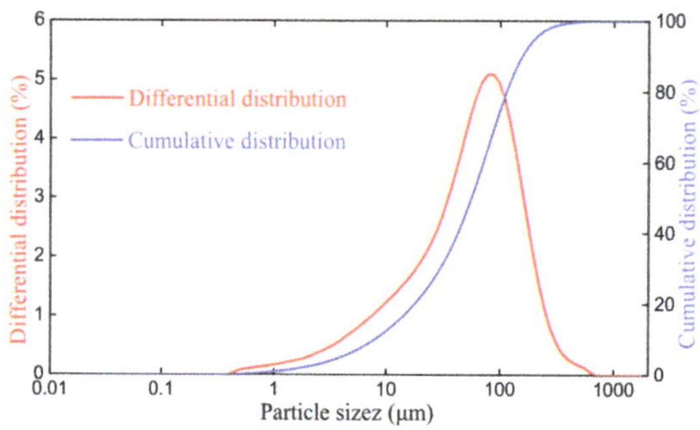

Figure 2. Particle size distribution of unclassified tailings.

Table 1. Main chemical elements in the tailings, cement and crushed mine waste rocks.

Component	Tailings (wt.%)	Crushed Mine Waste Rocks (wt.%)	Cement (wt.%)
SiO_2	36.41	37.31	21.5
Al_2O_3	7.77	3.39	4.5
Fe_2O_3	9.90	- [1]	2.0
CaO	3.09	3.86	63.5
MgO	27.79	28.15	4.0
SO_3	1.63	- [1]	2.5
Others	13.41	10.38	2.0

[1] "-" indicates that the component was not measured.

The pumping agent used, JKJ-NF, was produced by Jinchang Building Materials Chemical Plant, Jinchang, China. This product was selected according to the preliminary comparison of slumping results. JKJ-NF pumping water reducer is a cement concrete pumping agent with naphthalene sulfonated formaldehyde condensate superplasticizer as the main component, which is composed of naphthalene sulfonated naphthalene formaldehyde superplasticizer, air entraining agent, retarder, water retaining agent, slump inhibitor, etc. It is a brown-yellow powder, with a fineness of less than 1.25 mm, and a solid content of more than 94%, and it is non-toxic and non-combustible. Other characteristics include: no corrosion to reinforcement; pH 5–6, water-insoluble matter $\leq 1\%$, sodium sulfate content $\leq 10\%$, and chloride ion content $\leq 0.5\%$. The main features include a high water-reducing rate, an obvious improvement of mixture workability, good pumpability at low and room temperature, and good compatibility with cement.

Tap water, at a pH value of 7.9, was used, which met the experimental requirements for CPB preparation.

2.2. Testing Procedures

2.2.1. Fluidity Tests

The fluidity of CPB refers to its performance in terms of of how it flows by itself or under pump pressure, and its capability to maintain a certain flowing state uniformly and stably. Generally, the slump test can reflect the fluidity of paste and a desired slump value is around 23~26 cm [27].

The CPB samples were prepared at a cement content of 280 kg/m^3, a solid concentration of 79 wt.%, and the ratio of tailings to waste rocks was 1:1, with various pumping agent dosages of 0, 2%, 3%, 4% and 5% relative to cement weight. The dimensions of the slump cone are shown in Figure 3a. In the slumping tests, the CPB was filled into the slump cone and compacted before the cone was lifted at a constant rate, which was completed within 5~10 s. The maximum drop of the paste was the slump and the diameter of the slumping was the divergence, as shown in Figure 3b. The latter is generally an average value of the longest and shortest diameters. To ensure the reliability of the experimental results, three tests were conducted for each CPB, and the average value was adopted.

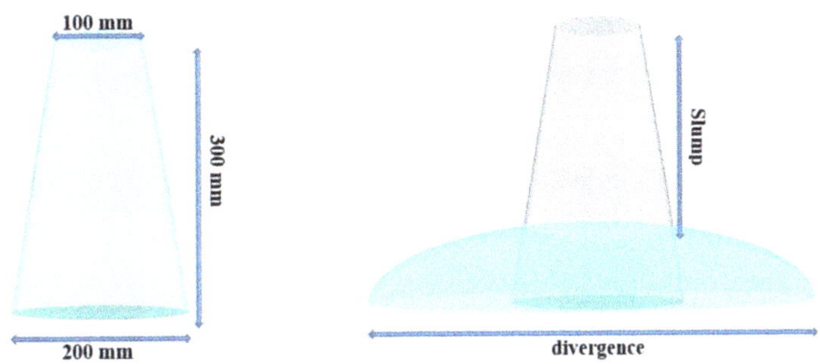

(a) (b)

Figure 3. Schematic of slump tests: (**a**) size of the slump cone and (**b**) measurement of slump and divergence.

2.2.2. Rheological Tests

The proportioning of CPB samples used in the rheological tests was consistent with that in the fluidity tests. The experiments were conducted with a Brookfield R/S plus four-blade vane rheometer, and the vane diameter and height were 20 mm and 40 mm, respectively. According to the characteristics of the paste, the yield stress was measured using a controlled shear rate mode (CSR) [28,29]. The freshly prepared paste was tested at room temperature. A volume of 480 mL CPB in a beaker was pre-sheared at a constant shear rate of 15 s^{-1} for 20 s. Thereafter, the CPB was left standing for 10 s before undergoing an increasing shear from a shear rate of 0 to 180 s^{-1}. From 30 s to 210 s, a data point was collected every second, and a total of 180 data points were collected, as shown in Figure 4.

The shear stress-shear rate curves obtained in the rheological tests were analyzed using the Bingham model to calculate the yield stress and plastic viscosity. The Bingham model is one of the classic models used for characterizing viscoplastic fluids with yield stress, and it describes the rheological properties of yielding fluids when their viscosity is independent of the shear rate (constant). The Bingham model is of good applicability to CPB and is widely used [30,31]. Its mathematical model expression is shown in formula (1):

$$\begin{cases} \dot{\gamma} = 0 & \tau < \tau_y \\ \tau = \tau_y + \eta_p \dot{\gamma} & \tau \geq \tau_y \end{cases} \quad (1)$$

where τ is the shear stress, τ_y is the yield stress, η_p is the plastic viscosity, and $\dot{\gamma}$ is the shear rate.

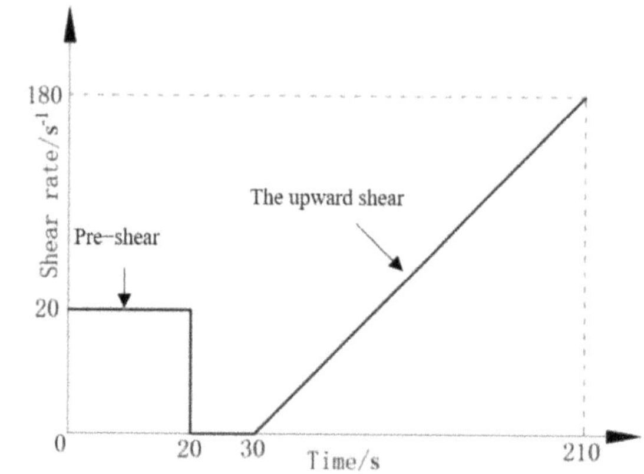

Figure 4. The test program.

2.2.3. ESEM Experiments

The microstructure of the paste was observed by using an environmental scanning electron microscope (FEI Quanta 200, FEI, Eindhoven, Netherlands). The CPB samples were prepared at a cement content of 280 kg/m^3 and a solid concentration of 79 wt.%, and the ratio of tailings to waste rocks was 1:1. Two conditions were set: no pumping agent and 5% pumping agent. Fresh samples were prepared at room temperature for the ESEM test. The samples without pumping agent and those with a proportion of 5% underwent observation to analyze their effects on the microstructure of the paste. In order to better reflect the changes in the paste, the scanning was carried out at 1000 times, 2000 times and 5000 times magnification, respectively, in order to achieve a relatively comprehensive comparison.

3. Results

3.1. Experimental Results

1. The results of the slump and divergence of paste with different contents of pumping agent are shown in Figure 5. The following trend was identified: the better the slump and divergence, the stronger the paste flow performance and the smoother the pipeline transportation flow. It can be seen that the two parameters were at a relatively low level before the pumping agent was added. As the dosage increased, the slump and divergence gradually increased. When the dosage reached 5%, the slump increased from 11 to about 24 cm, and the divergence increased from 25 to 60 cm. This shows that the pumping agent can effectively improve the fluidity of paste.
2. The yield stresses of paste at different dosages of pumping agent are shown in Figure 6. The paste with large yield stress had poor fluidity compared to the paste with small yield stress, which was found to be more likely to have greater pipeline friction in the transportation process and more likely to lead to pipe plugging. As shown in Figure 6, the yield stress of paste without pumping agent reached 600 Pa and did not have fluidity. With the increase in pumping agent, the yield stress gradually decreased. When the content reached 5%, the yield stress decreased to 200 Pa, about 1/3 of the original quantity, and the rheological properties of the paste were significantly improved.

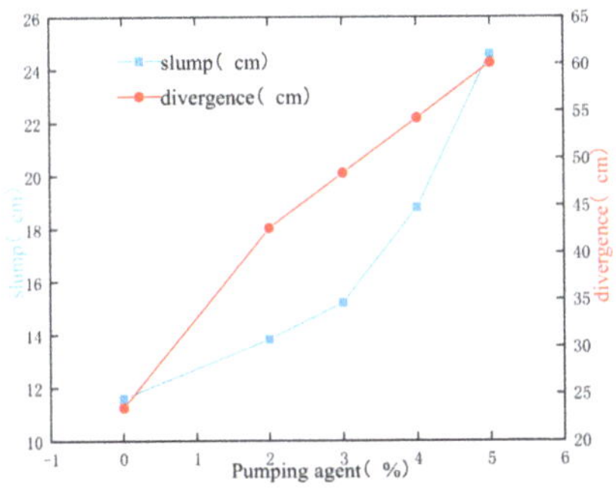

Figure 5. Amount of slump and expansion degree with different pumping agents added.

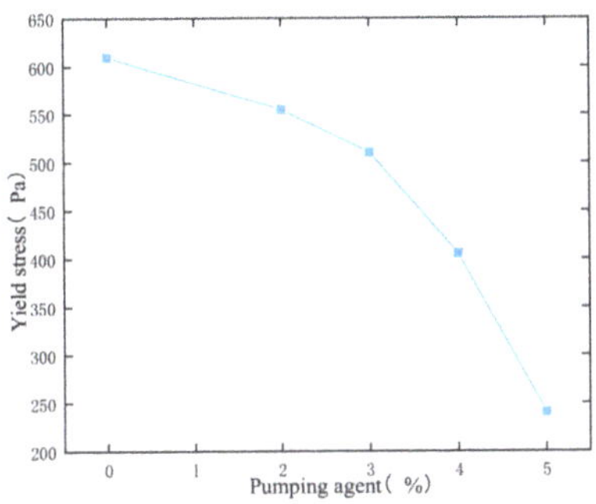

Figure 6. Yield stress curve for different pumping agent dosages.

3. The ESEM images facilitate an insight into the microstructure of freshly prepared paste. As shown in Figure 7, the particles clustered into a dense structure and the pores between the clusters had a small number (a single pore area is large) and a nonuniform distribution without pumping agent. For those pastes that included pumping agent, flocs were disintegrated into smaller and uniformly-distributed particles; although the number of pores increased, the area of individual pores between particles decreased and the pore distribution was relatively uniform.

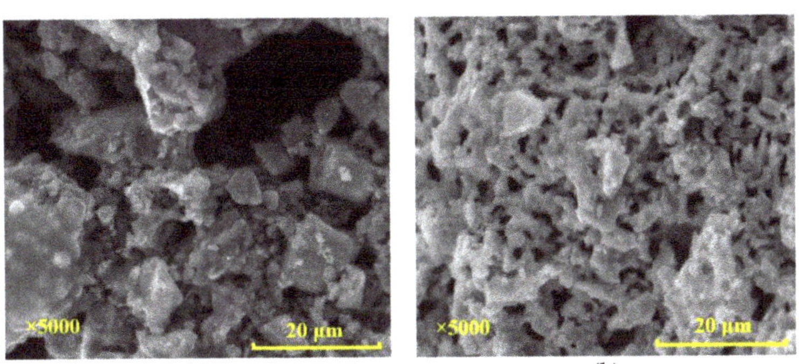

(**a**) (**b**)

Figure 7. SEM image at magnification of 5000: (**a**) no pumping agent added; (**b**) pumping agent added.

The results show that pumping agent can significantly reduce the yield stress of paste and improve its rheological properties. Moreover, it can enhance the fluidity of paste and, thus, the related issues in the pipeline transportation of paste, which occur as a result of excess yield stress and poor fluidity, can be tackled. The macroscopic performance is generally a result of the changes in the microstructure [32]. Next, the article will further study the mechanism of action of the pumping agent from the micro perspective.

3.2. Effects of Pumping Agent on Flocs

Solids in CPB, including cement, tailings and waste stones, have a wide range of particle sizes. In the initial mixing process, flocs formed among particles, mainly as a result of interparticle force [33], as shown in Figure 8 [23,34].

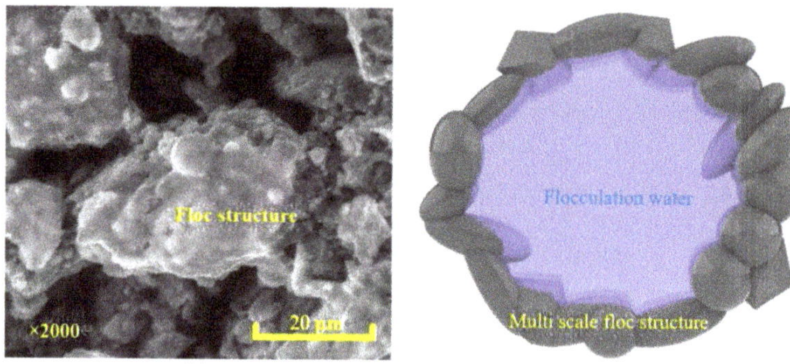

(a) (b)

Figure 8. The floc structure: (**a**) SEM image; (**b**) schematic diagram.

The flocculation partially turned the free water into flocculated water, leading to a decrease in the mass fraction of free water, and the latter had an evident influence on the paste's fluidity and rheology. The friction resistance between particles increased with the reduced lubrication of the water, thus increasing the yield stress and plastic viscosity of the paste. Free water could provide hydrodynamic force to drive the particles in the paste to move forward, and less free water led to the weakening of the hydrodynamic force and also to poor fluidity. A number of particles that could not move forward in a steady state due to lack of force were likely to cause problems such as settlement, segregation and pipe blockage.

The pumping agent was able to break down the floc structure as a result of adsorption and dispersion effects, and thereafter, the water enclosed in the flocs could escape and turned to be free water. The increase in the mass fraction of free water could improve the rheology and fluidity of the paste. It can be observed in SEM images that the floc structure was destroyed and a concave structure formed after the escape of flocculated water, as shown in Figure 9.

Furthermore, the flocs in the paste were generally larger than particles, particularly those huge ones, which caused a number of particles to be closely arranged and, thu, to have a nonuniform distribution. This led to a higher collision probability and larger contact area between the particles, thus increasing the shearing between flocs and leading to a larger yield stress. The diameters of typical flocs and particles were calculated by using ImageJ, and the results are shown in Figure 10. It can be seen that the diameter of a typical floc was larger than 10 microns and the pumping agent could break the flocs into small flocs and particles of several microns, thus obtaining a more uniform microstructure of CPB.

There were many flocs in the paste as shown in Figure 11a, which shows that the paste had many dense flocs that were coagulated together before the pumping agent was added. Flocculation could produce flocs of large diameter, reduce the proportion of free water, and cause the particles to be unevenly distributed and reduce their porosity, which significantly reduced the paste's fluidity. In order to improve the fluidity and rheology of the paste, it was necessary to destroy these flocs, and the pumping agent could break most of them. As shown in Figure 11b, the particles became clearly smaller and uniformly distributed.

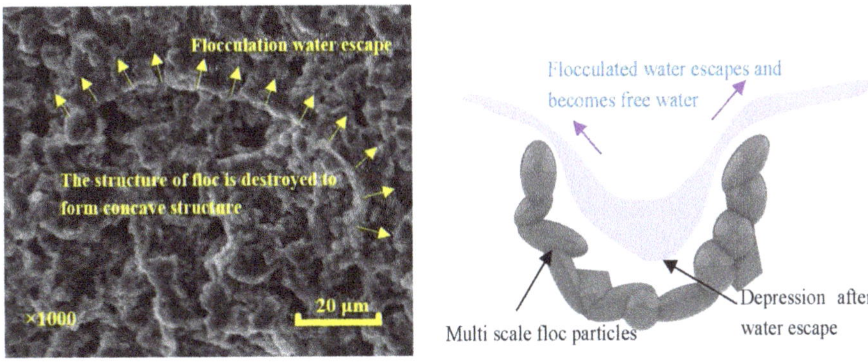

(**a**) (**b**)

Figure 9. The concave structure formed after the breakage of flocs: (**a**) SEM image; (**b**) schematic diagram.

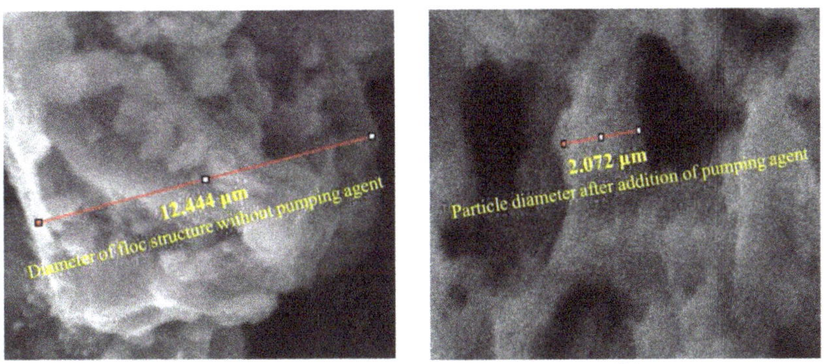

(**a**) (**b**)

Figure 10. Particle diameter measurement: (**a**) diameter of typical floc structure without pumping agent; (**b**) particle diameter after the addition of pumping agent.

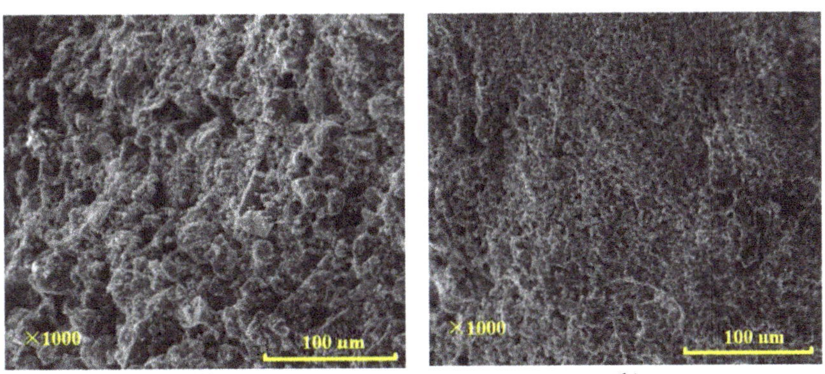

(**a**) (**b**)

Figure 11. SEM pictures at a magnification of 1000 denoting the destructive effect of pumping agent on flocs: (**a**) CPB without pumping agent; (**b**) CPB with pumping agent.

3.3. Evolution of Force Chains under the Effect of Pumping Agent

The flocs were destroyed under the action of pumping agent to different extents. The yield stress of the paste was reduced and its fluidity and rheology were improved. The force chain structure in the microstructure was constructed on the basis of SEM images, as shown in Figures 12 and 13, which can reflect the changes of force chains under the action of the pumping agent. The changes in force chains can explain the changes in energy reserve and yield stress.

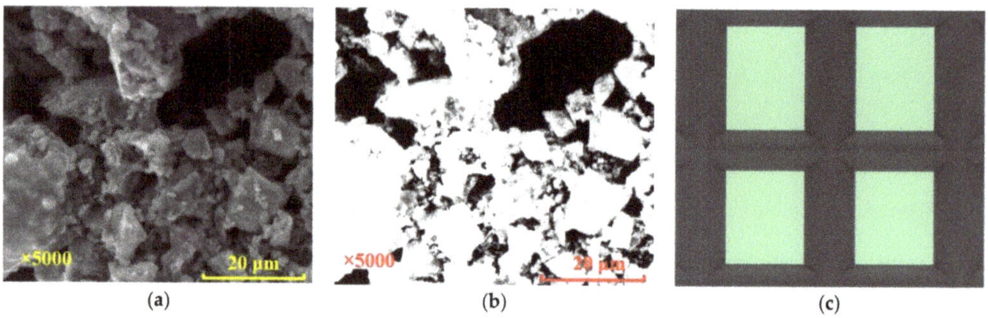

Figure 12. Binary analysis of a microscopic image of the paste without pumping agent and the extraction process of the force chain: (**a**) SEM image at a magnification of 5000; (**b**) binary image; (**c**) idealized sketch of force chain.

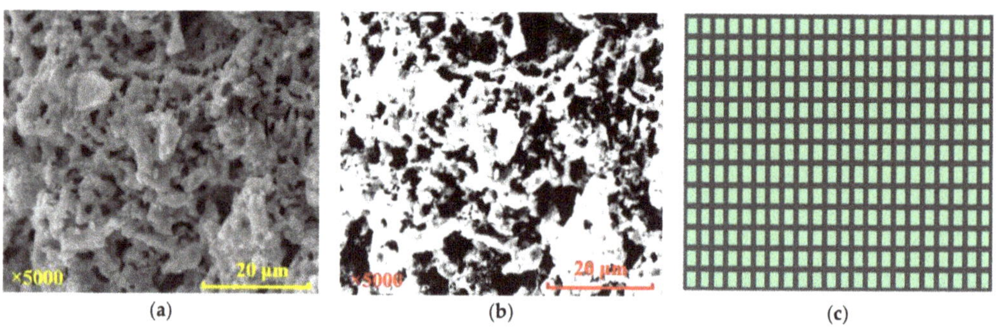

Figure 13. Binary analysis of a microscopic image of paste with pumping agent and the extraction process of the force chain: (**a**) SEM image at a magnification of 5000; (**b**) binary image; (**c**) idealized sketch of force chain.

Figure 12 shows that the force chain was thick and the voids between force chains were large even over a small number, forming an enhancement effect for the paste without the pumping agent. The fully dispersed force chain was defined as the unit force chain. The force chain of the paste without the pumping agent is shown in Figure 12c. The shear stress required to overcome the yield stress increased linearly, and the growth was related to the number of flocs and the tightness of bonding.

For the paste that included pumping agent, the force chains were thinner and evenly distributed, and were similar to the unit force chain, while the voids were smaller and occurred in larger numbers, as shown in Figure 13c. When the unit force chain was damaged by a small shear stress, the fracture coefficient of force chain increased and the shear stress required to move the flocs increased nonlinearly due to the influence of the action time and the recombination coefficient. The force chain structure model is shown in Formula (2) [18]. The pumping agent could change the structure of force chains in paste, thus reducing the stress required to start a flow and improving the fluidity and rheology of the paste.

$$\tau = a \times n \times \tau_0 / f\left(t, \frac{1}{a}\right) \quad (2)$$

where τ is the shear stress for breakage, τ_0 is the yield stress of the unit force chain, a is the recombination coefficient, indicating the ratio of the force chain to the unit force chain, n is the degree of freedom for the linkage, indicating the number of force chains linked, and $f\left(t, \frac{1}{a}\right)$ is the fracture coefficient, which is positively correlated with the action time and the reciprocal of the recombination coefficient.

3.4. Evolution of Liquid Network

Liquid networks are composed of connected pores and free water, which is central to the lubricating and pushing of particles. A fully developed liquid network that has more branches and sufficient contact with particles will present a better performance in terms of lubrication and promotion, corresponding to better fluidity and rheology properties. The abundance of liquid networks can be quantitatively evaluated by the boundary line between pores and particles, that is, the wetted perimeter. In this study, this was measured using ImageJ, on the basis of Figures 11b and 12b, and the results are shown in Figure 14.

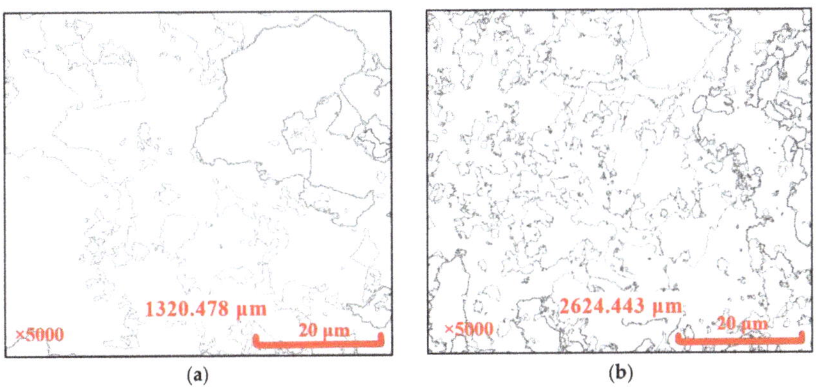

Figure 14. Quantitative analysis of boundaries between particles and pores: (**a**) no pumping agent added; (**b**) pumping agent added.

The length of boundaries for the paste without pumping agent was measured to be 1320.478 um, as shown in Figure 14a, and this value was 2624.443 um for the paste that had pumping agent added, which was nearly twice as long as the former, as shown in Figure 14b. The addition of pumping agent caused the particles to more fully make contact with the free water, and the liquid network was more developed. This was more favorable for lubrication and promotion on particles. Therefore, the pumping agent was able to enhance the liquid network, thus improving the fluidity and rheology of the paste.

3.5. Analysis of Pore Area and Fractal Characteristics

A quantitative evaluation of the area of pores in the images was conducted to provide a direct illustration of the effect of the pumping agent on the porosity and distribution of pores, as well as their dimensions, and the minimum and maximum pore area. The processed images are shown in Figure 15a,b with a magnification of 2000 and the obtained binary pictures are shown in Figure 15c,d.

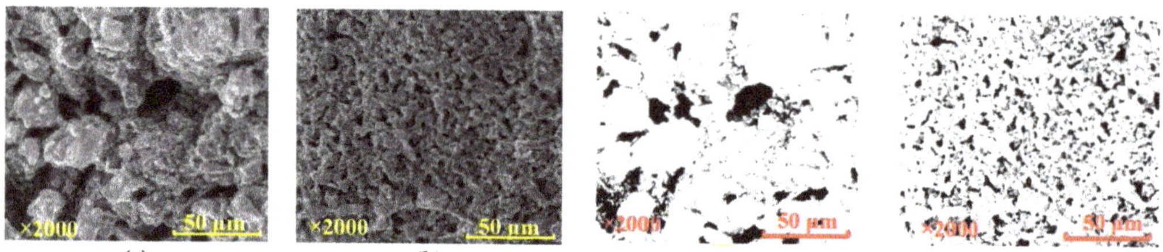

(a) (b) (c) (d)

Figure 15. SEM images at 2000 magnification and binary pictures: (**a**) SEM without pumping agent; (**b**) SEM with pumping agent; (**c**) binary picture without pumping agent; (**d**) binary picture with pumping agent.

The area of pores, as shown in Figure 15c,d, were measured using ImageJ. Prior to measurement, the threshold had been set and binarized. The part where the particles were located is white and the part where the pores were located is black. The measurement and analysis module of ImageJ was used to statistically analyze the number and area of black areas, to number and calculate each closed black pore area, and to analyze their areas. With the pore distribution data obtained, the pore distribution was illustrated and is shown in Figure 16. Figure 16a shows that when no pumping agent was added, the number of extracted pores was 109, the maximum pore area was 411.301 μm^2, there were two pores of more than 300 μm^2, there were also nine pores of more than 50 μm^2, and the pore area span was from several square microns to 400 square microns. Figure 16b shows that after adding the pumping agent, the number of pores was 422, the maximum pore area was only 79.887 μm^2, there were only two pores above 50 μm^2, and the pore area span was between a few square microns and 80 square microns. The porosities were 10.89% and 12.58% for samples without and with pumping agent, respectively. This shows that the use of the pumping agent led to a higher number of pores and a uniform pore distribution.

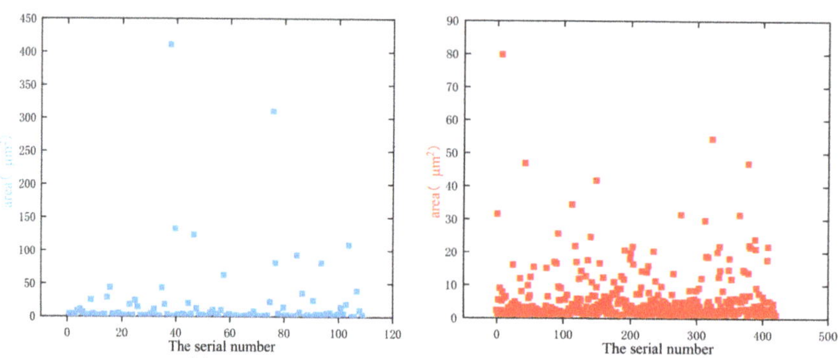

(a) (b)

Figure 16. Distribution of pores: (**a**) no pumping agent added; (**b**) pumping agent added.

ImageJ has a built-in box counting dimension analysis module, which can directly process binary images. The box counting dimension reflects the occupancy of the fractal body in the space. Let $F \in R^n$ be any non-empty bounded subset, $N_x(F)$ represent the minimum number of covering f sets, and the maximum diameter of the set be x, then the upper and lower fractal box dimensions of F can be obtained as shown in Equations (3) and (4), respectively. When the two equations are equal, take this value as the fractal box dimension of F.

$$\mathrm{Dim_B}(F) = \lim_{x\to\infty} \sup \frac{\lg N_x(F)}{-\lg x} \quad (3)$$

$$\mathrm{Dim_B}(F) = \lim_{x\to\infty} \inf \frac{\lg N_x(F)}{-\lg x} \quad (4)$$

The binary images, shown in Figure 15c,d, were processed using ImageJ based on the fractal theory of the box-counting dimension, and the results are shown in Figure 17. The fractal dimension of the pores was 1.6431 for the paste without pumping agent, with a correlation coefficient of 0.99789. This figure became 1.7559 after adding pumping agent, with a correlation coefficient of 0.99561. The increase in the fractal dimension indicated that the self-similarity at the micro level was higher [35]. Furthermore, the pore structure became more complicated [36], the pore distribution became more uniform, and the proportion of pores was higher. These changes improved the development of the liquid network in the paste and enhanced its fluidity and rheology. The following pattern was observed: the larger the fractal dimension of the pores, the better the fluidity and rheology of the paste.

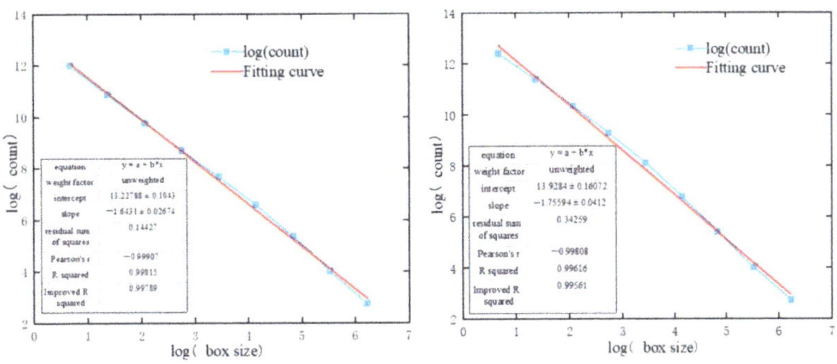

(**a**) (**b**)

Figure 17. Calculation of fractal dimension: (**a**) Without pumping agent; (**b**) with pumping agent.

4. Conclusions

The characteristics of floc networks exert a dominant effect on the fluidity and rheology of paste. A closely connected and complex floc network tends to correspond to strong force chains and underdeveloped channels for free water. In this case, the paste needs to overcome a higher yield stress in order for the flow to begin. In contrast, the pumping agent will break the floc network and a liquid network will form, which will play a dominant role; as such, the lubrication and promotion of multi-scale particles will be enhanced and, thus, the fluidity and rheology of the paste will be improved. The main conclusions are as follows:

1. The pumping agent could effectively improve the fluidity of paste and led to a larger slump and divergence. Flocs of more than 10 microns were, to a significant extent, broken into particles of several microns. In addition, the pumping agent caused the force chains to become fragile and scattered and, as a result, caused the yield stress to decrease.
2. The pumping agent could facilitate the development of liquid networks. The boundaries between the pores and the particles increased from 1324.478 to 2624.443 um after using the pumping agent, allowing the particles to fully make contact with the free water. This allowed the lubrication and promotion of particles.
3. There were 109 pores nonuniformly distributed in the specified area of the paste without pumping agent. The maximum number of pores reached 411.301 um^2 and the proportion of pores was 10.89%. In contrast, for paste with pumping agent added, the number of pores increased to 422, and these were of relatively uniform distribution, with a maximum pore area of 79.887 um^2 and a pore proportion of 12.58%. The addition of pumping agent contributed to the development of a liquid network with more pores and more uniform pore distribution.
4. The fractal dimensions of pores in specified areas before and after the addition of pumping agent were 1.6431 and 1.7559, respectively. The increase in fractal dimension indicated that the self-similarity at the micro level was higher, the pore structure

became more complicated, and the more complicated pore structure promoted the free water flow and the development of the liquid network. The enhancement of the liquid network made it possible to better promote and lubricate the migration of particles. The changes in the microstructure enabled the enhancement of the fluidity and rheological properties of the paste.

Author Contributions: Conceptualization, H.C. and J.Z.; methodology, S.W.; software, H.C. and J.Z.; validation, H.C. and J.Z. and S.W.; formal analysis, X.G.; investigation, J.L.; resources, H.C.; data curation, J.Z.; writing—original draft preparation, J.Z.; writing—review and editing, S.W.; visualization, J.L.; supervision, S.W.; project administration, H.C.; funding acquisition, H.C. All authors have read and agreed to the pblished version of the manuscript.

Funding: This research was funded by National Natural Science Foundation of China, grant number 52074137; the China Postdoctoral Science Foundation, grant number 2020T130272; the Yunnan Province Science Youth Experts Foundation, grant number 202001AU070036; the Yunnan Innovation Team, grant number 202105AE160023.

Data Availability Statement: Data sharing is not applicable.

Acknowledgments: This work was financially supported by the National Natural Science Foundation of China (52074137), the China Postdoctoral Science Foundation (2020T130272), the Yunnan Province Science Youth Experts Foundation (202001AU070036), and the Yunnan Innovation Team (202105AE160023).

Conflicts of Interest: The authors declare no conflict of interest.

References

1. Jiao, H.; Wang, S.; Yang, Y.; Chen, X. Water recovery improvement by shearing of gravity-thickened tailings for cemented paste backfill. *J. Clean. Prod.* **2020**, *245*, 118882.
2. Cheng, H.; Wu, S.; Li, H.; Zhang, X. Influence of time and temperature on rheology and flow performance of cemented paste backfill. *Constr. Build. Mater.* **2020**, *231*, 117117. [CrossRef]
3. Yang, Y.; Zhao, T.; Jiao, H.; Wang, Y.; Li, H. Potential Effect of Porosity Evolution of Cemented Paste Backfill on Selective Solidification of Heavy Metal Ions. *Int. J. Environ. Res. Public Health* **2020**, *17*, 814. [CrossRef]
4. Cheng, H.; Wu, S.; Zhang, X.; Wu, A. Effect of particle gradation characteristics on yield stress of cemented paste backfill. *Int. J. Miner. Metall. Mater.* **2020**, *27*, 10–17. [CrossRef]
5. Mascaro, I.; Benvenuti, M.; Corsini, F.; Costagliola, P.; Lattanzi, P.; Parrini, P.; Tanelli, G. Mine wastes at the polymetallic deposit of Fenice Capanne (southern Tuscany, Italy). Mineralogy, geochemistry, and environmental impact. *Environ. Geol.* **2001**, *41*, 417–429.
6. Wu, Y.Y.A.; Cheng, H.; Chen, S.; Han, Y. Status and prospects of paste technology in China. *Chin. J. Eng.* **2018**, *40*, 517–525.
7. Henriquez, J.; Simms, P. Dynamic imaging and modelling of multilayer deposition of gold paste tailings. *Miner. Eng.* **2009**, *22*, 128–139. [CrossRef]
8. Deschamps, T.; Benzaazoua, M.; Bussière, B.; Aubertin, M. Laboratory study of surface paste disposal for sulfidic tailings: Physical model testing. *Miner. Eng.* **2011**, *24*, 794–806. [CrossRef]
9. Xue, Z.; Gan, D.; Zhang, Y.; Liu, Z. Rheological behavior of ultrafine-tailings cemented paste backfill in high-temperature mining conditions. *Constr. Build. Mater.* **2020**, *253*, 119212. [CrossRef]
10. Cheng, H.; Wu, S.; Wu, A.; Cheng, W. Grading characterization and yield stress prediction based on paste stability coefficient. *Chin. J. Eng.* **2018**, *40*, 1168–1176.
11. Lim, S.; Kim, S.; Ahn, K.; Lee, S. The effect of binders on the rheological properties and the microstructure formation of lithium-ion battery anode slurries. *J. Power Sources* **2015**, *299*, 221–230. [CrossRef]
12. Uchikawa, H.; Sawaki, D.; Hanehara, S. Influence of kind and added timing of organic admixture on the composition, structure and property of fresh cement paste. *Cem. Concr. Res.* **1995**, *25*, 353–364. [CrossRef]
13. Huynh, L.; Beattie, D.A.; Fornasiero, D.; Ralston, J. Effect of polyphosphate and naphthalene sulfonate formaldehyde condensate on the rheological properties of dewatered tailings and cemented paste backfill. *Miner. Eng.* **2006**, *19*, 28–36. [CrossRef]
14. Liu, S.; Wang, H.; Wu, A.; Peng, N.; Yang, P. Study on the Rheological Properties of the Unclassified-tailings Paste Mixed with Pumping Agent. *J. Wuhan Univ. Technol.* **2014**, *38*, 919–922.
15. Ercikdi, B.; Cihangir, F.; Kesimal, A.; Deveci, H.; Alp, I. Utilization of water-reducing admixtures in cemented paste backfill of sulphide-rich mill tailings. *J. Hazard. Mater.* **2010**, *179*, 940–946. [CrossRef]
16. Yang, L.; Wang, H.; Wu, A.; Pan, G.; Liu, S.; Li, H. The Effect of Pumping Aid on Paste Pipeline Transportation. *Metal Mine.* **2014**, *11*, 22–26.
17. Wu, A.; Ruan, Z.; Wang, Y.; Yin, S.; Wang, S.; Wang, Y.; Wang, J. Simulation of long-distance pipeline transportation properties of whole-tailings paste with high sliming. *J. Cent. South Univ.* **2018**, *25*, 141–150. [CrossRef]

18. Wu, A.; Liu, X.; Wang, H.; Wang, Y.; Jiao, H.; Liu, S. Resistance characteristics of structure fluid backfilling slurry in pipeline transort. *J. Cent. South Univ.* **2014**, *45*, 4325–4330.
19. Li, G.; Wang, H.; Wu, A.; Yang, P. Experimental study and mechanism analysis on the effect of pumping agents on the compressive strength of paste slurries. *Chin. J. Eng.* **2016**, *38*, 595–601.
20. Zheng, J.; Gu, D. Study on influence of water-reducing agent on properties of total tailings cemented paste backfill. *Nonferrous Met.* **2014**, *66*, 60–65.
21. Zuo, Y.; Wang, D.; Li, W.; Song, S. Review of interaction between superplasticizer and cement. *Concrete* **2007**, *12*, 79–83.
22. Zheng, J.; Zhao, Z.; Lu, S. Effect of Superplasticizers on the Fluidity of Microfine Slurry. *Met. Mine* **2014**, *01*, 154–157.
23. Yang, P.; Wu, A.; Wang, H.; Li, G.; Peng, N.; Chen, H. Microstructure model of paste slurry rheological properties with pumping admixture. *Nonferrous Met.* **2015**, *67*, 59–64.
24. Wu, A.; Ai, C.; Wang, Y.; Yang, X.; Zhou, F. Test and mechanism analysis on improving rheological property of paste with pumping agent. *J. Cent. South Univ.* **2016**, *47*, 2752–2758.
25. Zhang, L.; Wu, A.; Wang, H. Effects and mechanism of pumping agent on rheological properties of highly muddy paste. *Chin. J. Eng.* **2018**, *40*, 918–924.
26. Hoang, Q.G.; Kaci, A.; Kadri, E.H.; Gallias, J.L. A new methodology for characterizing segregation of cement grouts during rheological tests. *Constr. Build. Mater.* **2015**, *96*, 119–126. [CrossRef]
27. Liu, T.; Zhou, P.; Jin, M. *Backfill Mining Technology and Application*, 1st ed.; Metallurgical Industry Press: Beijing, China, 2001.
28. Wu, A.X.; Jiao, H.Z.; Wang, H.J.; Li, H.; Yi, H.; Liu, X.; Liu, S. Yield stress measurements and optimization of paste tailings. *J. Cent. South Univ.* **2013**, *44*, 3371–3376.
29. Boger, D.V. Rheology forming process. *Int. J. Mech. Sci.* **2009**, *64*, 4225–4536.
30. Wu, A.; Li, H.; Cheng, H.; Wang, Y.; Li, C.; Ruan, Z.-E. Status and prospects of researches on rheology of paste backfill using unclassified-tailings (Part 1): Concepts, characteristics and models. *Chin. J. Eng.* **2020**, *42*, 803–813.
31. Adeyinka, O.B.; Samiei, S.; Xu, Z.; Masliyah, J.H. Effect of particle size on the rheology of Athabasca clay suspensions. *Can. J. Chem. Eng.* **2009**, *87*, 422–434. [CrossRef]
32. Cheng, H.; Wu, S.; Zhang, X.; Li, J. A novel prediction model of strength of paste backfill prepared from waste-unclassified tailings. *Adv. Mater. Sci. Eng.* **2019**, *2019*, 357419. [CrossRef]
33. Tombácz, E.; Szekeres, M. Surface charge heterogeneity of kaolinite in aqueous suspension in comparison with montmorillonite. *Appl. Clay Sci.* **2006**, *34*, 105–124. [CrossRef]
34. McFarlane, A.; Bremmell, K.; Addai-Mensah, J. Improved dewatering behavior of clay minerals dispersions via interfacial chemistry and particle interactions optimization. *J. Colloid Interface Sci.* **2006**, *293*, 116–127. [CrossRef] [PubMed]
35. Cao, S.; Xue, G.; Song, W.; Teng, Q. Strain rate effect on dynamic mechanical properties and microstructure of cemented tailings composites. *Constr. Build. Mater.* **2020**, *247*, 118537. [CrossRef]
36. Sun, W.; Hou, K.; Yang, Z.; Wen, Y. X-ray CT three-dimensional reconstruction and discrete element analysis of the cement paste backfill pore structure under uniaxial compression. *Constr. Build. Mater.* **2017**, *138*, 69–78. [CrossRef]

metals

MDPI

Article

Tailings Settlement Velocity Identification Based on Unsupervised Learning

Jincheng Xie, Dengpan Qiao *, Runsheng Han and Jun Wang

Lianhua Campus, College of Land and Resources Engineering, Kunming University of Science and Technology, Kunming 650093, China; 20202012@kust.edu.cn (J.X.); 20030033@kust.edu.cn (R.H.); 20190077@kust.edu.cn (J.W.)
* Correspondence: danielshak1@hotmail.com

Abstract: In order to reasonably and accurately acquire the settlement interface and velocity of tailings, an identification model of tailing settlement velocity, based on gray images of the settlement process and unsupervised learning, is constructed. Unsupervised learning is used to classify stabilized tailing mortar, and the gray value range of overflow water is determined. Through the identification of overflow water in the settlement process, the interface can be determined, and the settlement velocity of tailings can be calculated. Taking the tailings from a copper mine as an example, the identification of tailings settling velocity was determined. The results show that the identification model of tailing settlement speed based on unsupervised learning can identify the settlement interface, which cannot be manually determined in the initial stage of settlement, effectively avoiding the subjectivity and randomness of manual identification, and provide a more scientific and accurate judgment. For interfaces that can be manually recognized, the model has high recognition accuracy, has a rapid and efficient recognition process, and the relative error can be controlled within 3%. It can be used as a new technology for measuring the settling velocity of tailings.

Keywords: settlement velocity measurement; K-means; tailings backfill; unsupervised learning

Citation: Xie, J.; Qiao, D.; Han, R.; Wang, J. Tailings Settlement Velocity Identification Based on Unsupervised Learning. *Metals* **2021**, *11*, 1903. https://doi.org/10.3390/met11121903

Academic Editors: Lijie Guo and Antoni Roca

Received: 20 October 2021
Accepted: 23 November 2021
Published: 26 November 2021

1. Introduction

In the 14th Five-Year Plan, China explicitly listed "carbon emissions after peaking, steady and declining" as its long-term goal for 2035. General Secretary Xi Jinping has put forward greater requirements to address climate change, for green and low-carbon development and the construction of an ecological society in the present and future, and throughout in the middle of the century. An ideal solution for green and low-carbon development of the mining industry is the use of the backfilling mining method, which is widely used in many mines because of its clean and efficient qualities and its ability to solve the problem of surface tailing waste storage [1–3].

One of the essential processes in the backfilling mining method is to concentrate tailing slurry; however, with the development of grinding and separation technologies, the particle sizes of the tailings are very small, and the proportion of mud is relatively large. Therefore, determining what kind of settling equipment and method should be used in the process of tailing settlement is the main research direction for tailings slurry settlement and concentration, at home and abroad [4]. In practice, due to the different properties of tailings from different mines, the selection of tailing slurry concentration parameters (such as feed concentration and flocculant parameters) cannot completely copy the qualities of other mines; thus, a large number of tailing settlement tests and large-scale prediction analyses are needed. For example, Jiao [5] used tailings from a mine and polyacrylamide (PAM) as experimental raw materials to carry out static flocculation sedimentation tests. Zhang and colleagues [6] analyzed the settlement law of solid particles filled with aggregates. Shi et al. [7] selected different flocculants for tail mortar liquids with different sand supply concentrations, and they carried out flocculation settlement tests for the tail mortar liquid of a vertical sand silo. In tests, it is necessary to measure the settling velocity of tailings, and the method of measuring the settling velocity of tailings is performed by marking the solid–liquid interface between the tailings and water [8,9]. The actual experiments show that the solid–liquid interface does exist in the initial stage of sedimentation, but it is very difficult to determine its position using the naked eye. However, as the sedimentation time goes on, the solid–liquid interface will become more and more obvious. Therefore, in many works in the literature, the settling velocity of tailings is the average velocity after the solid–liquid interface becomes clear and can be discriminated. However, this method obviously cannot fully reflect the settlement process of tailings and has a significant impact on the calculation of subsequent settlement data.

In order to overcome the issues with settling velocity measurements, it is convenient and to accurately identify the position of the solid–liquid interface. In this paper, an unsupervised learning method is proposed to realize automatic tracking and identification of the solid–liquid interface, as well as cluster analysis of sediment and overflow water, so as to accurately and quickly measure the settling velocity of tailings.

2. Identification Model of Tailings Settlement Velocity Based on Unsupervised Learning

2.1. Unsupervised Learning

In the measurements of the settling velocity of tailings, because the particle sizes of tailings are fine, Table 1 shows the particle size distribution and average particle size of tailings from some mines. The particle sizes of the tailings from most copper and iron mines is less than 250 μm, ranging from 250 μm to 20 μm, and the average particle sizes are 80 μm to 140 μm. Compared with the sand used in the concrete industry, these particle sizes are very fine. There is a solid–liquid interface in the early stage of settling, but its specific position is very difficult to determine.

Table 1. Particle size distribution of different mine tailings.

Particle size	Liuju Copper Mine	Dayao Copper Mine	Jiangfeng Iron Mine	Lala Copper Mine	Dahongshan Copper Mine
500.00 μm	0.00%	0.00%	4.27%	0.00%	0.00%
250.00 μm	1.79%	6.98%	24.22%	7.98%	15.25%
185.00 μm	19.27%	36.95%	22.33%	21.50%	32.25%
97.50 μm	21.37%	21.37%	11.84%	32.45%	27.03%
61.50 μm	23.90%	18.97%	10.33%	24.18%	15.77%
43.00 μm	7.70%	5.38%	4.35%	4.72%	2.77%
34.50 μm	6.37%	3.29%	4.40%	3.12%	1.97%
30.00 μm	13.40%	6.01%	10.22%	4.84%	4.97%
20.00 μm	6.21%	1.05%	8.05%	1.21%	0.00%
Average particle size (μm)	86.43 μm	123.77 μm	138.50 μm	111.04 μm	137.20 μm

As shown in Figure 1, although a solid–liquid interface exists in the first measuring cylinder, the specific position is difficult to determine with the naked eye, which is due to a lack of sufficient prior knowledge, which makes it difficult to manually mark. The unsupervised learning algorithm is generally used for the recognition and classification of information in such pictures.

Commonly used unsupervised learning algorithms mainly include principal component analysis (PCA), local linear embedding, and Laplacian Eigenmaps. PCA is a multivariable statistical method, and it is one of the most commonly used dimensionality reduction methods. Using orthogonal transformation, a group of variable data, which may be correlated, is transformed into a group of linearly unrelated variables, and the transformed variables are called principal components. There are two PCA methods, namely feature decomposition and singular value decomposition (SVD). The PCA algorithm is widely used in speech recognition and facial recognition [10]. Locally linear embedding (LLE) is a nonlinear dimension reduction method, proposed by Sam T. Roweis and colleagues in 2000 and published in Science magazine. LLE tries to preserve the local properties of the original high-dimensional data; by assuming that local original data are approximately located on a hyperplane, some local data can be linearly represented by their neighborhood data [11]. Laplacian Eigenmaps (LEs) are used to construct the relationship between the data from the perspective of local approximation. An LE is a graph-based dimension reduction algorithm that constructs the data to be reduced into a graph, and each node in the graph establishes an edge relationship with the nearest k nodes. Then, it attempts to ensure that connected points in the graph (points close to each other in the original space) are as close as possible in the space after dimensional reduction, so that the original local structural relationship can be maintained [12].

Figure 1. Schematic diagram of the settlement process.

In principle, PCA and other data dimensionality reduction algorithms are also used in deep learning, but these data dimensionality reduction methods are complicated and cause abstracted low-dimensional data to lack secondary information, which may be the main factor in distinguishing data at higher levels [13]. Therefore, the unsupervised learning methods presently used in deep learning usually adopt relatively simple algorithms and intuitive evaluation criteria. A typical example of unsupervised learning is clustering. The purpose of clustering is to bring similar object together [14]. In a tailing settlement test, it is necessary to separate overflow water from the tail mortar using a clustering algorithm. Because the particle sizes of tailings are different, and the concentrations of the tail mortar are different, the color and transparency of the overflow water in each test are different; as such, it is obviously not feasible to adopt a method of manual calibration or big data learning. Therefore, the unsupervised learning clustering algorithm can be used to divide the solid–liquid interface. Generally, there are five clustering methods [15–17], the most important of which are the partition method and the hierarchical method. The clustering algorithm divides a dataset into k parts by optimizing the evaluation function, which requires k as an input parameter. Typical clustering algorithms are K-means, K-medoids, and CLARANS. Hierarchical clustering is composed of different levels of segmentation clustering, and the segmentation between levels has a nested relationship. This method does not need input parameters, which is an obvious advantage over the segmentation clustering algorithm, but its disadvantage is that termination conditions must be specified.

The K-means clustering algorithm is the most commonly used clustering algorithm and is based on Euclidean distance, which holds that the less distance there is between two objects, the greater the similarity. The steps of the K-means algorithm are as follows:

1. Select k initialized samples as initial clustering centers.

$$a = a_1, a_2, \ldots a_k. \tag{1}$$

2. For each sample, x_i in the dataset, calculate its distance to the k-many cluster centers and classify it into the class corresponding to the cluster center with the smallest distance.

$$label_i = \underset{1 \le j \le k}{\mathrm{arg}min} \parallel x_i - \mu_j \parallel . \tag{2}$$

3. For each category, a_j, recalculate its cluster center.

$$a_j = \frac{1}{|c_i|} \sum_{x \in c_i} x_i \tag{3}$$

4. Repeat the second and third steps above until a certain suspension condition (iteration times, minimum error change, etc.) is reached. Finally, the samples are divided into K classes, and the centroid is determined to ensure the minimum distance between each class of samples and the corresponding centroid.

In the measurement of the settling velocity of tailings, the whole slurry should be divided into two categories: overflow water and tail mortar. On the basis of a clear classification, the position of the solid–liquid interface can be determined, and the settling velocity of tailings can be determined by moving the position of solid–liquid interface at different times. Therefore, K-means clustering analysis in unsupervised learning can be used to determine the solid–liquid interface.

2.2. Identification Model of Settling Velocity of Tailings

K-means clustering analysis can be used to classify a sample set (data) based on mathematical principles. For the application of settlement velocity measurements, it is necessary that the classification set have physical significance; that is, the results of unsupervised learning should have clear objectives; otherwise, classification results will be unsatisfactory [18].

The process of tailings settlement can be divided into two stages. The first stage is the process of free settlement of solid particles in water, and the second stage is the gradual densification process of solid particles after reaching a certain concentration. The time required by the second stage is much longer than that of the first stage, which is also called the natural settlement process, and the second stage is also called the compression compaction process. When the natural settling process is completed, the slurry is divided into two parts, as shown in the upper left corner of Figure 2; the part with a light color and low concentration is called the supernatant, and the supernatant continues to contain some tailing particles with very small particle sizes that do not settle easily. In the industrial application of tailing mortar concentrations (sand silo concentration or deep cone concentration), supernatant will flow out of the overflow port, also known as overflow water. In the identification model of tailing settlement velocity, the image after natural settlement is selected as the benchmark. At that time, the image is divided into two parts, namely the overflow water (supernatant) and the concentrated tailings mortar. Because the gray values of the overflow water and the concentrated tailings mortar are not continuous at the edges of their images, and the gray value ranges of these two parts of the graphics are obviously different, the K-means classification method is used to divide the settlement images into two categories, according to gray values, and, thus, the gray ranges of the overflow water and the concentrated tailing mortar are obtained. The gray range of the overflow water is [a, b]. Because the color of the overflow water is lighter, the range of [a, b] must be closer to 255. Because tailing particles are opaque, the less tailing particles in the slurry, the lighter the color. If a gray image is used, the lower the concentration in the tailing slurry, the larger the average gray value. The gray scale range of the overflow water is extended to [a, 255] by using the relationship trend between the gray scale value and the tailing mortar concentration. Using this condition, the gray images in the settlement process are compared, and the scope of the overflow water is determined, so as to draw the interface in the settlement process.

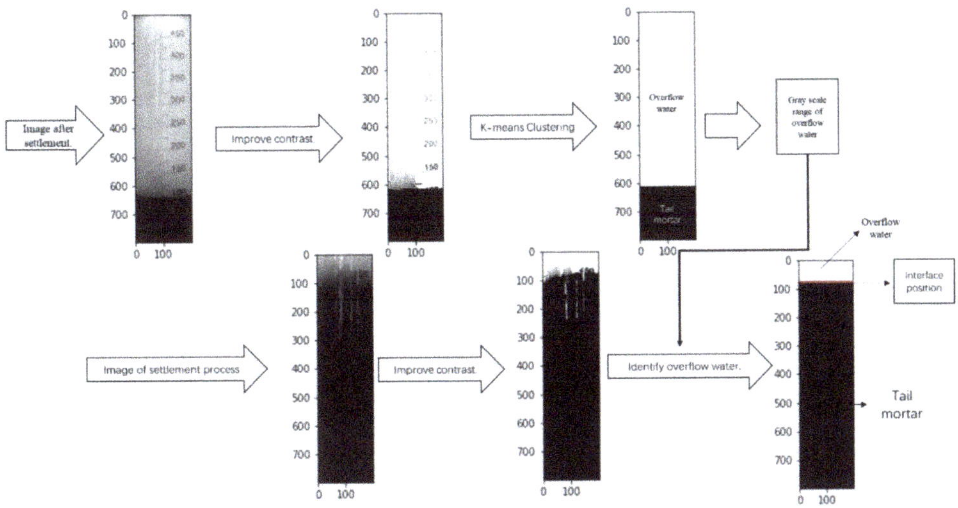

Figure 2. Principle of unsupervised learning to identify settling velocity of tailings.

The principle of unsupervised learning recognition of the settling velocity is shown in Figure 2 (the horizontal and vertical coordinates in the figure are the pixel scale/dpi). The specific identification steps are as follows:

1. Import high-definition video of the tailing sedimentation process, and take a final stable image as the classification basis. The upper liquid is generally regarded as the overflow water, which contains tailing particles that cannot settle.
2. Import the stable grayscale image. The image is processed into an m × n two-dimensional matrix, where each element in the matrix represents the meeting value of the corresponding pixel, and the gray value varies from 0 (black) to 255 (white).
3. Using the histogram equalization method, the contrast is improved, and the image is enhanced, so that the objects and shapes in the image are more prominent. Each row of pixels in the image is regarded as an object for K-means clustering analysis, and all objects are divided into two groups, namely the overflow water group and the tailings group. At the same time, center point A of the classification, that is, the basis point of the classification, can be obtained. In the gray image, the smaller the gray value, the darker the image; the more solid particles contained in the tail mortar, the darker the image will be, so the range of the gray value from 255 to "a" indicates the gray range of the tail mortar that cannot settle, that is, the gray range of the overflow water group.
4. Import the gray image in the process of settlement, and use histogram equalization to improve the contrast and enhance the image. Use the gray scale range of the overflow water group, obtained in Step 3, to identify the overflow water group in the image. Finally, the position of the solid–liquid interface is obtained, from which the settling velocity can be calculated.

The settling process of the tailings is affected by factors, such as the type of tailing, physical and mechanical properties, particle size, concentration, and flocculant, which will lead to drastic changes in the concentration and color of overflow water. Taking the gray scale range of the overflow water in a stable state as a reference, the deviation caused by the above reasons can be effectively avoided, and the accuracy of the settlement interface identification is better ensured.

3. Application Example

Taking the tailings from a copper mine as an example, unsupervised learning is used as the identification method to determine the settlement interface of tailings and calculate the settlement velocity. The mine tailings are small in terms of granularity, and the tailings of a vertical sand silo are used for a cement filling. The key technology lies in the rapid settlement of tailings. The realization of rapid settlement depends on the selection of tailing settlement parameters (feed concentration, flocculant, etc.), and the basis of the selection of settlement parameters is used to measure the settlement speed of tailing slurry at different concentrations.

3.1. Physical and Mechanical Properties of Tailings

The physical and mechanical properties of the mine tailings are shown in Table 2, and the gradation distribution of the tailings is shown in Table 3. It can be seen from Table 3 that 19.61% of the total tailings are smaller than 500 mesh. The average particle size of the tailings is 87.22 μm, and the median particle size is 58 μm.

Table 2. Physical properties of tailings.

Number of Tests	Mass (g)	Volume (mm^3)	Apparent Density (t/m^3)	Loose Density (t/m^3)	Packing Compactness	Void Ratio
Test 1	53.76	20.40	2.64	1.39	-	-
Test 2	53.75	20.40	2.64	1.37	-	-
Test 3	56.61	21.39	2.65	1.38	-	-
Average	-	-	2.68	1.38	51.50%	48.50%

Table 3. Gradation distribution of tailings.

Particle Size (μm)	Distribution (%)
250	1.79
185	19.27
97.5	21.37
61.5	23.90
43	7.70
34.5	6.37
30	19.61

From Table 4, showing the mineral composition analysis, it can be seen that the main components of the tailings are calcite and mica, accounting for 40–50% and 23–35%, respectively. The chemical properties of the tailings are relatively stable, and there is no chemical reaction with water. The PH of the tailings slurry is shown in Table 5, The PH of the tailings slurry is about 7; the slurry is neutral, and the PH remains basically unchanged with the increase in concentration.

Table 4. Analysis results of mineral composition of tailings.

Mineral Name	Mineral Content (%)
Calcite	40–50
Mica	25–35
Fluorite	5–10
Dolomite	5–10
Amphibole	1–5

Table 5. PH value of tailing slurry with different concentrations.

Mass Concentration	PH
15%	7.15
20%	7.15
25%	7.15
30%	7.15
35%	7.16
40%	7.16

3.2. Experimental Study

The K-means unsupervised learning classification method can classify tailings and overflow water after settlement stabilization and distinguish them in images in order to judge the interface in the settlement process and calculate the settlement speed. Because of visual recognition, the influence of illumination on environmental variables is very important; therefore, the test device (shown in Figure 3) was designed.

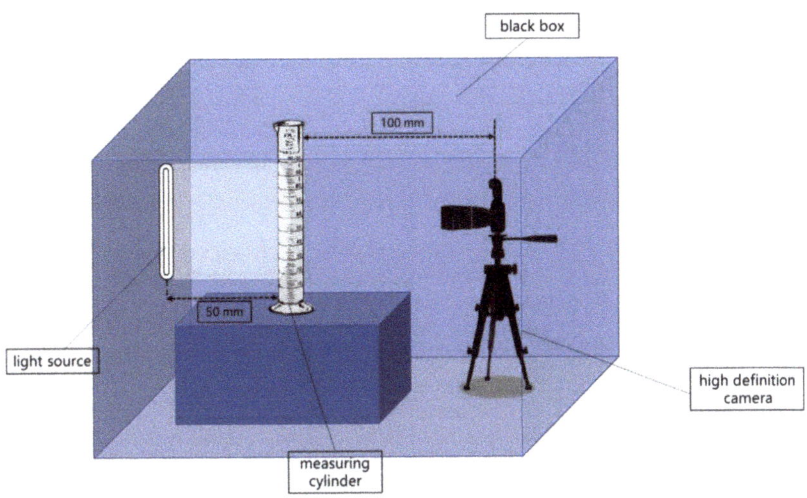

Figure 3. Schematic diagram of experiment device.

The experimental device consists of a black box, a measuring cylinder, a light source, and a high-definition camera. The experiment is performed in the black box and is not affected by external light source to ensure that the influence of external light sources on visual recognition is minimized. The light source is a 300-mm-long LED strip. There are 30 SMD LED beads on the lamp belt. The luminous brightness of the beads is 300 LM, the power of the lamp belt is 7.2 W/m, the voltage is 3–3.2 V, and the luminous angle is 120°. The light source is located behind the measuring cylinder, 50 mm away from the measuring cylinder, which provides the illumination conditions for the experimental process. The experiment was recorded using a high-definition camera. The camera used a Canon EOS

200D optical viewfinder, The manufacturer is Japan Canon Company, manufactured in Oita Prefecture, Japan. and the photosensitive element is CMOS. When recording video, it uses the 1080P short film shooting mode, with the sensitivity of ISO100–ISO12800, and the automatic exposure function is turned off. The optical viewfinder is installed on a tripod, and its highest point is level with the center of the measuring cylinder. The distance between the optical viewfinder and the measuring cylinder is 100 mm, focusing on the measuring cylinder. After enough settling time, the black box is opened to obtain the image data taken using the high-definition camera.

Tailing settlement is not only affected by physical and mechanical properties and particle size distribution of tailings but also by the slurry concentration. Therefore, different concentrations of tailings, from 15% to 40%, were designed. The unsupervised learning method is compared with manual recognition to verify the accuracy of the unsupervised learning results.

3.3. Experimental Results and Analysis

Screenshots of the shooting results of the HD camera at certain time intervals are shown in Figure 4. In Figure 4, on the left, is the tailing slurry with mass concentration of 15%, and the axis is the interval time of each image; the time scale of the axis is 125 s. On the right is the tailing slurry with mass concentration of 20%; the time scale of several axes is 160 s.

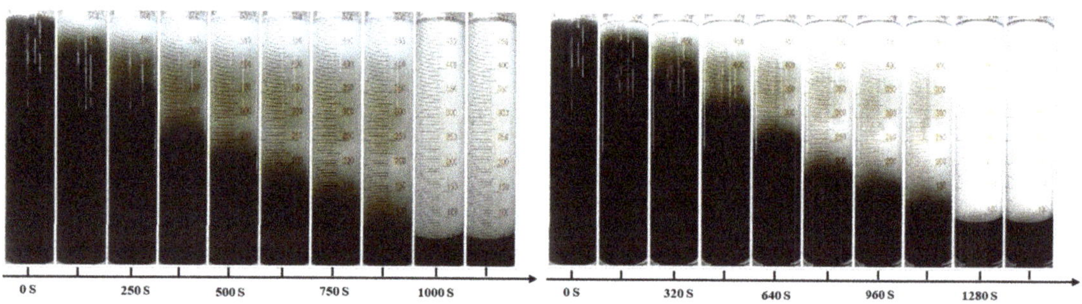

Figure 4. Settlement experiment: (**left**) mass concentration 15%, (**right**) mass concentration 20%.

It can be seen that the settling speed of particles is very fast in the early stage of the experiment, but it is difficult to manually determine the interface position due to the influence of fine particles. In the later stage, the settlement velocity decreases, the interface tends to be stable, and the interface position remains unchanged for a long period of time. Therefore, the gray scale interval of the overflow water can be obtained using the image after sedimentation stabilization.

The process of determining the gray range of the overflow water is shown in Figure 5. The grayscale mode is used to import the image after settlement, and the range of the grayscale values is [0, 255], as shown in Figure 5a. (In this example, the pixels of the image are adjusted to 800 $\times$ 200 dpi, and the more pixels there are, the higher the accuracy is, and the more computing resources are occupied.) In order to conveniently explain the principle of judging the gray value of overflow water, each row of pixels in the picture is averaged and expanded into a gray statistical chart, as shown in Figure 5b. Each row of pixels will be in different positions according to the average gray value, and its positions can be roughly divided into two categories. One is the point with a larger gray value, where the gray value is greater than 100 and is visually lighter in color. The other type is the point with a smaller gray value, with a gray value below 50, and the visual expression of the color is darker.

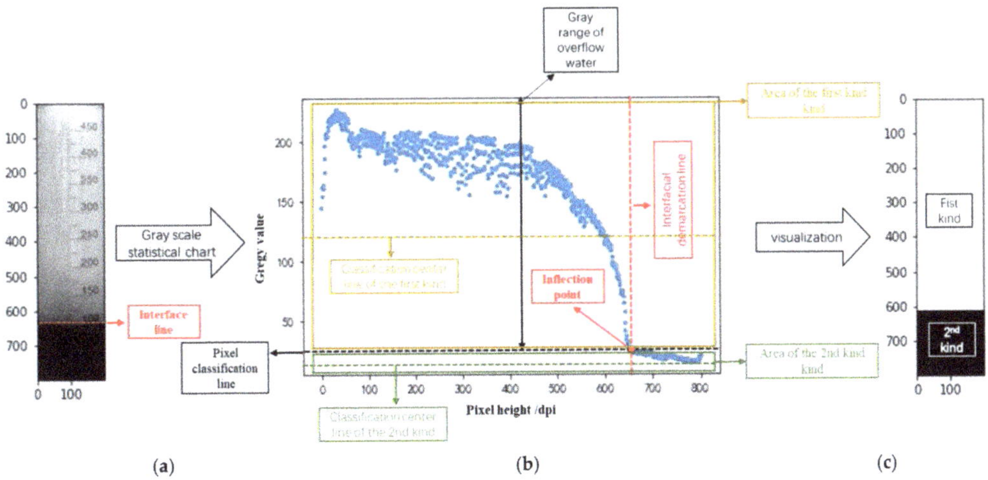

Figure 5. Determination of gray scale range of overflow water. (**a**) original image; (**b**) gray statistical chart; (**c**) visualization of classification results.

As can be seen from Figure 5b, there are obvious differences in the gray values at different heights. Because the imported image is a settled image, its interface is visible to the naked eye (as shown in Figure 5a). The gray value is discontinuous at the interface, and there is an inflection point (as shown in Figure 5b). The inflection point is the point where the gray value of the overflow water has the smallest (because the larger the gray value, the lighter the color, and the larger the gray value, the lower the concentration). Therefore, as long as the coordinates of the inflection point are obtained, the minimum gray value of the overflow water can be obtained (Assuming that the value is a, then, the interval [a, 255] is the gray range of the overflow water). With different concentrations and physical characteristics of tailing slurry, the gray image of the tailings after settlement is different. The only similarity is that they all have inflection points that can determine the gray value of the interface. That is to say, as long as gray value a of the inflection point is obtained in each video, the gray range of the overflow water [a, 255] can be determined, and the overflow water and the interface position of each image in the video can be judged based on this. Averaging the gray values of each row of pixels is only for convenience of explaining the principle and visualization of coordinates. In actual operations, the gray values of each row of pixels are represented by a 200-dimension matrix (hypersurface.) Therefore, the K-means clustering method is used to classify the gray values of each row. They are fully divided into two categories, as shown in Figure 5b. Each category has a center line (plane), which indicates that each element in the category has the shortest distance from the center line. The first category is the overflow water. As long as the classification center of the overflow water is determined, the minimum gray value of the overflow water can be obtained. In actual operations, the gray value boundary of the overflow water can be obtained through iterative comparison, etc. With the solution with water as medium, a larger gray value means clearer water, and the lower the slurry concentration, the range of the minimum gray value of the overflow water to 255 is the gray value range of the overflow water. It is important to point out that the gray value range of the overflow water is affected by the physical characteristics of the tailings, settling concentration, water quality, pH, etc., so, the gray value range of the overflow water only represents the range of the gray value of the overflow water of tailing slurry in this video, and it cannot be applied to other concentrations and other tailings.

Taking the gray scale range of the overflow water as the judging condition of the overflow water, we can determine the image area of the overflow water by introducing it into each frame of the settlement process. Because the settlement process is caused by gravity, the area of the overflow water must expand from the top to the bottom. Therefore, in each image, the part excluding the overflow water is regarded as part of the tailing slurry, and the dividing line between the overflow water and the tailing slurry on the pixel scale is the solid–liquid interface to be identified. Therefore, the interface of each frame of the images in the settlement process can be determined. Based on the above principle, the images of the interface changes over time under different concentrations are shown in Figure 6, and the calculated settlement speed is shown in Table 6.

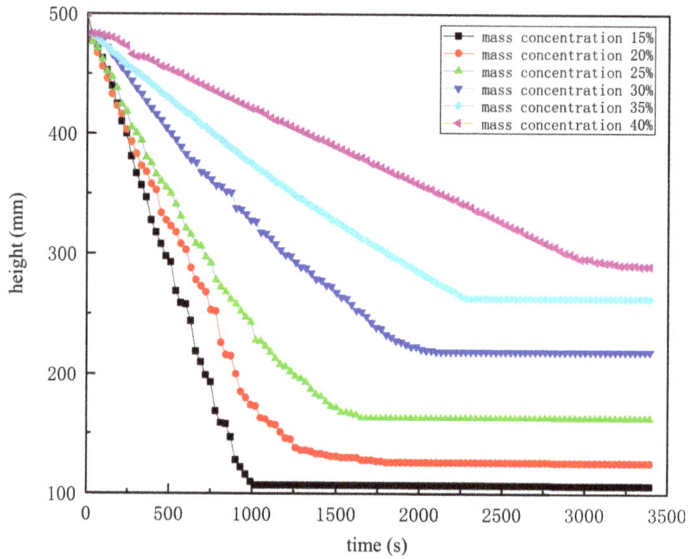

Figure 6. Unsupervised learning recognition result of interface settlement.

Table 6. Tailings settling velocity.

Mass Concentration (%)	Volume Concentration (%)	Sedimentation Rate (m/s)
15	5.68	8.333×10^{-4}
20	7.86	7.658×10^{-4}
25	10.21	6.982×10^{-4}
30	12.76	6.270×10^{-4}
35	15.52	6.091×10^{-4}
40	18.54	5.466×10^{-4}

As shown in Figure 6, the relationship between the height and time of the settlement interface, identified using the unsupervised learning method, agrees with the actual situation; that is, the settlement interface will gradually decrease with the passage of time, and after a certain period of settlement, the interface will remain stable and not decrease. Comparing the settling velocity of the tailing slurry at different concentrations, it can be seen that the settling velocity will decrease with the increase in tailings slurry concentration, and the decreasing range is smaller and smaller, which is also consistent with the relationship between the settling velocity and the concentration that was observed manually. After settlement, the interface height basically does not change.

Comparing the results of K-means unsupervised learning with those of manual recognition (taking a concentration of 15% as an example), as shown in Figure 7, we can clearly see the advantages of unsupervised learning: it can read the interface data that cannot be distinguished through manual recognition; that is, the data that cannot be recognized through manual recognition because the interface is not obvious at the beginning of the settlement process. Secondly, there is little difference between the results of K-means unsupervised learning and those of manual recognition, and the maximum error does not exceed 3%. Moreover, using K-means unsupervised learning to identify the settlement speed is very fast, and the identification process is completed in a few seconds (60 images), which saves a great deal of time compared with manual identification and, thus, greatly improves work efficiency.

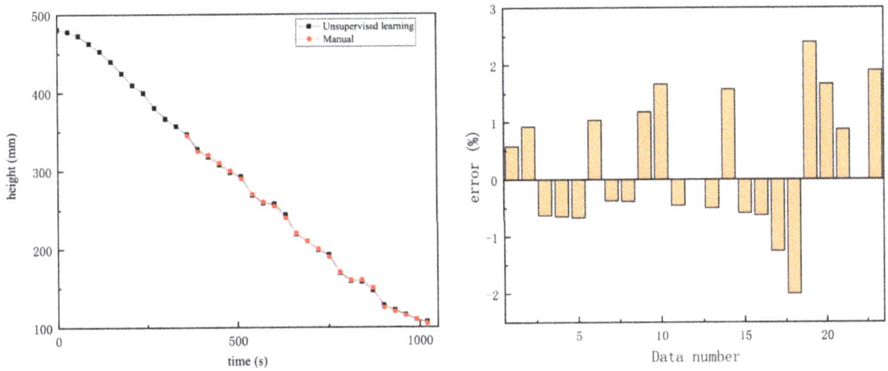

Figure 7. Comparison between the unsupervised learning recognition results and the manual recognition results.

4. Conclusions

Tailing settling velocity is important information to be obtained for filling slurry making and filling system optimization. Because of the small particle size of tailings, the interface is not obvious in the initial stage of sedimentation. Unsupervised learning can effectively avoid the subjectivity and randomness of settlement interface judgments and make more scientific, accurate, and evidence-based judgments.

1. The identification model of the tailing settlement speed adopts the K-means unsupervised classification method to identify the stabilized settlement image and judge the gray value interval of the overflow water. By identifying the overflow water, the interface in the settlement process can be judged, and the settlement speed can be calculated. The model has the characteristics of a high recognition accuracy and high speed. Because of the unsupervised learning method, the recognition accuracy of the model is independent of the amount of learning data, and only stable images are needed for recognition and analysis. Moreover, the model has a wide range of applications and has a high recognition accuracy for the settlement interface of tailing mortar with different particle size distributions, physical and mechanical properties, and flocculant addition or not.
2. Taking the tailing settlement of a copper mine as an example, the unsupervised learning model is applied to the tailing settlement process. Based on images of tailings at different concentrations in a final stable state, the interface position in the process of tailing settlement is identified, and the settlement speed of the tailings is obtained. The experimental results show that the model can identify the interface position, which cannot be distinguished manually, and gives the settlement velocity of the initial settlement. In addition, when the position of the interface can be determined, the accuracy of the model recognition is high, and the error rate is small.

The research results provide a new idea and reliable technical support for the determination of the settlement velocity of mine tailings and the optimization of parameters.

Author Contributions: Conceptualization, J.X.; methodology, J.X. and D.Q.; software, J.X.; validation, D.Q., R.H. and J.W.; formal analysis, J.X.; investigation, J.X. and D.Q.; resources, J.X., D.Q., R.H. and J.W.; data curation, J.X.; writing—original draft preparation, J.X.; writing—review and editing,

J.X. and D.Q.; visualization, J.X.; supervision, D.Q.; project administration, J.X. and D.Q.; funding acquisition, J.W. All authors have read and agreed to the published version of the manuscript.

Funding: This research was funded by Basic Research Program of Yunnan Province—Youth Program, grant number NO.202001AU070062 and The APC was funded by Talent Training Fund of Kunming University of Science and Technology, grant number No. KKZ3202021040.

Institutional Review Board Statement: Not applicable.

Informed Consent Statement: Not applicable.

Data Availability Statement: The data presented in this study are available on request from the corresponding author.

Conflicts of Interest: The authors declare no conflict of interest.

References

1. Wu, A.; Yang, Y.; Cheng, H.; Chen, S.; Han, Y. Development status and trend of paste technology in China. *J. Eng. Sci.* **2018**, *40*, 517–525.
2. Wu, A.; Wang, Y.; Wang, H. Present situation and trend of paste filling technology. *Metal. Mine* **2016**, *7*, 1–9.
3. Qiao, D.; Cheng, W.; Zhang, L.; Yao, W.; Wang, X.; Wang, H. Modern mining concept and filling mining. *Sci. Eng. Nonferr. Met.* **2011**, *2*, 7–14.
4. Rulyov, N.; Laskowski, J.S.; Concha, F. The use of ultra-flocculation in optimization of the experimental flocculation procedures. *Physicochem. Probl. Miner. Proc.* **2011**, *1*, 5–16.
5. Jiao, H.; Wu, A.; Wang, H.; Liu, X.; Yang, S.; Xiao, Y. Experiment study on the flocculation settlement characteristic of unclassified tailings. *J. Beijing Univ. Sci. Technol.* **2011**, *33*, 1437–1441. [CrossRef]
6. Zhang, Q.; Xie, S.; Zheng, J.; Wang, X. Study on settlement law of filling slurry and feasibility analysis of transportation. *J. Chongqing Univ.* **2011**, *34*, 105–107.
7. Shi, X.; Hu, H.; Du, X.; Li, M.; Wang, H. Experimental study on flocculation settlement of mortar liquid at the tail of vertical sand silo. *Min. Metall. Eng.* **2010**, *30*, 1–11.
8. Caifu, U.; Cuiping, L.; Bingheng, Y.; Gezhong, C. Effect of water reducer on rheological properties of fine tailings paste. *Trans. Nonferr. Met. Soc. China* **2021**, *10–13*, 1–17.
9. Wen, Z.; Yang, X.; Li, L.; Gao, Q.; Wang, Z. Selection and optimization of flocculation settlement parameters of whole mortar based on RSM-BBD. *Trans. Nonferr. Met. Soc. China* **2020**, *30*, 1437–1445.
10. Xi, W.; Chao, Y.; Qu, W.; Kun, G. Research on speech emotion recognition based on PCA-RF classification. *Sci. Technol. Innov.* **2021**, *29*, 91–93.
11. Wu, X.; Yan, D. Analysis and research on data dimension reduction methods. *Res. Comput. Appl.* **2009**, *26*, 2832–2835.
12. Elkin, M.; Niyogi, P. Laplacian Eigenmaps and Spectral Techniques for Embedding and Clustering. *Adv. Neural Inf. Process. Syst.* **2001**, *14*, 585–591.
13. Yuting, K.; Fuxiang, T.; Xin, Z.; Zhenghang, Z.; Lu, B.; Yurong, Q. Review of K-means Algorithm Optimization Based on Differential Privacy. *Comput. Sci.* **2021**, 1–21. [CrossRef]
14. Lu, X.; Xu, L.; Zhang, M. Security hierarchical virtual network mapping method based on clustering. *Telecommun. Sci.* **2021**, *37*, 112–117.
15. Ji, Q.; Sun, Y.; Hu, Y.; Yin, B. A survey of deep clustering algorithms. *J. Beijing Univ. Technol.* **2021**, *47*, 912–924.
16. Deng, X.; Yu, L. Summary of Deep Clustering Algorithms. *Commun. Technol.* **2021**, *54*, 1807–1814.
17. Yin, F.; Cao, X.; Qi, X. Summary of recommendation algorithms based on clustering. *J. Southwest Minzu Univ.* **2021**, *47*, 303–309.
18. Zhou, Y.; Ren, Q.; Niu, H. A summary of research on training sample data selection methods. *Comput. Sci.* **2020**, *47*, 402–408.

metals MDPI

Article

Dispatch Optimization Model for Haulage Equipment between Stopes Based on Mine Short-Term Resource Planning

Ning Li [1,2,*], Shuzhao Feng [1,*], Haiwang Ye [1,2], Qizhou Wang [1,2,*], Mingtao Jia [3], Liguan Wang [3], Shugang Zhao [4] and Dongfang Chen [1,2]

1 School of Resource and Environment Engineering, Wuhan University of Technology, Wuhan 430070, China; yehaiwang@sina.com (H.Y.); chendongfang@whut.edu.cn (D.C.)
2 Hubei Key Laboratory of Mineral Resources Processing and Environment, Wuhan 430070, China
3 School of Resource and Safety Engineering, Central South University, Changsha 410083, China; mingtao_jia@163.com (M.J.); 13808478410@163.com (L.W.)
4 MCC Zhicheng (Wuhan) Engineering Technology Co., Ltd., Wuhan 430074, China; zhaoshugangkaka@outlook.com
* Correspondence: 13875910191@163.com (N.L.); fsz2224284697@163.com (S.F.); wqz@whut.edu.cn (Q.W.); Tel.: +86-13875910191 (N.L.)

Abstract: The working environment of underground mines is complicated, making it difficult to construct an underground mine production plan. In response to the requirements for the preparation of a short-term production plan for underground mines, an optimization model for short-term resource planning was constructed, with the goal of maximizing the total revenue during the planning period. The artificial bee colony optimization algorithm is used to solve the model using MATLAB. According to the basic requirements of underground mine ore haulage and ore hoisting, a haulage equipment inter-stopes dispatch plan model was constructed, with the primary goal of minimizing the haulage equipment wait time. A non-dominated sorting genetic algorithm is used to solve the optimization model. An underground mine is examined using the two models, and the optimization results are compared and verified with the scheme obtained by using traditional optimization algorithms. Results show that based on the improved optimization algorithm, the use of short-term production planning schemes to guide mine production operations can increase the haulage equipment utilization rate, thereby increasing mine production revenue.

Keywords: digital mine; mine short-term production planning; haulage equipment dispatch plan; ABCA; NSGA

Citation: Li, N.; Feng, S.; Ye, H.; Wang, Q.; Jia, M.; Wang, L.; Zhao, S.; Chen, D. Dispatch Optimization Model for Haulage Equipment between Stopes Based on Mine Short-Term Resource Planning. *Metals* **2021**, *11*, 1848. https://doi.org/10.3390/met11111848

Academic Editors: Jean François Blais and Lijie Guo

Received: 20 October 2021
Accepted: 16 November 2021
Published: 17 November 2021

1. Introduction

The complete mining cycle of a mine can be divided into two categories: the development period and production period. During development, a series of tunnels from the surface to the ore body is excavated to establish a passage between the surface and ore body for transporting personnel, equipment, materials, ore and waste rock, as well as creating appropriate conditions for mining and forming an independent system for hoisting, transportation, ventilation, drainage, power supply, and water supply, within these tunnels. The main material to be transported is generally rock, which has a low economic value. Investments during this period are primarily used to purchase equipment and hire manpower; therefore, the mine operates at a negative profit during the development period. After mine development and any necessary planning are completed, the mine enters the production period. The haulage is primarily composed of ore with high economic value. The production period is divided into three periods: ramp-up, steady state and tailing-off periods. Early revenue is to offset the costs from the development period, and mines begin to make profit during the middle-term; however, in the last term, a portion of the profit must be used to maintain the stability of the ore body. In order to maximize profit during the production period, Topal [1] and Sandanayake et al. [2] considered mining process allocation, comprehensive scheduling, stope design, and construction of system planning.

Different mining cycles correspond to different mining plans. Wu and Li [3] pointed out that mine production planning can be divided into long-term, medium-term, and short-term planning according to the length of the planning time. The length of the time period for each planning type is different, leading to different optimization goals. Long-term planning defines the company's long-term goals, emphasizing timeliness and strategy, and fully considers market changes; medium-term planning is guided by the long-term goals to complete the annual plan; and short-term planning can be divided into four plan types according to the length of time: seasonal planning, monthly planning, weekly planning, and daily planning, including product output, operation arrangement, labor allocation, grade control, product sales, and haulage cost. Planning the mine location and scheduling the mining operations is conducted during the production period. In short-term planning, resource planning is aligned to the mining cycle, and daily planning is generally aimed at equipment dispatch planning.

Previous scholars' research on short-term production in underground mines only divided the mining areas of the mine at each period, and did not involve the actual mining process. The production dispatching planning of underground mining vehicles is a short-term production planning process that is different from short-term resource planning. Only after short-term resource planning is completed, can production dispatching planning begin. However, most short-term production planning does not involve precise equipment dispatch path optimization research. This article combines resource planning and dispatching path optimization to guide the actual short-term production of a mine.

Newman et al. [4] pointed out that an underground mine contains more influencing factors than an open-pit mine, and discussed the advantages of mine production optimization. Based on their work, this article presents two mathematical models for optimizing different types of short-term production planning in underground metal mines and presents different optimization algorithms to obtain better short-term production planning schemes.

In the following sections, the data set used to test the model calculation results is described. Next, the mathematical models used in the construction of different production plans and their corresponding solving algorithms are introduced. The results are compared to the operational plan obtained from the traditional optimization algorithm.

1.1. Terms and Concepts

Every industry has its own unique terminology to describe its operating status and industry characteristics, and the mining industry is no exception. This section briefly introduces technical terms and concepts mentioned in this article.

In short-term production planning, weekly planning generally refers to resource arrangement, and daily planning generally refers to dispatching planning of production equipment. The excavation can start only after vehicle dispatching planning is completed.

During resource planning, the overall profit of the mine is maximized as much as possible while satisfying the constraints and medium-term planning objectives. The purpose of constructing a haulage equipment dispatch plan is to complete the resource planning scheme more efficiently. This process does not consider the benefits and objectives. Making the equipment utilization rate as high as possible while meeting the haulage conditions is necessary. Short-term production planning combines resource planning with equipment dispatch planning and can plan daily production tasks more accurately as well as enable the mine to perform production operations efficiently.

1.2. Dataset

In order to test the validity of the mathematical model and optimization algorithm, actual data from an iron mine (Figure 1) in Handan City, Hebei Province, People's Republic of China is analyzed. The ore body of the mine is approximately 1400 m long and 300 m wide. Here, we will briefly introduce some mining work and equipment involved in short-term production operations of underground mines.

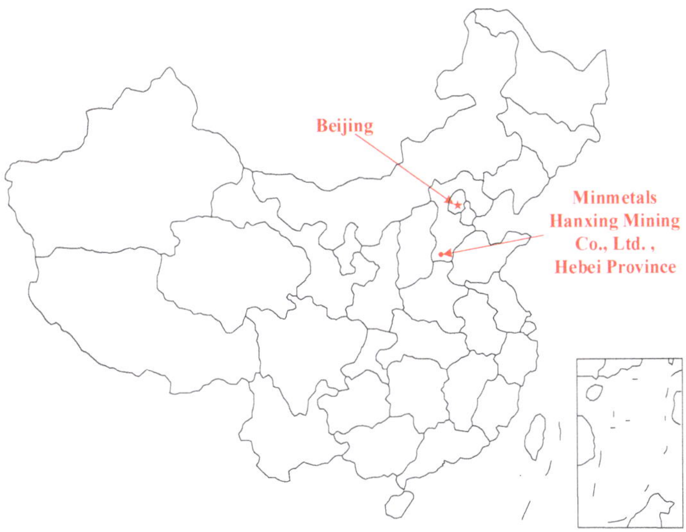

Figure 1. Location map.

1.2.1. Mine Layout

After the mine ground construction and road repairs are completed, ore extraction begins with the excavation of the main shaft. When the depth of the main shaft reaches the depth of the ore body, a shaft station connecting the transport roadway and the beginning of the main shaft is excavated. Generally, the shaft station chambers have storage lines, traffic lines, shunting lines, water pump chambers, substations, dispatching rooms, and repair warehouses. The transportation tunnel is placed between the ore body and the shaft station and is used for transporting ore and ventilation. Cross-cuts are excavated horizontally as links between the levels and the main shaft. The transportation tunnels that are parallel to the run of the ore vein are called drifts, and those vertical to the run of the ore vein are called crosscuts (Figure 2).

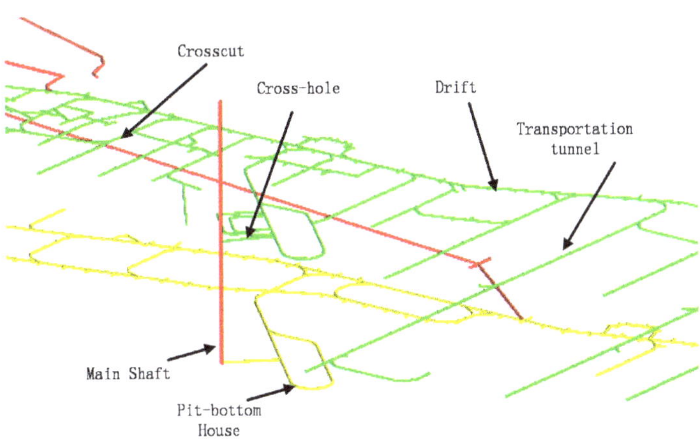

Figure 2. Mine layout.

1.2.2. Mining Methods

All or part of the mineral deposits mined by a mining enterprise is called mine fields. Within the mine fields, the main haulage tunnels that are consistent with the strike of the ore body are excavated within certain vertical distance, and the mine fields are divided vertically into several mine sections. These mine sections are called levels, which are divided into a number of independent mining units according to a certain size, called ore blocks.

The selection of mining methods for underground mines should consider multiple factors, such as location, thickness, shape, occurrence, and geological conditions of the ore body. Different mining methods have significant differences in mining rules, ore output, and production capacity of mining equipment. Based on the above factors, the mine chose to adopt the cut-and-fill mining method.

The cut-and-fill mining method commonly divides an ore block into an ore pillar and chamber, and the mining order is the chamber first, then the ore pillar. The chamber is excavated from bottom to top by sublevels, then filled sequentially to maintain the surrounding rock of the upper and lower walls to create continuous upward mining operation conditions. Top filling is carried out when the chamber reaches the last sublevel. After several chambers or whole levels have been mined, the ore pillar begins to be mined. This method is a working face cycle operation. After one cycle of rock drilling and blasting, ore extraction, backfilling, and roof protection is completed, the next sublevel cycle is carried out.

During the medium-term planning of the mine, mining two levels at the same time are necessary, each level has two sublevels, and each sublevel has 36 sites, 12 along the X-axis direction and three along the Y-axis direction. Short-term resource planning needs to determine which sites within the two levels can be used as stopes for production operations, and haulage equipment dispatch planning needs to reasonably arrange vehicles for haulage operations in different levels. The short-term planning period is seven days. During the planning period, the output ore volume is approximately 40,000 tons, the market metal price of iron is 488 dollars/t, the mining cost is 38.6 dollars/t, the ore recovery ratio is 91%, and the mineral processing recovery ratio is 81%.

1.2.3. Spatial Relations

There is no restriction on the dimensions of each mine point in an open-pit mine. Due to the special mining environment of underground mines, the selection of sites needs to be determined based on the mining method as well as the temporal and spatial development of the ore block.

(1) During the same planning period, multiple levels are mined simultaneously. Each level is divided into multiple ore blocks and subsequently divided into chambers and ore pillars. The chambers are stoped in sublevels from bottom to top. Ore blocks in each sublevel have unique attributes such as grade and total volume of ore and waste. (2) During the early mining stage, in order to make the mine recuperate costs as soon as possible, the ratio of ore mining in the upper levels to the lower levels should be greater than one. (3) In order to ensure safe mining, during the same mining period, the ore blocks at the same horizontal position in each level are stoped within a given sublevel from bottom to top. The mining conditions are classified into three types: (1) The lower sublevel ore block is not mined, and mining of the upper sublevel ore block is prohibited; (2) the lower sublevel ore block is being mined, and mining of the upper sublevel ore block is prohibited; and (3) the lower sublevel ore block has been mined, and the upper sublevel ore block can be mined. (4) As shown in Figure 3, in the horizontal direction within each sublevel, only one of the four adjacent ore blocks can be mined in the same mining period. This is to avoid excessively large exposed areas of the roof between sublevels, which may cause stress concentrations and accidents.

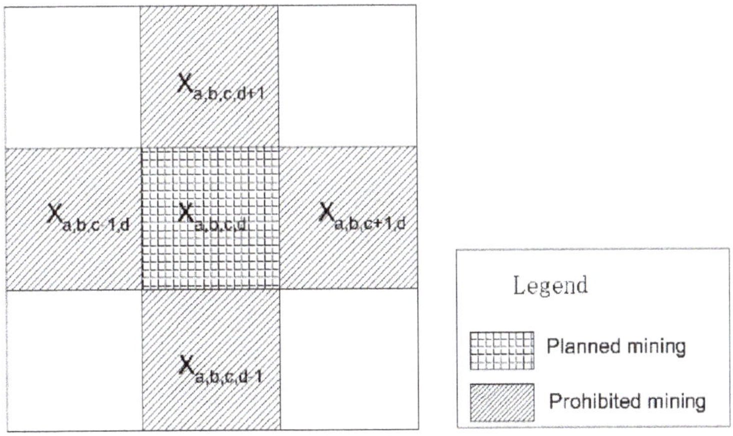

Figure 3. Horizontal direction mining rule.

1.2.4. Haulage Equipment

Currently, the production equipment of underground mines is gradually becoming trackless. Zhan [5] analyzed the current research and application status of trackless mining equipment as well as comprehensively reviewed the main problems of trackless mining equipment in the People's Republic of China and the development trend of trackless mining equipment technology. The primary equipment for underground mining operations includes rock drills, scrapers, electric locomotives, and bogies. With the exception of rock drills, the main purpose of the other equipment is to transport ore. Yi [6] pointed out that the scraper has become a core piece of trackless equipment. The scraper is responsible for transporting ore from the stope to the ore pass on each sublevel. An electric locomotive is used to transport the ore from the lowest sublevel of each level to the main shaft for hoisting. In the mine mentioned in this article, six scrapers with the same specifications and two electric locomotives with the same specifications are used for production operations during each planning period. The load capacity of the scraper is 3 t, and each electric locomotive can pull 13 mine bogies with a load capacity of 4 t each.

Compared to rock drills, scrapers and electric locomotives are more flexible to use. They consistently engage in irregular movements, making the operation time and route difficult to predict. A reasonable dispatching plan for haulage equipment is required. In addition, the operational planning of scraper and the electric locomotive is affected by a variety of constraints. The most basic constraint is the waiting time of the scraper, followed by the mining volume requirements of the mining and unloading sites, as well as the scraper's running route and speed limit within the complex roadway. Under these constraints, the scraper and electric locomotive are not restricted to only move back and forth between the fixed mining and unloading sites, allowing for a variety of dispatching options in each day. The purpose of this article to optimize the dispatching planning of haulage equipment by selecting a scheme that can minimize the total equipment waiting time, improve the equipment utilization of haulage equipment, and further reduce production costs. The simple dispatching model of haulage equipment is shown in Figure 4.

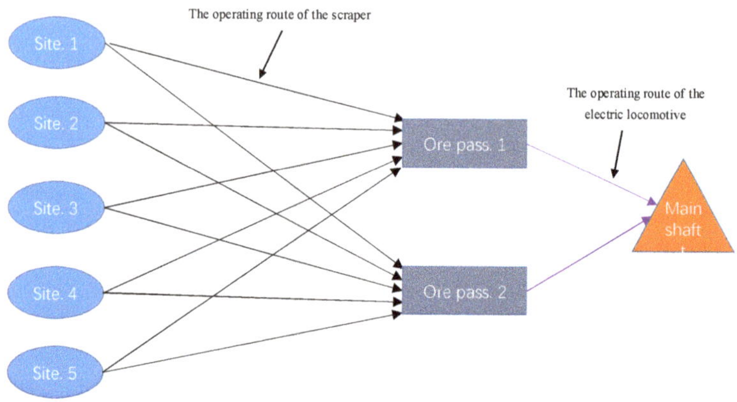

Figure 4. Simple dispatching model.

2. Relevant Literature

2.1. Short-Term Resource Planning

For underground mines, long-term and medium-term planning goals are set by the enterprise or decision-makers, with a wide range of goals and relatively few constraints. However, short-term planning is closely related to actual mine operations, which requires precise targets and tight plans. It is difficult to realize short-term planning by relying solely on decision support systems based on experience. Therefore, a mathematical model is needed to solve the short-term production planning problem by considering it as two constrained optimization problems that are long-term and medium-term planning goals, as well as constraints of the mine itself. The mathematical model is solved using optimization algorithms to quickly and effectively obtain a better short-term production scheme.

Li and Tong [7] proposed that the main evaluation methods for underground mine planning are optimization and simulation. The simulation method has strong applicability, while the optimization method has higher theoretical and application value. In the optimization method, scholars visualize the resource planning problem as a corresponding mathematical model or ore allocation problem, accounting for the various constraints using linear programming, multi-objective programming, and dynamic programming in Operational Research methods to determine the solution. Ming et al. [8] established a mineral product market demand planning model based on the production characteristics and operation of underground metal mines. In addition to the methods mentioned above, Nehring et al. [9] adopted a single mathematical optimization model combining the short-term goal of minimizing the deviation of the feed grade of the target concentrator with the medium-term goal of maximizing the net present value, to solve the short-term and medium-term integrated production planning tasks. O'Sullivan and Newman [10] established a mathematical model with both resource constraints and priority constraints, and then used a heuristic method to reduce the model specifications. They also constructed an optimization-based decomposition heuristic method to solve the model.

Based on operational research methods, in order to speed up the solution process, intelligence algorithms have gradually become the mainstream solution for solving mathematical models. Zhou and Gu [11] proposed the application of genetic algorithms to optimize the mining sequence for underground mines based on numerical simulations of the mining sequence. Hou et al. [12] considered the technical and economic requirements and the spatial sequence relationship in the mining process, constructed a dynamic optimization model for the production planning of polymetallic underground mines with the goal of maximizing profit, and gave a solution algorithm based on the artificial bee colony model. Foroughi et al. [13] constructed a multi-objective integer programming model that introduced a non-dominated sorting genetic algorithm to solve the target model. The algorithm showed good convergence and diversity, and the solution time was significantly reduced. Gligoric et al. [14] expressed the ore body as a set of minable blocks based on the establishment of a production planning model, and applied a multi-objective iterative greedy algorithm to define a set of minable blocks each year to make the deviation from the target less than or equal to the given minimum error.

With the continuous development of computer hardware, three-dimensional visualization technology has been gradually applied to resource planning. Jiang et al. [15] proposed a construction method for three-dimensional visualization production plan based on simulated mining technology by using DEMINE. In order to resolve the issue that operations research methods or optimization algorithms cannot be connected with 3D visualization technology, Liu et al. [16] used multi-objective planning, combined with logical constraints, business constraints, and spatial constraints, as well as established a multi-objective planning model using three-dimensional visualization technology to show the spatial logical relationship of mine engineering.

Other methods can be also used to solve production planning problems. Sarin and West-Hansen [17] proposed a model to optimize the start-up time of different parts of underground coal mines. Newman and Kuchta [18] constructed a mixed integer program to plan ore production over multiple time periods. Riff et al. [19] constructed a "greedy random adaptive search" program to speed up the model solving process for copper mines. Little et al. [20] showed the value of optimizing the shape of the stope by providing a model to optimize the shape of the two stopes. Mousavi and Sellers [21] integrated in mine recovery (IMR) into conventional mining operations, which can significantly increase the net present value of the project by recycling low-grade material from conventional mining that is commonly left as waste rock. Campeau and Gamache [22] proposed an optimization model for short-term planning, taking into account the various working points from the development and production phases, as well as the specific equipment and worker restrictions using a mixed integer program with priority. Gligoric et al. [23] proposed a long-term mine planning method for underground lead-zinc mines based on fuzzy logic aimed at the production plan of lead-zinc mines under uncertain conditions. A fuzzy stochastic inventory control model was established.

2.2. Haulage Equipment Dispatch Planning

Compared with research on the short-term production planning of underground mines, there are relatively few studies on the production dispatching planning of underground mine vehicles. Gamache et al. [24] proposed a solution based on the shortest-path algorithm, in which each decision of the solution accounts for the current state of the mine. With automation comes the possibility for optimization. Saayman [25] looked at possible solutions to the problem of optimizing the autonomous vehicle dispatch system in an underground mine, and evaluated possible optimization strategies using a simulated environment. Nehring et al. [26] proposed a classic mixed integer programming model to optimize the production dispatching of the sublevel shutdown process and proposed a new model formula that can significantly reduce the solution time and maintain all constraints without changing the results.

The optimization problem of the vehicle path can be regarded as a non-polynomial problem. The optimal solution cannot be derived directly. It can only be verified by polynomials whether the proposed solution is a feasible solution, and then find the optimal solution. Sun and Lian [27] used the ant colony algorithm to solve this NP problem to optimize the vehicle haulage route of a certain shift in underground mine production dispatching, and this algorithm first used adaptive strategy to control its convergence speed, which can improve search performance and then optimize performance indicators based on the traditional ant colony algorithm during each iteration. Åstrand et al. [28] proposed a constraint programming method that can automatically realize the short-term dispatching process of cut-and-fill mines. This method builds on previous work by considering the running time of the fleet.

3. Model

Since the short-term production plan of the underground mine mentioned in this article includes a short-term resource plan and a dispatching plan for haulage equipment, two different models are needed to construct the different plans. The following will focus on the selection of objective functions and constraints of the two models. According to the time span used by most mines and planners, the short-term resource plan cycle in this paper is one week, and the haulage equipment dispatch plan cycle is 8 h in a shift.

3.1. Model of Short-Term Resource Planning

The underground mine short-term resource plan is based on the medium-term and long-term plans. First, determine the time span of the short-term plan, then determine the planned mining volume within the time span of the short-term plan, and finally determine the short-term mining scheme based on the medium-term and long-term plan. The purpose of a short-term resource plan is to determine the mining sequence in most levels. The optimization model established by the traditional 0-1 integer programming method can solve this problem.

3.1.1. Sets

$\mathcal{A}$: Set of levels mined simultaneously during the planning period.

$$\mathcal{A} = \{1, \ldots, A\},\ a \in \mathcal{A}$$

$\mathcal{B}$: Set of sublevels divided during the mining of the whole block.

$$\mathcal{B} = \{1, \ldots, B\},\ b \in \mathcal{B}$$

$\mathcal{C}$: Set of ore blocks in the x direction on each sublevel.

$$\mathcal{C} = \{1, \ldots, C\},\ c \in \mathcal{C}$$

$\mathcal{D}$: Set of ore blocks in the y direction on each sublevel.

$$\mathcal{D} = \{1, \ldots, D\},\ d \in \mathcal{D}$$

3.1.2. Parameters

P: Market price of iron;
$W_{a,b,c,d}$: Total tonnage of ore and waste in the (a, b, c, d) ore block;
$G_{a,b,c,d}$: Metal grade in the (a, b, c, d) ore block;
R: Ore recovery ratio;
ε: Dressing recovery ratio of the metal;
$C_{a,b,c,d}$: Mining cost of the (a, b, c, d) ore block;
$maxG$: The upper limit of the grade required by the dressing plant of the metal mining enterprise;
$minG$: The lower limit of the grade required by the dressing plant of the metal mining enterprise;
Q_{max}: The planned maximum tonnage of ore and waste during the plan period;
Q_{min}: The planned minimum tonnage of ore and waste during the plan period;
$N_{a,b}$: The number of scrapers and rock drills in b sublevel, a level.

3.1.3. Variables

$X_{a,b,c,d}$ is a binary variable, indicating whether the (a, b, c, d) ore block will be mined in the short-term plan.

$$X_{a,b,c,d} = \begin{cases} 1, & \text{The ore blocks are mined during the plan period} \\ 0, & \text{The ore blocks will not be mined during the plan period} \end{cases}$$

3.1.4. Objective

When constructing the short-term resource plan for the mine, the layout of the block should be rationally planned according to the current metal market price, combined with the grade of each mining ore block. Then, the difference between the income and cost of all mining blocks during the planning period is taken as an objective function to maximize the total production income of the mine in each short-term plan period. The first part of Equation (1) is the sum of the income of all mining blocks during this planning period. The second part of the equation is the sum of the mining costs during the plan period.

$$maxP = \sum_{a\in\mathcal{A}}\sum_{b\in\mathcal{B}}\sum_{c\in\mathcal{C}}\sum_{d\in\mathcal{D}} PW_{a,b,c,d}G_{a,b,c,d}R\varepsilon X_{a,b,c,d} - \sum_{a\in\mathcal{A}}\sum_{b\in\mathcal{B}}\sum_{c\in\mathcal{C}}\sum_{d\in\mathcal{D}} C_{a,b,c,d}X_{a,b,c,d} \quad (1)$$

Under the premise of meeting the mine production requirements, the mine can obtain greater profits during each plan period. This is the goal pursued by every planner, to maximize the economic benefits of the mine and provide funds for later production or purchase advanced equipment, thereby improving the efficiency and economic benefits of subsequent production.

3.1.5. Constraints

Constraint (2) determines the tonnage of ore produced during the plan period so that it cannot be greater or less than a predetermined tonnage, which would affect the production process of the mine's medium- and long-term plan. Constraint (3) is the restriction on the dressing grade. In order to meet the grade requirements for the dressing plant, the grade of the mined ore should be guaranteed within a certain range; otherwise, the dressing recovery rate will decrease. Due to the current way of work, constraint (4) determines the number of ore blocks mined in each sublevel during the plan period, which should be greater than the total number of scrapers and rock drills in that sublevel. This constraint can effectively prevent production equipment from waiting in the stope and improve equipment time utilization ratio. Constraint (5) restricts the mined total tonnage of ore and waste between the levels, making the mined total tonnage of ore and waste in the upper level in the early stage greater than that in the lower level. Next, Constraints (6) and (7) are spatial relationship constraints in the vertical direction. Constraint (6) means that during the same plan period, the lower sublevel ore blocks are not mined, and the ore blocks with the same horizontal position in the upper sublevel are not mined. Once mining of the lower sublevel ore blocks is completed, the ore blocks with the same horizontal position of the upper sublevel can be mined. Constraint (7) restricts simultaneous mining of ore blocks in adjacent sublevels with the same horizontal position. Constraint (8) is a horizontal spatial constraint that specifically states that in the same plan period, only one of the four adjacent ore blocks in the same sublevel can be mined.

$$Q_{min} \le \sum_{a\in\mathcal{A}}\sum_{b\in\mathcal{B}}\sum_{c\in\mathcal{C}}\sum_{d\in\mathcal{D}} W_{a,b,c,d}X_{a,b,c,d} \le Q_{max} \quad (2)$$

$$minG \le \frac{\sum_{a\in\mathcal{A}}\sum_{b\in\mathcal{B}}\sum_{c\in\mathcal{C}}\sum_{d\in\mathcal{D}} W_{a,b,c,d}G_{a,b,c,d}X_{a,b,c,d}R}{\sum_{a\in\mathcal{A}}\sum_{b\in\mathcal{B}}\sum_{c\in\mathcal{C}}\sum_{d\in\mathcal{D}} W_{a,b,c,d}X_{a,b,c,d}} \le maxG \quad (3)$$

$$\sum_{c\in\mathcal{C}}\sum_{d\in\mathcal{D}} X_{a,b,c,d} \ge N_{a,b} \quad (4)$$

$$\sum_{b\in\mathcal{B}}\sum_{c\in\mathcal{C}}\sum_{d\in\mathcal{D}} W_{a,b,c,d}X_{a,b,c,d} \ge \sum_{b\in\mathcal{B}}\sum_{c\in\mathcal{C}}\sum_{d\in\mathcal{D}} W_{a+1,b,c,d}X_{a+1,b,c,d} \quad (5)$$

$$X_{a,b,c,d} - X_{a,b+1,c,d} \ge 0, a \in [1,A], b \in [1,B-1], c \in [1,C], d \in [1,D] \quad (6)$$

$$X_{a,b,c,d} + X_{a,b+1,c,d} \le 1, a \in [1,A], b \in [1,B-1], c \in [1,C], d \in [1,D] \quad (7)$$

$$X_{a,b,c,d} + X_{a,b,c-1,d} + X_{a,b,c+1,d} + X_{a,b,c,d-1} + X_{a,b,c,d+1} \leq 1, a \in [1, A], b \in [1, B], c \in [2, C - 1], d \in [2, D - 1] \quad (8)$$

3.2. Haulage Equipment Dispatch Plan Model

Haulage equipment in the mine under study includes scrapers and electric locomotives. The scrapers are responsible for hauling the ore from the stope of each sublevel to the ore pass. The electric locomotive hauls the ore from the ore pass of each level to the main shaft for hoisting. After the short-term resource plan is created, the scrapers will alternately operate among the sites. It is not possible to make a scraper only run back and forth between specific stopes and ore passes, which will affect the production balance and efficiency of the short-term plan.

The most important parameter in the dispatch plan model is time, including the loaded and empty travel time of the scraper as well as the loaded and unloaded travel time of the electric locomotive. Only by determining each operating time can a haulage equipment dispatch plan be constructed.

3.2.1. Sets

$\mathcal{A}$: Set of levels mined simultaneously during the plan period.

$$\mathcal{A} = \{1, \ldots, \mathrm{A}\}, a \in \mathcal{A}$$

$\mathcal{B}$: Set of sublevels divided during the mining of the whole block.

$$\mathcal{B} = \{1, \ldots, \mathrm{B}\}, b \in \mathcal{B}$$

$\mathcal{C}_{ab}$: Set of scrapers in b sublevel, a level.

$$\mathcal{C}_{ab} = \{1, \ldots, C_{ab}\}, c \in \mathcal{C}_{ab}$$

$\mathcal{D}_a$: Set of electric locomotives in a level.

$$\mathcal{D}_a = \{1, \ldots, D_a\}, d \in \mathcal{D}_a$$

$\mathcal{N}_{ab}$: Set of stopes in b sublevel, a level.

$$\mathcal{N}_{ab} = \{1, \ldots, N_{ab}\}, i \in \mathcal{N}_{ab}$$

$\mathcal{M}_a$: Set of ore passes in a level.

$$\mathcal{M}_a = \{1, \ldots, M_a\}, j \in \mathcal{M}_a$$

3.2.2. Parameters

T_{limit}: Working hours in a shift, generally 8 h;
z_{abcij}: The loaded travel time of c scraper from i stope to j ore pass in b sublevel, a level;
q_{abcij}: The empty travel time of c scraper from i stope to j ore pass in b sublevel, a level;
z_{adj}: The loaded travel time of d electric locomotive from j ore pass to main shaft in a level;
q_{adj}: The empty travel time of d electric locomotive from j ore pass to main shaft in a level;
R_{abc}: The capacity of c scraper in b sublevel, a level;
g_{abj}: The total tonnage of ore and waste of i stope in b sublevel, a level;
h_{abj}: The maximum number of loading vehicles in one shift at i stope in b sublevel, a level;

J_{aj}: The maximum ore draw of j ore pass in a level, that is, the tonnage of ore drawn when the ore has been drawn in one shift;

R_{ad}: The capacity of d electric locomotive in a level;

F_{aj}: The drawing capacity of j ore pass in a level, that is, the general tonnage of ore drawn obtained from a statistical analysis of the general tonnage of ore drawn from multiple ore passes in a shift;

Q_{ajmax}: Maximum ore storage capacity of j ore pass in a level;

Q_{ajmin}: Minimum ore storage capacity of j ore pass in a level;

Q_{min}: The minimum tonnage of ore and waste scheduled to be mined during the given shift;

Q_{max}: The maximum tonnage of ore and waste scheduled to be mined during the given shift.

3.2.3. Variables

x_{abcij}: The number of trips of c scraper from i stope to j ore pass in b sublevel, a level;

y_{abcij}: The number of trips of c scraper from j ore pass to i stope in b sublevel, a level;

x_{adj}: The number of trips of d electric locomotive from j ore pass to the main shaft in a level;

y_{adj}: The number of trips of d electric locomotive from the main shaft to j ore pass in a level.

3.2.4. Objective

For mine production equipment, long waiting times reduce the production capacity of the mine and increase mining cost. In the production of underground mines, the haulage equipment has the most variable time; therefore, increasing the equipment utilization of the scraper and electric locomotive will indirectly increase the production income of the mine. Equation (9) gives the total waiting time of the scrapers, and Equation (10) gives the total waiting time of the electric locomotives.

$$F_1 = \sum_{a\in\mathcal{A}} \sum_{b\in\mathcal{B}} \sum_{c\in\mathcal{C}_{ab}} \left(T_{limit} - \sum_{i\in\mathcal{N}_{ab}} \sum_{j\in\mathcal{M}_a} z_{abcij}x_{abcij} - \sum_{i\in\mathcal{N}_{ab}} \sum_{j\in\mathcal{M}_a} q_{abcij}y_{abcij} \right) \tag{9}$$

$$F_2 = \sum_{a\in\mathcal{A}} \sum_{d\in\mathcal{D}_a} \left(T_{limit} - \sum_{j\in\mathcal{M}_a} z_{adj}x_{adj} - \sum_{j\in\mathcal{M}_a} q_{adj}y_{adj} \right) \tag{10}$$

When the constraints are met, the model is optimized using the objective function, which can minimize the total waiting time of the scrapers and electric locomotives in the haulage equipment dispatch scheme. The scheme can meet the daily production balance and high-efficiency mining requirements of the mine as well as maximize the utilization rate of the equipment as much as possible.

3.2.5. Constraints

Constraints (11) and (12) are the logical limits of the operating time of the scrapers and electric locomotives, respectively. The total operating time of each scraper and electric locomotive during a shift cannot exceed the length of the shift. Constraint (13) ensures that the tonnage of ore hauled from each stope does not exceed the total tonnage of ore in the stope. Constraint (14) ensures that the loading times of any mining site are less than the maximum loading times of the stope within a shift. Constraint (15) ensures that the total ore input in each ore pass does not exceed the maximum tonnage of ore drawn from the ore pass in a shift. Constraint (16) ensures that the ore hauled by the electric locomotive from each ore pass does not exceed the general tonnage of ore drawn from the ore pass in each level. Constraint (17) requires that the tonnage of ore in each ore pass is always within the upper and lower limits of the ore storage of the ore pass. Constraint (18) ensures that the tonnage of ore hoisted by the main shaft does not exceed the total ore output of all scrapers in a shift. Constraint (19) determines the volume of ore and waste scheduled to be mined during each shift based on the number of shift hours in the plan period determined by the short-term planning period. Constraints (20) and (21) require that the total number of round trips of the scrapers or electric locomotives on each line be equal.

$$\sum_{i\in\mathcal{N}_{ab}}\sum_{j\in\mathcal{M}_a} z_{abcij}x_{abcij} + \sum_{i\in\mathcal{N}_{ab}}\sum_{j\in\mathcal{M}_a} q_{abcij}y_{abcij} \le T_{limit}, \forall a,b,c \tag{11}$$

$$\sum_{j\in\mathcal{M}_a} z_{adj}x_{adj} + \sum_{j\in\mathcal{M}_a} q_{adj}y_{adj} \le T_{limit}, \forall a,d \tag{12}$$

$$\sum_{c\in\mathcal{C}_{ab}}\sum_{j\in\mathcal{M}_a} R_{abc}x_{abcij} - g_{abj} \le 0, \forall a,b,i \tag{13}$$

$$\sum_{c\in\mathcal{C}_{ab}}\sum_{j\in\mathcal{M}_a} x_{abcij} - h_{abj} \le 0, \forall a,b,i \tag{14}$$

$$\sum_{b\in\mathcal{B}}\sum_{c\in\mathcal{C}_{ab}}\sum_{i\in\mathcal{N}_{ab}} R_{abc}x_{abcij} - J_{aj} \le 0, \forall a,j \tag{15}$$

$$\sum_{d\in\mathcal{D}_a} R_{ad}x_{adj} - F_{aj} \le 0, \forall a,j \tag{16}$$

$$Q_{ajmin} \le \sum_{b\in\mathcal{B}}\sum_{c\in\mathcal{C}_{ab}}\sum_{i\in\mathcal{N}_{ab}} R_{abc}x_{abcij} - \sum_{d\in\mathcal{D}_a} R_{ad}x_{adj} \le Q_{ajmax}, \forall a,j \tag{17}$$

$$\sum_{a\in\mathcal{A}}\sum_{b\in\mathcal{B}}\sum_{c\in\mathcal{C}_{ab}}\sum_{i\in\mathcal{N}_{ab}}\sum_{j\in\mathcal{M}_a} R_{abc}x_{abcij} \ge \sum_{a\in\mathcal{A}}\sum_{d\in\mathcal{D}_a}\sum_{j\in\mathcal{M}_a} R_{ad}x_{adj} \tag{18}$$

$$Q_{min} \le \sum_{a\in\mathcal{A}}\sum_{d\in\mathcal{D}_a}\sum_{j\in\mathcal{M}_a} R_{ad}x_{adj} \le Q_{max} \tag{19}$$

$$\sum_{c\in\mathcal{C}_{ab}} x_{abcij} = \sum_{c\in\mathcal{C}_{ab}} y_{abcij}, \forall a,b,i,j \tag{20}$$

$$\sum_{d\in\mathcal{D}_a} x_{adj} = \sum_{d\in\mathcal{D}_a} y_{adj}, \forall a,j \tag{21}$$

4. Optimization Algorithm

Currently, optimization algorithms have become a common method of solving mathematical models, and their development speed is also increasing. According to the variable characteristics of different mathematical models, this paper chooses different optimization algorithms to optimize and solve the corresponding plan. For the short-term resource planning model with binary variables, this article intends to improve the artificial bee colony algorithm for model optimization. The artificial bee colony algorithm has a fast convergence speed. Through the individual's local optimization behavior, the global optimal value will finally emerge in the group. For the haulage equipment dispatch model with multiple continuous integer variables, this paper intends to integrate the non-dominated sorting algorithm with genetic algorithm to optimize the scheduling plan. The chromosome encoding method in genetic algorithm is very suitable for route planning problems. Then, the optimal solution can be found for the multi-objective problem through the non-dominated solution.

4.1. Improved Artificial Bee Colony Optimization Algorithm

An artificial bee colony optimization algorithm is a swarm intelligence optimization algorithm inspired by bee colony foraging behavior. This algorithm introduces three types of bees: picking bees, following bees, and scout bees. Different bees perform different tasks in the process of finding an optimal nectar source. The task of picking bees is to extensively search for nectar sources, perform a neighborhood search for better nectar sources, and determine whether to replace the nectar source according to the comparison of fitness. Following bees select the nectar source after neighborhood search using the roulette method and determine whether to replace the nectar source according to the comparison of fitness. When the nectar source location of the picking bee and the following bee meets the nectar source abandonment condition, they will become the scout bee, and the scout bee will randomly search for a new nectar source at the abandoned nectar source. The specific implementation process of the algorithm is as follows:

(1) Determine the fitness value of the objective function and initialize the parameters, including the nectar population N, the maximum evolutionary generation t, and the custom generation limit;
(2) The coding rules of the nectar source location, the nectar source population adopts binary coding are expressed as $\begin{bmatrix} a_{11} & \cdots & a_{1N} \\ \vdots & \ddots & \vdots \\ a_{m1} & \cdots & a_{mN} \end{bmatrix}$ where m represents the sum of all variable elements of a single individual;
(3) Initialize the nectar population, find a feasible solution according to the constraints of the optimization model, and randomly generate feasible solutions in the surrounding area of the feasible solution. All the generated feasible solutions form the initial nectar population;
(4) Calculate the fitness value of the initial nectar source population, compare the fitness value of the current population, record the best individual value in the current population, and position the honeybees at the half of the nectar source in the population where the fitness value is better. The number of following bees is the same as the number of picking bees;
(5) Picking bees are used to search the neighborhood at the current nectar source location. When the binary code of discrete variables is used, the neighborhood search becomes a value change 0 and 1. After the value is changed, it is judged whether it satisfies the constraint condition. If the constraint condition is not met, the variable is reselected near the value of the variable for transformation until the constraint condition is met, at which point, it can be used as a new nectar location. Then, calculate the fitness value and compare the fitness value of the new nectar source with the original nectar source. If the nectar source quality of the new nectar source is better, replace the original location with the new nectar source to update the nectar source population;
(6) Compare the fitness value of the current population. Compare the current optimal individual with the recorded optimal individual, if it is better than the recorded optimal individual, replace it; otherwise, replace the recorded optimal individual back to the original one. Then, continue subsequent operations at the location;
(7) According to the roulette selection method which is $p_i = fit_i / \sum_{n=1}^{N} fit_n$ following bees choose a better position in the current nectar population and go through the neighborhood search method of step (5) and generate a new nectar source location around this location. Then, the following bee calculates the fitness value, compares the fitness value of the new nectar source with the original nectar source, selects the best to form a new nectar source population, and proceeds to step (6);
(8) Determine whether part of the nectar source in the current nectar source population meets the abandonment condition. If a nectar source has not been replaced after the limit generation neighborhood search, then go to step (9), otherwise go to step (10);
(9) If the nectar source is the best nectar source in the current population, do not abandon it; otherwise, the current nectar source is abandoned, and the picking or following bees at the current nectar source location become the scout bees. The nectar source randomly changes its position to form a new nectar source and updates the nectar source population;

(10) Judge whether the end condition is reached, if the maximum number of iterations is not reached, the current nectar source population is used as the initial nectar source population, and steps (5), (6), (7), (8), and (9) are repeated. If the maximum evolutionary generation is reached, the optimal nectar source position in the current population is considered to be the optimal solution.

4.2. Non-Dominated Sorting Genetic Algorithm (NSGA-III)

The elite-level retention strategy is added to the traditional non-dominated sorting genetic algorithm, and the current optimal individual is retained in each iteration. Rapid non-dominated ranking of individuals in the mixed population is used to divide the non-dominated level. The selection of individuals at the same non-dominated level no longer uses crowding degree distance, but uses the constraint dominance relationship based on the reference point to select the elites with the mixed population. In this way, the elite individual in the mixed population is selected to enter the next generation, retaining the superior genes of the parent. The specific description is as follows:

(1) Initialization parameters. The maximum evolutionary generation is G_{max} the reference point size is H, the population size is N + 1, the crossover and mutation rates are P_c and P_m, respectively, and the evolutionary generation is t, which is set as 0.
(2) Initialize the P_t population. The scale is N. Each individual in the population consists of N+M chromosomes. The individual chromosomes are independent of each other. The maximum and minimum lengths of chromosomes are set. The length of each chromosome is randomly generated between the maximum and minimum values. The chromosomes are coded using characters, with lowercase for stopes, capital letters for ore passes, and the letter 'Z' for the main shaft. For example: $[aAbBcC]$ represents the operating route of a scraper, while $[AZBZAZ]$ represents the operating route of an electric locomotive.
(3) Determine the objective function. The Pareto ranking hierarchical comparison method is used for multiple objective functions.
(4) Using rapid non-dominated sorting based on Pareto dominance, divide the current P_t population into several dominance layers, select the best individuals in the first dominance layer based on the reference point-based constraint dominance relationship method, and extract them as a single population that does not participate in genetic manipulation. Then, subpopulation B_t is generated through the crossover, mutation, and breaking of the genetic algorithm, with a scale of N. Crossover: randomly select chromosomes of the same nature for different individuals in the population to cross over random gene positions. Mutation: Chromosomal genetic properties mutate at random. Breaking: Each chromosome randomly chooses whether to perform the break operation, if so, cut two genes to the last position of the chromosome. The progeny population is generated through the operation sequence of crossover, mutation, and breaking.
(5) Combine the B_t offspring population and the P_t parent population to form the R_t population, use the non-dominated sort based on Pareto dominance to divide R_t into several different non-dominated layers, and select N higher-level individuals as the next P_{t+1} parent population. Individuals in the same level are selected using a reference point-based constraint dominance relationship method, also the best individual in the first dominance level of the P_{t+1} parent population is selected using the same method.
(6) Compare and select the optimal individual produced by the t + 1 generation with the optimal individual of the t generation. Use the method based on the reference point to select the superior individual among the two adjacent generations of optimal individuals. If the optimal individual of the t + 1 generation is superior, it will be placed in a separate population, and the original t generation optimal individual will be placed in t, where the best individual of the t + 1 generation is located, and form a new P_{t+1} parent group.

(7) Judge whether the P_{t+1} parent population meets the termination conditions. If not, then t = t + 1, repeat steps (4), (5), and (6). If the termination conditions are met, then output the best individual.

5. Computational Study

In the following sections, the optimization scheme is compared and verified with the scheme obtained using the traditional optimization algorithm. The mathematical model, as well as the algorithm's high efficiency and feasibility, is described.

5.1. Optimization Results

First, the actual production data of the mine are brought into the bee colony algorithm to solve the short-term resource plan. Then, according to the spatial location of the stopes and ore passes to be mined in the short-term resource plan, the time parameters in the haulage equipment dispatch plan model are obtained, when are subsequently brought into non-dominated sorting genetic algorithm to calculate the haulage equipment scheduling scheme.

5.1.1. Resource Plan

The calculation results of the resource planning model are shown in Table 1. The number '1' indicates that the site can be mined during the plan period; on the contrary, the number '0' indicates that the site cannot be mined during the plan period. The number of the ore blocks in Table 1 are shown in Figure 5. For example, (1, 1, 4, 3) means the ore block numbered (4, 3) in the plan view in (1, 1) sublevel.

Table 1. Calculation results.

Number	Result	Total Storage/t	Remaining Storage/t	Grade/%
(1, 1, 4, 3)	1	2883.92	2883.92	24.03
(1, 1, 8, 3)	1	2951.56	2951.56	61.74
(1, 1, 11, 3)	1	2631.26	2631.26	44.26
(1, 1, 1, 2)	1	2951.37	2951.37	36.46
(1, 1, 6, 2)	1	2601.16	2601.16	49.15
(1, 1, 9, 1)	1	2667.56	2667.56	49.28
(1, 1, 12, 1)	1	2916.49	2916.49	56.74
(2, 1, 1, 3)	1	2643.29	2643.29	35.77
(2, 1, 4, 3)	1	2708.67	2708.67	42.90
(2, 1, 10, 3)	1	2947.29	2947.29	31.48
(2, 1, 2, 1)	1	2938.19	2938.19	52.25
(2, 1, 5, 1)	1	2646.24	2646.24	54.64
(2, 1, 8, 1)	1	2951.56	2951.56	61.74
(2, 1, 11, 1)	1	2969.06	2969.06	44.26
(2, 1, 7, 3)	1	2671.67	2671.67	48.00
Others	0	/	/	/

5.1.2. Haulage Equipment Dispatch Plan

The short-term resource plan indicates that there are 15 stopes in the plan period, seven stopes in (1, 1) sublevel, and eight stopes in (2, 1) sublevel. According to the production arrangement of the mine, there are two ore passes in the first level and three ore passes in the second level during the plan period. The travel time parameters of the scrapers and electric locomotives measured by the mine are shown in Tables 2–7.

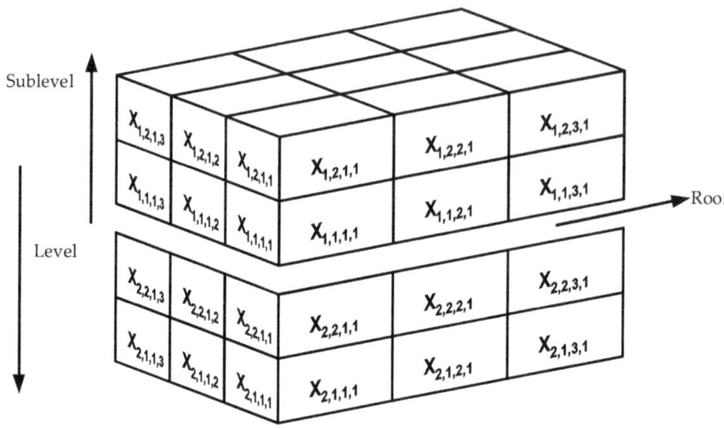

Figure 5. Simple spatial model of ore blocks.

The total income during the plan period is 5,325,966 dollars.

Table 2. Loaded travel time of the scraper between the stope and ore pass in *(1, 1)* sublevel/s (Units: s).

Ore Pass	Stope/(Number)						
	a	b	c	d	e	f	g
	(1, 1, 4, 3)	(1, 1, 8, 3)	(1, 1, 11, 3)	(1, 1, 1, 2)	(1, 1, 6, 2)	(1, 1, 9, 1)	(1, 1, 12, 1)
A	75.5	149.3	179.2	97.8	100.5	145.6	210.3
B	154.2	85.6	96.2	205.2	137.3	79.7	100.5

Table 3. Empty travel time of the scraper between the stope and ore pass in *(1, 1)* sublevel/s (Units: s).

Ore Pass	Stope/(Number)						
	a	b	c	d	e	f	g
	(1, 1, 4, 3)	(1, 1, 8, 3)	(1, 1, 11, 3)	(1, 1, 1, 2)	(1, 1, 6, 2)	(1, 1, 9, 1)	(1, 1, 12, 1)
A	52.49	105.7	147.5	64.5	67.8	104.3	172.6
B	110.7	57.5	61.3	144.2	84.6	54.0	68.2

Table 4. Loaded travel time of the scraper between the stope and ore pass in (2, *1*) sublevel/s (Units: s).

Ore Pass	Stope/(Number)							
	s	t	u	v	w	x	y	z
	(2, 1, 1, 3)	(2, 1, 4, 3)	(2, 1, 7, 3)	(2, 1, 10, 3)	(2, 1, 2, 1)	(2, 1, 5, 1)	(2, 1, 8, 1)	(2, 1, 11, 1)
C	87.3	86.7	144.7	173.7	91.8	114.7	152.9	177.4
D	157.6	133.3	92.3	132.7	145.7	87.2	67.1	134.7
E	214.7	195.7	113.8	56.4	215.3	187.3	105.4	80.5

Table 5. Empty travel time of the scraper between the stope and ore pass in (2, *1*) sublevel/s (Units: s).

Ore Pass	Stope/(Number)							
	s	t	u	v	w	x	y	z
	(2, 1, 1, 3)	(2, 1, 4, 3)	(2, 1, 7, 3)	(2, 1, 10, 3)	(2, 1, 2, 1)	(2, 1, 5, 1)	(2, 1, 8, 1)	(2, 1, 11, 1)
C	54.5	54.2	106.7	129.8	56.3	73.2	111.4	132.3
D	110.3	89.6	56.7	90.5	107.3	55.1	43.8	92.3
E	162.3	141.4	72.7	34.3	160.9	137.9	69.6	51.7

Table 6. Loaded travel time of the electric locomotive between ore pass and the main shaft/s (Units: s).

Main Shaft	Ore Pass				
	A	B	C	D	E
Z	900	963	935	973	892

Table 7. Empty travel time of the electric locomotive between the ore pass and main shaft/s (Units: s).

Main Shaft	Ore Pass				
	A	B	C	D	E
Z	725	772	741	694	711

The short-term resource planning scheme and running time parameters of the above scrapers and electric locomotives are brought into MATLAB to create the haulage equipment scheduling plan. The change curve of the target optimization value with the evolutionary generation is shown in Figure 6, where Object 1 represents F1, Object 2 represents F2, the x-axis is generations, and the y-axis is fitness value. The calculation results of variables in the final haulage equipment dispatch plan model are shown in Tables 8 and 9.

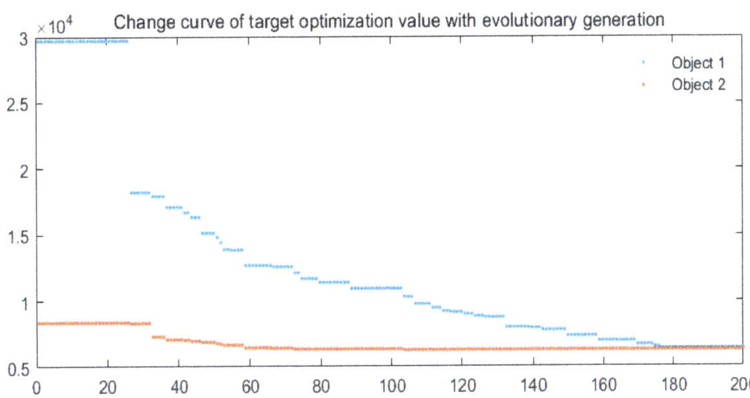

Figure 6. Change curve of the target optimization value with evolutionary generation.

Table 8. Loaded haulage number of times of the scraper (Units: 1).

Ore Pass	Stope													
	a	b	c	d	e	f	g	s	t	u	v	w	x	y	z
A	19	20	26	20	26	30	35	/	/	/	/	/	/	/	/
B	21	13	16	28	15	18	15	/	/	/	/	/	/	/	/
C	/	/	/	/	/	/	/	9	14	7	17	9	16	19	17
D	/	/	/	/	/	/	/	13	9	6	13	8	12	6	7
E	/	/	/	/	/	/	/	14	4	10	6	15	13	7	6

Table 9. Unloaded haulage number of times of electric locomotives (Units: 1).

Main Shaft	Ore Pass				
	A	B	C	D	E
Z	6	7	6	5	5

In order to prove the superiority of the non-dominated sorting genetic algorithm, the calculation results of the non-dominated sorting genetic algorithm are compared with the results of the traditional genetic algorithm. Solving a multi-objective problem using the traditional genetic algorithm method involves transforming the multi-objective problem into a single objective by weighting to select the superior individual and generating the superior scheme. Since the single running time of the electric locomotive is longer than that of the scraper, the proportion of F2 is relatively low, at 10%, while the proportion of F1 is 90%. The change curve of the target value in the calculation process with the evolutionary generation is shown in Figure 7. The results of comparison between traditional scheme and the scheme in Tables 8 and 9 are shown in Table 10.

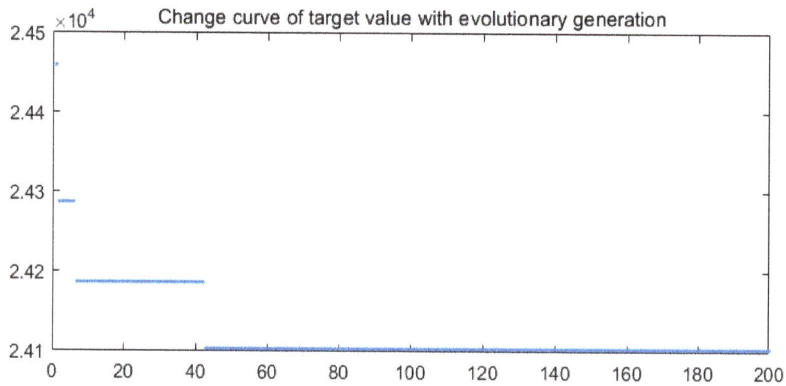

Figure 7. Change curve of the target value with evolutionary generation.

Table 10. Comparison of the two schemes.

Method	The Tonnage of Ore Transported by the Scraper/t	The Tonnage of Ore Transported by the Electric Locomotive/t	Total Waiting Time/h
NSGA-III	1677	1508	3.5
GA	1118	1005	11

5.2. Result Analysis

The change curve in Figure 5 reflects the change trend of the multi-target value of the optimal individual in the evolution process. The multi-target value of the optimal individual in each generation changes significantly, and when it evolves to approximately 180 generations, the curve tends to stable, at which point the optimal solution of the multi-objective optimization problem is obtained. The optimal target value curve of the traditional genetic algorithm in Figure 6 is relatively simple and hierarchical. The curve tends to be stable when it evolves to approximately 40 generations, indicating that the optimization intensity is insufficient in the calculation process, and the optimization result is relatively rough.

In the scheme calculated by the NSGA-III-algorithm, the tonnage of ore hauled by the scraper from the stope is 1677 t, and the tonnage of ore hauled by the electric locomotive to the main shaft is 1508 t during a shift (Table 10). The wait time is 3.5 h. In the scheme calculated using the traditional genetic algorithm, the total wait time is 11 h, the tonnage of ore hauled from the stope is 1118 t, and the tonnage of ore hauled to the main shaft is 1005 t. In the better scheme, the total wait time of the scraper and electric locomotive was reduced by 7.5 h, the ore hauled from the stope increased by 559 t, and the ore hauled to the main shaft for hoisting increased by 503 t. The non-dominated sorting genetic algorithm has better performance than the traditional genetic algorithm and has higher practical value.

The operation route of the scraper is not limited to a single stope and ore pass (Figure 8). The simultaneous digging and loading operation of multiple stopes can improve the smoothness of the succession of each process. Through this operation mode, the U-turn and wait time of the scraper at the ore pass are reduced.

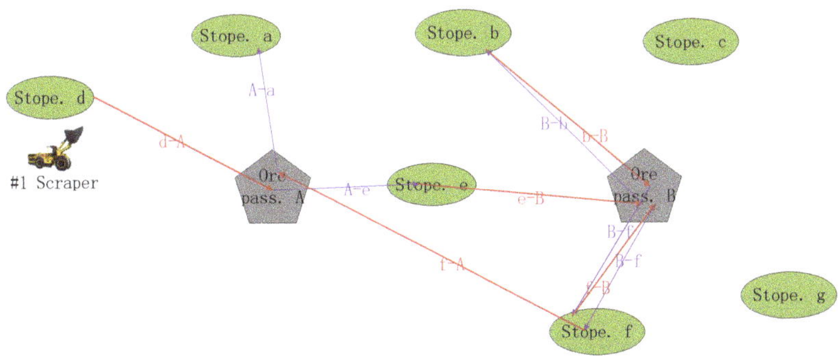

Figure 8. Operation route of the #1 scraper during a certain period of time.

For the studied mine, the constraints considered are insufficient when formulating a short-term resource plan. The constructed resource plan barely meets the mine's production needs and fails to maximize the mine's production benefits. The haulage equipment dispatch plan prepared on this basis is inefficient. In this paper, the equipment dispatch plan is based on the short-term resource plan. From the optimization results, the short-term resource plan is used to guide the dispatching plan of the scraper and the electric locomotive, which can efficiently complete the short-term production tasks. Therefore, as far as the traditional scheduling scheme of the mine is concerned, this research can provide a comprehensive and excellent production decision-making scheme based on meeting the production requirements of the mine, which can improve the utilization rate of equipment, and is suitable for production and haulage in the mine under study.

6. Discussion

Through the mine short-term resource planning optimization model, the operating location of the scraper is obtained. Then, the dispatching plan of the scraper is prepared using the dispatching model. This article only arranges the dispatching plan of the scraper in a certain period of time, and does not realize the preparation of the dispatching plan of the scraper for the entire short-term planning cycle. Therefore, the follow-up research can be conducted to further improve the entire short-term dynamic dispatching plan of the scraper. In addition, the size of the population limits its global search, so the population size can be increased appropriately, and the parameters in the algorithm flow can also be appropriately optimized.

7. Conclusions

Based on 0-1 integer programming, combined with the production needs of underground mines under study, a short-term resource plan optimization model for underground mines is constructed with the maximum profit as the primary objective. The constraints of this model conform to the actual conditions of the mine. The artificial bee colony optimization algorithm is used to solve the model, and the requirements of the artificial bee colony for the nectar source are restricted by the constraints. The neighborhood search method can be used to find the optimal nectar source over a wide range. The addition of the following bee and scout bee can improve the optimization results. Then, based on the short-term resource plan, considering the constraints of each link in the underground mine ore haulage process, the shortest total equipment wait time is used as the objective function to construct an optimization model for haulage equipment between stopes, which can more completely describe the process of dynamic allocation of the scraper and electric locomotives. Using a non-dominated sorting genetic algorithm, a more accurate Pareto optimal solution set can be obtained by introducing elite selection strategy and selection methods based on reference points. Finally, verified by examples, this paper can provide a basis for the formulation of a short-term plan and equipment dispatch decisions for underground mines, and improve the mine's production revenue and equipment utilization. Future research directions should focus on the preparation of dispatching planning for multiple pieces of equipment and should not be limited to haulage equipment.

Author Contributions: Conceptualization, N.L.; Data curation, H.Y.; Formal analysis, S.F. and D.C.; Funding acquisition, N.L., Q.W. and M.J.; Investigation, Q.W.; Methodology, N.L. and M.J.; Resources, H.Y., S.Z. and L.W.; Supervision, N.L.; Validation, S.F. and S.Z.; Writing—original draft, S.F.; Writing—review & editing, H.Y., Q.W., S.Z., M.J., L.W. and D.C. All authors have read and agreed to the published version of the manuscript.

Funding: This research is supported by the National Key R&D Program of China (Grant No. 2019YFC0605304), National Natural Science Foundation of China (Grant No. 51704218), and Special Fund Project of the Basic Scientific Research Operation Fee of the Central University (Grant No. 2019III086CG and 2018IVB054).

Data Availability Statement: Not applicable.

Acknowledgments: Thanks to the relevant departments for funding and the mining companies that provided the experiment.

Conflicts of Interest: The authors declare no conflict of interest.

References

1. Topal, E. Early start and late start algorithms to improve the solution time for long-term underground mine production scheduling. *J. South. Afr. Inst. Min. Metall.* **2008**, *108*, 99–107.
2. Sandanayake, D.S.S.; Topal, E.; Ali Asad, M.W. A heuristic approach to optimal design of an underground mine stope layout. *Appl. Soft Comput.* **2015**, *30*, 595–603. [CrossRef]
3. Wu, H.; Li, J. Open-pit Mine Production Planning: The Current, Problem& Strategies. *Met. Mine* **2005**, *4*, 4–6+42.
4. Newman, A.M.; Rubio, E.; Caro, R.; Weintraub, A.; Eurek, K. A Review of Operations Research in Mine Planning. *Interfaces* **2010**, *40*, 222–245. [CrossRef]
5. Zhan, K. Development trend of trackless mining equipment in underground metal mines. *Min. Technol.* **2006**, *3*, 34–38.
6. Yi, X. Development Trend of Trackless Equipment in Mines in my country. *Min. Equip.* **2019**, *1*, 10–15.
7. Li, Y.; Tong, G. General situation of the development of mine production planning method. *Met. Mine* **1994**, *12*, 11–16.
8. Ming, J.; Li, G.; Hu, N. Optimization of Production Planning for Underground Metal Mines Based on Market. *J. Univ. Sci. Technol. Beijing* **2013**, *35*, 1215–1220.
9. Nehring, M.; Topal, E.; Kizil, M.; Knights, P. Integrated short- and medium-term underground mine production scheduling. *J. South. Afr. Inst. Min. Metall.* **2012**, *112*, 365–378.
10. O'sullivan, D.; Newman, A. Optimisation-based heuristics for underground mine scheduling. *Eur. J. Oper. Res.* **2015**, *241*, 248–259. [CrossRef]
11. Zhou, K.; Gu, D. Optimization of Stoping Sequence in Underground Mining by Using Genetic Algorithm Method. *China Min. Mag.* **2001**, *5*, 52–56.
12. Hou, J.; Hu, N.; Li, G.; Ma, Z.; Li, D.; Gong, J. Dynamic Optimization of Production Plans for Multi-Metal Underground Mines. *Chin. J. Eng.* **2016**, *38*, 453–460.
13. Foroughi, S.; Hamidi, J.K.; Monjezi, M.; Nehring, M. The integrated optimization of underground stope layout designing and production scheduling incorporating a non-dominated sorting genetic algorithm (NSGA-II). *Resour. Policy* **2019**, *63*, 101408. [CrossRef]
14. Gligoric, Z.; Gligoric, M.; Dimitrijevic, B.; Grozdanovic, I.; Milutinovic, A.; Ganic, A.; Gojkovic, Z. Model of room and pillar production planning in small scale underground mines with metal price and operating cost uncertainty. *Resour. Policy* **2020**, *65*, 101235. [CrossRef]
15. Jiang, C.; Peng, P.; Wang, L. 3D Visualization Production Planning in Underground Ming Based on Simulating Technology. *China Min. Mag.* **2015**, *24*, 152–156.
16. Liu, D.; Wang, L.; Chen, X.; Zhong, D.; Xu, Z. Study on Multi Objective Optimization and Application of Medium and Long Term Plan for Underground Mine. *Gold Sci. Technol.* **2018**, *26*, 228–233.
17. Sarin, S.C.; West-Hansen, J. The long-term mine production scheduling problem. *IIE Trans.* **2005**, *37*, 109–121. [CrossRef]

18. Newman, A.M.; Kuchta, M. Using aggregation to optimize long-term production planning at an underground mine. *Eur. J. Oper. Res.* **2007**, *176*, 1205–1218. [CrossRef]
19. Riff, M.-C.; Otto, E.; Bonnaire, X. A New Strategy Based on GRASP to Solve a Macro Mine Planning. In Proceedings of the International Symposium on Foundations of Intelligent Systems, Prague, Czech Republic, 14–17 September 2009; pp. 483–492.
20. Little, J.; Knights, P.; Topal, E. Integrated optimization of underground mine design and scheduling. *J. South. Afr. Inst. Min. Metall.* **2012**, *113*, 775–785.
21. Mousavi, A.; Sellers, E. Optimisation of production planning for an innovative hybrid underground mining method. *Resour. Policy* **2019**, *62*, 184–192. [CrossRef]
22. Campeau, L.-P.; Gamache, M. Short-term planning optimization model for underground mines. *Comput. Oper. Res.* **2020**, *115*, 104642. [CrossRef]
23. Gligoric, Z.; Beljic, C.; Gluscevic, B.; Cvijovic, C. Underground Lead-Zinc Mine Production Planning Using Fuzzy Stochastic Inventory Policy. *Arch. Min. Sci.* **2015**, *60*, 73–92.
24. Gamache, M.; Grimard, R.; Cohen, P. A shortest-path algorithm for solving the fleet management problem in underground mines. *Eur. J. Oper. Res.* **2005**, *166*, 497–506. [CrossRef]
25. Saayman, P.; Craig, I.K.; Camisani-Calzolari, F.R. Optimization of an autonomous vehicle dispatch system in an underground mine. *J. South Afr. Inst. Min. Metall.* **2006**, *106*, 77–86.
26. Nehring, M.; Topal, E.; Little, J. A new mathematical programming model for production schedule optimization in underground mining operations. *J. South. Afr. Inst. Min. Metall.* **2010**, *110*, 437–446.
27. Sun, Y.; Lian, M. Research on Vehicle Routing and Scheduling Problems Based on Improved ACA in Underground Mine. *Met. Min.* **2010**, *2*, 51–54.
28. Åstrand, M.; Johansson, M.; Zanarini, A. Underground Mine Scheduling of Mobile Machines using Constraint Programming and Large Neighborhood Search. *Comput. Oper. Res.* **2020**, *123*, 105036. [CrossRef]

metals

Article

Propagation Laws of Reclamation Risk in Tailings Ponds Using Complex Network Theory

Zhixin Zhen, Ying Zhang * and Mengrong Hu

School of Civil and Resource Engineering, University of Science and Technology Beijing, Beijing 100083, China; zzxzack@126.com (Z.Z.); humengrong0331@163.com (M.H.)
* Correspondence: zhangyingmail@ustb.edu.cn

Abstract: Accidents have occurred periodically in the tailings ponds where mine solid waste is stored in recent years, and thus their safety has become one of the constraints restricting the sustainable development of the mining industry. Reclamation is an important way to treat tailings ponds, but improper reclamation methods and measures not only cannot reduce the accident risk of tailings ponds, but will further increase the pollution to the surrounding environment. The influencing factors of reclamation accidents in tailings ponds are complex, and the existing models cannot characterize them. In order to study the propagation process of tailings pond reclamation risk, this paper proposes a three-dimensional identification framework for accident hazards based on evidence (TDIFAHE) to identify all potential hazards that may occur during the reclamation stage, and obtain a list of hazards. Based on the complex network theory, this paper uses identified hazards as network nodes and the correlation between hazards as the edges of the network. Based on the identified hazard data, the evolution network of reclamation risk in tailings ponds (ENRRTP) is constructed. By analyzing the statistical characteristics of ENRRTP, it can be found that ENRRTP has small world and scale-free characteristics. The above characteristics show that the reclamation risk of tailings ponds is coupled with multiple factors and the disaster path is short. Giving priority to those hub hazards that have a dominant impact on the reclamation risk can significantly reduce the reclamation risk of the tailings pond.

Keywords: tailings; reclamation risk; hazard identification; complex network; hazard management

Citation: Zhen, Z.; Zhang, Y.; Hu, M. Propagation Laws of Reclamation Risk in Tailings Ponds Using Complex Network Theory. *Metals* **2021**, *11*, 1789. https://doi.org/10.3390/met11111789

Academic Editor: Chris Aldrich

Received: 31 August 2021
Accepted: 3 November 2021
Published: 6 November 2021

1. Introduction

Reclamation of a tailings pond, also known as tailings reclamation, is one of the important methods for comprehensive management and utilization of mine tailings. Tailings reclamation is generally to cover the beach and slopes of the tailings pond after the pond area is closed, while plant crops or cash crops protect the tailings from being carried away by wind and rain under natural conditions [1]. Tailings reclamation would simultaneously promote both economic development and environmental protection [2]. However, if the mining enterprise adopts improper reclamation methods, the result will not only be unable to improve the environment of the pond area, but may even damage the stability of the dam, causing more serious safety accidents and environmental damage. For example, mine workers carried out tailings reclamation in violation of regulations, which may cause damage to drainage pipes. When heavy rainfall occurs in the pond area, the water level rises rapidly because the rainwater cannot be drained in time, which in turn may cause overtopping or dam break accidents. In order to avoid the occurrence of such accidents, it is important to objectively and systematically identify potential accident hazards in the reclamation of tailings ponds, master the propagation laws of reclamation risks, and then take targeted measures to reduce the occurrence of reclamation accidents.

Hazard identification is also known as risk identification, and its purpose is to discover, identify, and describe the risks [3]. Hazard identification is the basis and premise of underlying risk management in tailings ponds [4]. Before clarifying the hazards of the tailings pond, rushing to carry out reclamation work may cause serious consequences. Commonly used hazard identification methods include Safety Checklists, Accident Statistics Analyses, Preliminary Hazard Analyses, Failure Models and Effects Analyses, Hazard and Operability Analyses, Cause–Consequence Analyses, Fault Tree Analyses, and Event Tree Analyses [4]. To mitigate tailing-water interactions and support reclamation, Divine et al. inject applied tracers into legacy tailings, and observe the transport behavior of the tracers [5]. Liu et al. used the Root-State Hazard Identification (RSHI) method to identify the hazards in underground coal mine risk management [4]. The above methods provide theoretical support for the identification of hazards, and promote the development of risk management in tailings ponds. However, the composition of the tailings pond reclamation system is very complex, and the entire system is undergoing continuous changes with the progress of reclamation activities, so the existing methods are difficult to meet the needs of hazard identification and risk characterization.

The complex network theory is an appropriate tool for analyzing multidimensional hazards based on holistic thinking and understand the complex interactions among the hazards. The complexity of complex network is mainly reflected in two aspects: (i) the number of nodes in complex network is enough; (ii) the relationships between nodes are complicated and uncertain [6]. Based on the above characteristics, complex networks have been widely used in many fields. The network developed by Lam CY and Tai K. used a topological approach to modeling the causations and characteristics of railway incidents in Japan [7]. To analyze the influence factors of SOC stock, Zhang Z. used the complex network to characterize the complex relationship of SOC stock [6]. Xu W used the Minimum Cumulative Resistance model (MCR) to build a landscape ecological network to identify the important ecological elements, and the complex network method was used to analyze the structure of landscape ecological network [8]. In the field of tailings ponds, Qin et al. applied the complex network theory to the accident risk management of the tailings pond, and found that the influencing factors with larger betweenness centrality have a more important impact on the accident [9]. Zhen et al. used a complex network to characterize the evolution process of tailings dam break risk, and found that the propagation process of dam break risk has the characteristics of a small world and is scale-free [10–18]. Although complex networks have many advantages, no scholar has studied the propagation laws of accident risks during the reclamation of tailings ponds from the perspective of networks.

Based on the shortcomings of the above research, the paper proposes a three-dimensional identification framework for accident hazards based on evidence (TDIFAHE) that can systematically and completely identify the reclamation hazards of tailings pond. After that, this paper uses hazards to represent nodes, and the relationships between hazards represent the edges of complex networks. For the first time, an evolution network of reclamation risk in the tailings pond (ENRRTP) that intuitively characterizes the propagation process of reclamation risks is constructed. Through the analysis of the statistical characteristics of the ENRRTP, the propagation law of reclamation risk is researched from the network perspective. Finally, the method proposed in this paper is applied to a specific tailings pond case, and the network efficiency is used to verify whether the key hazards obtained by the method in this paper are reasonable. The above-mentioned research is helpful to discover the propagation laws of reclamation risk that has not been discovered in the past methods [19–21] from the network perspective, and can improve the safety management during the reclamation of the tailings pond.

2. Research Methods

2.1. Overview

In order to systematically and objectively identify the accident hazards during the tailings pond reclamation period, this paper proposes a three-dimensional identification framework for accident hazards based on evidence (TDIFAHE). After that, this paper uses the identified hazards as the nodes of the complex network, and the correlation between hazards as the edges of the network, and constructs the ENRRTP which characterizes the propagation process of the tailings pond reclamation risk. Based on the complex network theory, the statistical characteristics of ENRRTP are analyzed, and part of the propagation law of reclamation risk is derived from the network perspective. Finally, the paper adds an application case, and verifies the effectiveness of the method proposed through the index of global network efficiency. The entire research process and main results of this paper are shown in Figure 1.

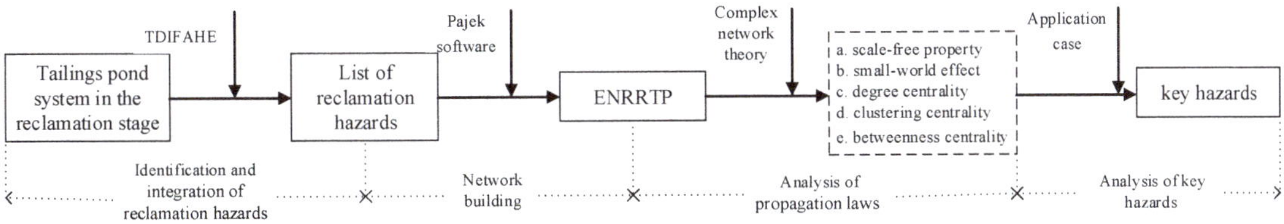

Figure 1. A flow chart of research methods and results.

2.2. Identification Method

2.2.1. The Definition of Hazard

The definition of hazard has many forms. For the purpose of characterizing the evolution process of tailings pond reclamation risk, combined with the structure and accident characteristics of the tailings pond, this paper divides the influencing factors, hidden dangers, accidents and consequences involved in risk propagation into four types of hazards. The first kind is the dormant hazard, which is the initial factor or event that causes the reclamation accident. Other factors or hazards cannot trigger it, and its state is stable. The second kind is the armed hazard, which refers to the intermediate state of hazard evolution [9], and may evolve from the dormant hazard or other armed hazards (armed hazard means the imminent accidents and disasters). The third hazard is the active hazard, which refers to the reclamation accidents that are happening [9]. The reclamation risk of tailings pond includes multiple types of accidents, such as dam break, seepage, overtopping, raise dust in the pond area, and leaks in transmission facilities. The fourth hazard is the consequence hazard, which is the type of disaster caused by the accident, including environment pollution, economic loss, personal injury and reputation loss. This hazard index helps to classify and evaluate the consequences caused by the reclamation accidents.

2.2.2. Hazard Identification and Data Extraction

The structure of the Tailings storage facilities is complex, and the facilities are constantly changing under the influence of the reclamation personnel and the external environment during reclamation process. This has caused many hazards in the reclamation process of the tailings pond, and the relationship and intensity between the hazards are uncertain over time. In order to solve the above problems, this paper proposes a three-dimensional identification framework for accident hazards based on evidence (TDIFAHE) to identify the reclamation hazards of tailing pond, as shown in Figure 2 [10,22]. Among them, the Z-axis represents the reclamation process, including five stages: pond area investigation, pond area remediation, reclamation design, reclamation construction and reclamation supervision, highlighting the temporal dynamics of reclamation work. The Y-axis contains eight subsystems of the tailings pond. They are a new division of the tailings pond system after considering the impact of the four influencing factors of personnel, material, environment, and management on the tailings pond reclamation work. In the process of identifying hazards, each subsystem can be further subdivided according to the roles played by each part of the system. The sub-division of the subsystems is not only conducive to detailing all the hazards involved in the reclamation process of the tailings pond, but also helps to ensure the independence of each hazard index. The Z-axis represents evidence such as laws and regulations, standards and norms, scientific and technological literature, and accident case. This evidence provides support for the identification of hazards in different subsystems and different stages of reclamation.

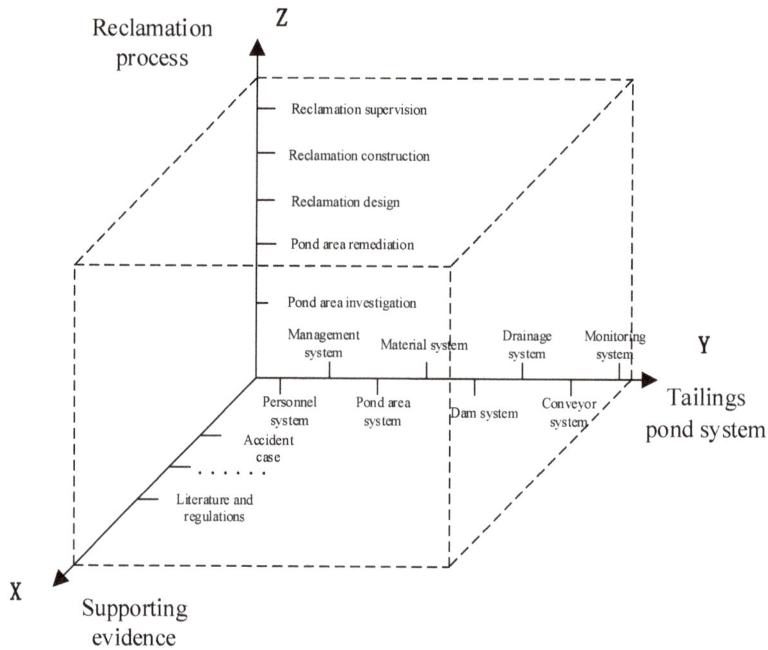

Figure 2. A three-dimensional identification framework for accident hazards based on evidence.

In Figure 2, the spatial nodes enclosed by three-dimensional coordinates are all initially identified reclamation hazards of tailings pond. Afterwards, by integrating the spatial nodes with the same meaning, such as both Safety Regulations for Tailings Pond (GB GB39496-2020) and Code for Construction of Tailings Facilities (AQ 2001–2018) can provide evidence to support for the hazard 65 (Dam deformation) in the dam subsystem during the reclamation construction stage. However, in Figure 2, the same hazard identified by these two supporting pieces of evidence is shown as two nodes. Therefore, when determining the final hazard index, it needs to be integrated. After identifying and integrating the reclamation hazard of tailings pond, a universal list of reclamation hazard of tailings pond can be obtained to construct a complex network. Therefore, this process can also be called data extraction.

2.3. Network Model

2.3.1. Model Building

Based on the identified reclamation hazards and the correlation between hazards of tailings pond, this paper constructs an evolution network of reclamation risk in the tailings pond (ENRRTP) that characterizes the propagation process of tailings pond reclamation risk. In ENRRTP, reclamation hazards represent nodes in a complex network, and the relationships between reclamation hazards represent edges between nodes. The ENRRTP has four hazard nodes (armed hazard, dormant hazard, active hazard/accident, consequence/disaster hazard), and three propagation stages (from armed hazard to dormant hazard, from dormant hazard to active hazard, active hazard to consequence hazard) [10].

2.3.2. Degree

The degree value of a hazard node refers to the number of hazard nodes directly connected to the node, reflecting the direct influence of the hazard [10]. Because the correlation between the reclamation hazards of tailings pond is directional, the ENRRTP is a directed network. The number of relationships from other hazard nodes to a hazard node is called the in-degree of this node, and the number of relationships emitted from a node is called the out-degree of this node. The average value of degree ki of node i in the ENRRTP is called the average degree of the network, which is denoted as K. For any network with M edges and N nodes, its average degree can be expressed as: $K = 2M/N$ [11].

The network density of ENRRTP is an index that characterizes the degree of connection between hazard nodes in the network. The greater the network density means that the relationship between hazards is closer, and the network may have a greater impact on the state or behavior of hazards. For a directed ENRRTP with N nodes and M edges, the theoretical maximum possible value of the total number of edges is $N(N-1)$, then the density of ENRRTP is [6]:

$$D = \frac{M}{N(N-1)}. \tag{1}$$

The degree distribution of ENRRTP represents the probability distribution function $P(K)$ of the degree of the hazard node, which refers to the probability that the hazard node has k edges connected. If there are a total of n nodes in ENRRTP, of which there are n_k nodes, and his degree is k, then [11]:

$$P(K) = \frac{n_k}{n}. \tag{2}$$

The cumulative degree distribution of ENRRTP is the probability of the occurrence of hazard nodes with degree greater than or equal to k in the network. When the cumulative degree distribution of ENRRTP can be approximated by Equation (3) in double logarithmic coordinates, it indicates that the network is scale-free [12,16,17].

$$Q(k) \sim ax^{-b}. \tag{3}$$

In Equation (3), a and b are constants greater than zero.

2.3.3. Average Path Length and Diameter

The distance of the ENRRTP refers to the step length from one hazard node to another hazard in the network, that is, the distance between hazard nodes i and j in the network, defined as d_{ij}, which represents the number of edges connecting these two hazards on the shortest path [23]. The average path length, also known as the characteristic path length, represents the average step length of the hazard node in the network to reach the non-neighboring node. The average path length of ENRRTP reflects the influence relationship between hazards in an average sense. The shorter the path length means the closer the correlation between hazards and the greater the mutual influence. Its mathematical expression is:

$$L = \frac{1}{\frac{1}{2}N(N-1)} \sum_{i \geq j} d_{ij}. \tag{4}$$

N is the total number of ENRRTP nodes [11].

The network diameter is also called the maximum path length of the network. The network diameter of ENRRTP represents the number of edges between the two furthest nodes in the network. The network diameter of ENRRTP is closely related to the risk transmission time and the stability of the entire network.

2.3.4. Clustering Coefficient

According to the definition of clustering coefficient, the clustering coefficient of ENRRTP refers to the degree of interconnection of a node with adjacent hazard nodes [23]. Assuming that a hazard node i in ENRRTP has k_i edges to connect it with other hazard nodes, there may be at most $k_i(k_i - 1)/2$ edges between these k_i hazard nodes. Then the ratio of the actual number of edges E_i between these k_i hazard nodes to the total possible number of $k_i(k_i - 1)/2$ is defined as the clustering coefficient of the hazard nodes, namely [24]:

$$C_i = \frac{2E_i}{k_i(k_i - 1)}. \tag{5}$$

Calculating the average value of the clustering coefficient of all hazard nodes can get the clustering coefficient of the entire ENRRTP, namely [24]:

$$C = \frac{1}{N}\sum_{i=1}^{N} C_i. \tag{6}$$

2.3.5. Betweenness Centrality

The node betweenness centrality of the ENRRTP reflects the control degree of the hazard node over the surrounding edges and hazards [10]. Specifically, if a hazard node is on the shortest path of many other hazard node pairs, it is said that the hazard node has a high betweenness centrality. Hazard nodes with high betweenness centrality play an intermediary role in the process of the spread of reclamation risks, and aggravate the spread of reclamation risks. The calculation formula of the betweenness centrality is [19]:

$$C_B = \sum_{j<k}\left[\frac{g_{jk}(i)}{g_{jk}}\right]. \tag{7}$$

The g_{jk} is the number of existing shortest paths between hazard nodes j and k, and g_{jk} (i) represents the number of shortest paths passing through hazard node i between hazard node j and hazard node k. For the betweenness centrality of the entire ENRRTP, the calculation formula is [24]:

$$C_B = \frac{2\sum_{i=1}^{n}\sum CB_{Bmax}}{(n-1)^2(n-2)} \tag{8}$$

3. Results

3.1. Hazard List

Based on the TDIFAHE, this paper has identified 191 potential accident hazards in the reclamation stage of tailings ponds and 1207 relationships among the hazards, as shown in Appendix A [25].

In Table A1, the fourth column is the reclamation hazards of tailings pond after identification and integration. The third column is the number of the hazard in the same row in the fourth column. The numbers in the topology model are conducive to the display of the logical relationship between hazards. The first column and the second column indicate, respectively, that the corresponding hazard belongs to which subsystem of the tailings pond and which element of the subsystem. The last column is the hazards caused by the hazard in the same row in the fourth column confirmed by laws and regulations, documents, accident cases and other evidence. These hazards are characterized by their unique numbers.

3.2. Mode of the ENRRTP

3.2.1. Network Model

Based on the identified reclamation hazards and relationship between hazards in Appendix A, the paper first constructs an adjacency matrix that characterizes the relationship between the reclamation hazards. After that, the adjacency matrix was imported into Pajek complex network software to construct the evolution network of reclamation risk in the tailings ponds (ENRRTP). The topological structure of the network is shown as in Figure 3.

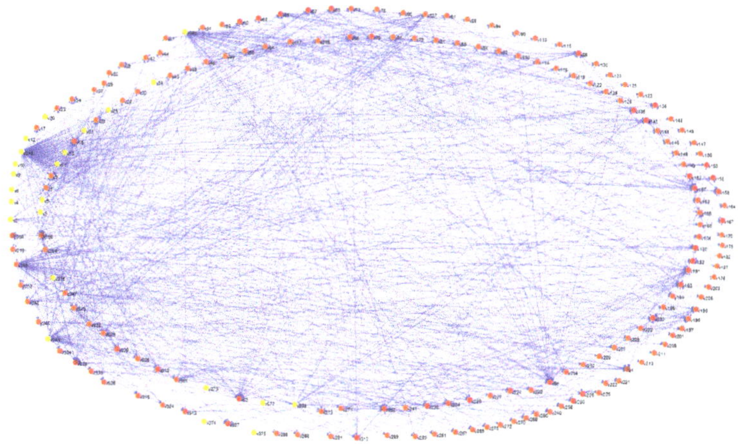

Figure 3. Propagation process of reclamation risk in the tailings ponds.

The ENRRTP systematically characterizes the four states and three transmission stages of the hazards of the tailings pond in the reclamation stage. In Figure 3, the yellow node represents the initial dormant hazard, including 23 hazards, with only out-degrees but no in-degrees; orange nodes reflect armed hazards, involving 127 hazards; red nodes reflect active hazards, involving 37 types of hazards, indicating 37 types of tailings pond reclamation accidents; red nodes reflect the consequence hazards of the accidents, including 4 types, and the in-degree value of these hazards is large, but the out-degree value is small.

3.2.2. Degree and Degree Distribution

The degree value of each node in ENRRTP can be obtained through Pajek complex network software, as shown in Figure 4. The average degree of the ENRRTP is 12.64, and the network density is 0.03, indicating that a hazard node is directly related to 11.81 hazards on average, but the overall density of ENRRTP is not large.

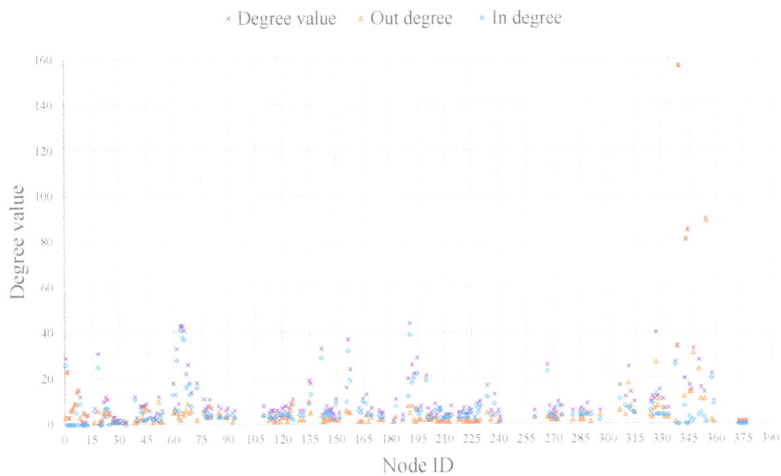

Figure 4. Node degree in the ENRRTP.

It can be seen from Figure 4 that among the 10 hazards with the largest degree value, hazard 340 (insufficient safety supervision) is the hazard node with the largest degree value in the ENRRTP, which directly affects the 157 reclamation hazards of tailing ponds. These hazards include the accident hazards of all subsystems in the tailings pond reclamation stage, indicating that safety supervision is the key to the safe operation of each subsystem in the reclamation stage of tailings ponds. Hazard 355 (Insufficient experience in personnel or organization qualification problems) is directly related to 90 hazards, which is the second largest hazard in ENRRTP and belongs to the personnel subsystem. The degree value of hazard 345 (Defects in safety production rules and regulations and operating procedures), 344 (Outdated specifications and standards for survey, design, construction and acceptance), 327 (Safety monitoring facilities cannot fully reflect the operating status of the tailings pond) are 85, 81 and 40 respectively. These hazards, and hazard 340, belong to the management subsystem, accounting for 40% of the top 10 hazards, highlighting that management factors play a leading role in the safety management of the reclamation stage of tailings ponds. The degree values of hazard 64 (Dam instability), 65 (Dam deformation), 66 (Dam cracks) and 157 (Filter failure) are 43, 43, 41 and 37, respectively, and these hazards belong to the dam body subsystem. Although the names and numbers of these hazards have appeared in the early stage, the meaning of these hazards has changed greatly during the reclamation stage. Hazard 191 (Fracture of drainage structure) is directly related to 44 types of accident hazards of tailings pond, which belong to the drainage subsystem, indicating that drainage problems are also an important factor affecting the safe operation of tailings ponds in the reclamation stage.

Hazards 340, 355, 345, 344, and 339 are the top five hazards, with an out-degree value of 157, 89, 85, 81, and 34, all of which are dormant hazards. Except for hazard 355, the other four hazards belong to the management subsystem. The above characteristics indicate that the personnel subsystem and the management subsystem are prone to produce hazards that induce the status changes of other hazards. These hazards need to be paid attention to and prevented in advance. In contrast, the five hazards with the largest in-degree are hazard 64 (Dam instability), 191 (Fracture of drainage structure), 65 (Dam deformation), 66 (Dam cracks) and 157 (Filter failure) (41, 39, 38, 37, and 32, respectively), and these hazards are all classified as armed hazards. With the exception of hazard 191, which belongs to the drainage subsystem, all four other hazards belong to the dam subsystem. The above characteristics indicate that the hazards of dams and drainage facilities generally require the use of external forces, and these external influences mainly involve management factors, human factors and environmental factors. The above characteristics indicate that hazards in dams and drainage facilities are generally caused by external forces, and these external influences mainly involve management factors, human factors and environmental factors.

In the double logarithmic coordinates, the cumulative degree distribution of the ENRRTP presents a power-law distribution, which approximates to $P(k) \sim 4.4265x^{-1.187}$ ($R^2 = 0.9061$), as shown in Figure 5 [12]. As the value of k gradually increases, the above results deviate from the power-law property, which indicates that the ENRRTP has scale-free characteristics. Therefore, in the process of controlling the reclamation risk, giving priority to those hazards with higher degree values can more effectively reduce the spread of reclamation risks and prevent reclamation accidents from occurring in tailings ponds.

3.2.3. Average Path Length and Diameter of the ENRRTP

After calculation, the average path length of the ENRRTP is 2.91, which represents the average step length of a node to a non-adjacent node; in the average sense, if a hazard wants to affect other hazards, it needs to pass 2.91 network edges in the reclamation stage, or a hazard in the reclamation stage can cause other hazards to change status within 3 steps.

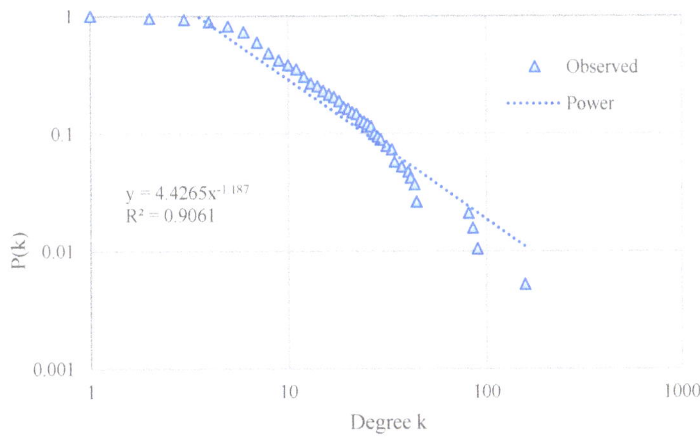

Figure 5. Cumulative degree distribution of the ENRRTP.

The network diameter of the ENRRTP is 11, which represents the number of edges between the two furthest nodes in the network. In other words, a hazard can spread to the entire network after 11 steps at most. The pair of nodes with the farthest distance in ENRRTP is node v93 to v142. Compared with some accident networks studied in the past [12–14], the diameter of ENRRTP is larger, and the evolution path of the network is complicated. Therefore, the reclamation hazards of tailings ponds have the characteristics of large workload and rapid risk transmission. To interrupt the transmission path of reclamation risks, it is necessary to find the key hazard nodes that affect risk transmission.

3.2.4. Clustering Coefficient and Small World Property

According to the definition of clustering coefficient, there is no clustering coefficient for nodes with a degree value of 1. Therefore, in the statistical calculation of the clustering coefficient of the ENRRTP, nodes with a degree value of 1 are excluded. The clustering coefficient of nodes in the ENRRTP is between 0 and 0.5, as shown in Figure 6. Among them, hazard 13 (Karst cave or existence of mine shafts) and 182 (Unqualified filter material) are the nodes with the largest clustering coefficient, with a value of 0.5. Hazard 13 is mainly caused by insufficient geological exploration when selecting the site of the tailings pond, which belongs to the pond area system. The hazard 182 belongs to the selection of anti-seepage materials and belongs to the dam subsystem. If an unqualified filter material is selected, it will not only cause seepage accidents but also damage the stability of the dam. In the ENRRTP, the clustering coefficients of most hazard nodes are mainly concentrated in the range of 0.05 to 0.35, which is relatively uniform and does not show clustering characteristics that are too strong.

This paper uses Pajek software to generate a random network with the same number of nodes and the same average degree value as ENRRTP, as shown in Figure 7. After calculation, the average clustering coefficient of the equal-sized random network of the reclamation risk for tailings pond is 0.07, which is smaller than the average clustering coefficient of the ENRRTP (0.1918). A small-world network refers to a network with characteristics of small path length and large clustering coefficient [15]. Therefore, through the above analysis of the two indicators, it can be concluded that the ENRRTP has a small-world effect, which means that the reclamation risk in the tailings pond has the characteristics of specific multi-factor coupling and short disaster path. Therefore, it is extremely important to find the key hazards that affect the spread of accident risk during the reclamation stage of the tailings pond through appropriate methods.

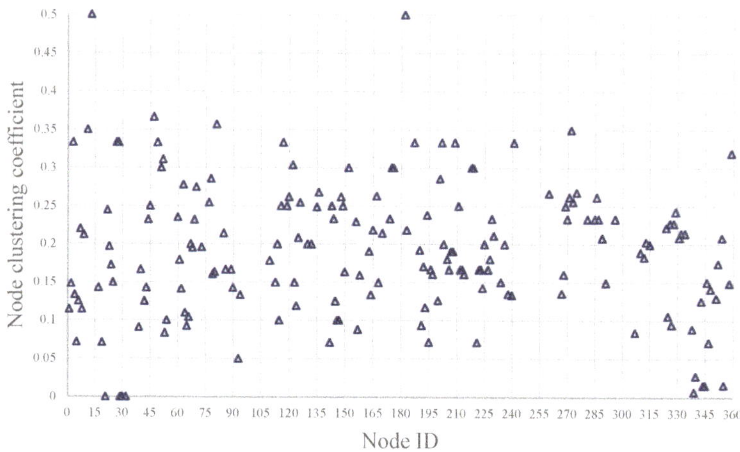

Figure 6. Clustering coefficient of the nodes in the ENRRTP.

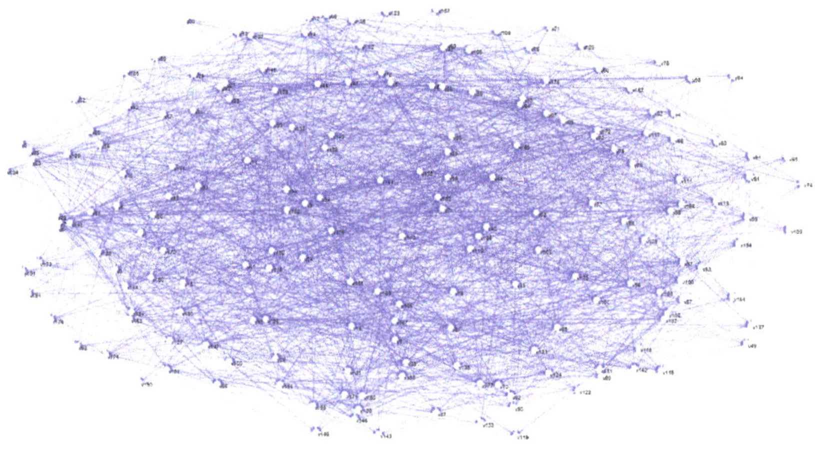

Figure 7. The equal-sized random network of the ENRRTP.

3.2.5. Betweenness Centrality and Key Hazards

The betweenness centrality of the hazard node in the ENRRTP is shown in Figure 8. The hazard with the largest betweenness centrality is hazard 65 (Dam deformation), with a value of 0.0677, indicating that the 'dam deformation' is the most important channel in the process of reclaiming risk transmission. In the ENRRTP, the minimum value of the betweenness centrality of the node is 0, the average value is 0.0021, and the standard deviation is 0.0051. In addition, it can be found that the maximum value of the betweenness centrality of the nodes is 32 times the average value, and the standard deviation is 2.4 times the average value, which proves that the betweenness centrality of the nodes of the ENRRTP is very heterogeneous. Under the effect of this heterogeneity, there are only a small number of hazards with a large betweenness centrality in the network, while most hazards have a small betweenness centrality. There are 25 hazards in the network with a betweenness centrality of 0; these hazards have only out-degree but no in-degree. They are located at the edge of the network and are the dormant hazards in the reclamation hazards of tailings pond. Heterogeneous networks usually have robustness against random attacks and vulnerability to specific attacks. Random management of hazards is difficult to quickly reduce the spread of reclamation risks. However, if we give priority to the management of those nodes with high betweenness centrality, we can make full use of the vulnerability of the ENRRTP to reduce the reclamation risk of the tailings pond.

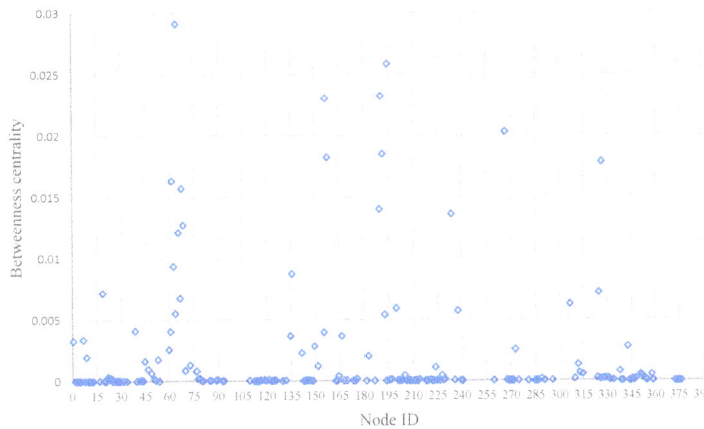

Figure 8. Betweenness centrality of nodes in the ENRRTP.

4. Application Case

To better demonstrate the method proposed in the paper, the Gaoxi tailings pond in Yun'an County, Guangdong was selected as an application case [26].

The stacking elevation of the tailings pond is 300~375 m. This area has a subtropical monsoon climate, with the maximum daily rainfall of 144.4 mm. The rainfall is mostly concentrated in April to September, accounting for 70% of the annual rainfall, and the instantaneous maximum wind speed is 23.3 m/s. The tailings of the tailing pond have poor physical structure, poor water and fertilizer retention capacity, and high levels of toxic heavy metals. The reclamation plan of the tailings pond mainly includes the stages of chemical passivation measures, engineering barrier measures and biological measures (establishment of vegetation growth layer or ecological restoration).

Based on the project overview information, management system and personnel qualifications of the Gaoxi tailings ponds during the reclamation, this paper excludes the hazards and the correlation between hazards that did not exist during the reclamation from Appendix A, and constructs the propagation network of reclamation risk for the Gaoxi tailings pond (PNRRGTP), as shown in Figure 9.

Global network efficiency refers to the average value of the reciprocal of the shortest path length between all pairs of hazard nodes in the network, reflecting the propagation speed of the reclamation risk on the network [10]. Therefore, this paper chooses the indicator to measure the spread efficiency of reclamation risks. Before remedying (deleting) the hazards (nodes), R software is used to calculate the global efficiency of the PNRRGTP as 0.1092.

In order to verify whether the hazards obtained based on the degree value and the betweenness centrality of the node are key hazards, the paper randomly deletes 10 nodes in the PNRRGTP and repeats it 10 times, and the average value of the network efficiency is 0.0906. For comparison, the paper first deletes the five hazards with the largest degree value of PNRRGTP (340, 355, 345, 344, and 191), and then deletes the five hazards with the largest betweenness centrality (65, 157, 195, 158, and 192), and the calculated network efficiency is 0.0541.

After comparison, we can find that after the remediation of some hazards, the propagation efficiency of reclamation risks has decreased, but remedying the hazards with the larger degree value and betweenness centrality can more effectively reduce the propagation efficiency of reclamation risks.

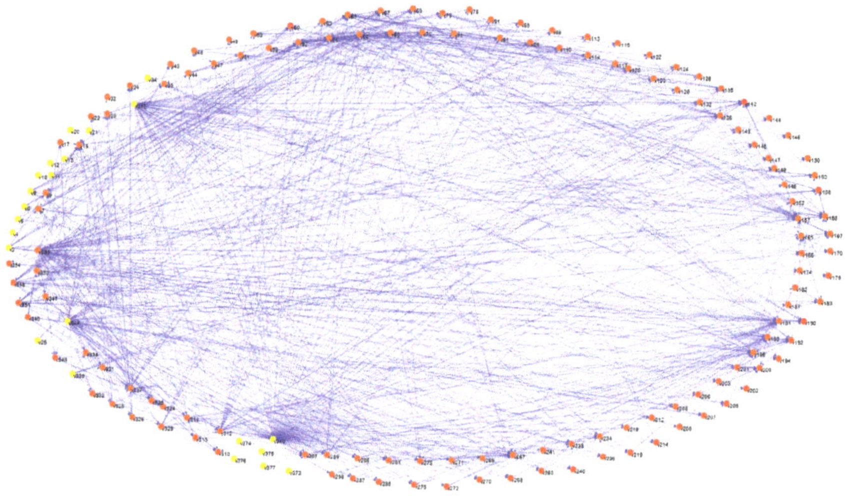

Figure 9. Mode of the PNRRGTP.

5. Discussion

The list of reclamation hazards of tailings pond is obtained by the TDIFAHE. The list covers the hazards that may exist during the reclamation of tailing ponds of different types and regions. When using the hazard list to identify the hazards in a specific tailings pond, it is only necessary to eliminate the hazards that do not exist in the hazard list based on the engineering geological information, management and personnel information of the tailings pond. In order to avoid the omission of the hazards and relationship between hazards, this paper analyzes the content one by one in the Code for Design of Tailings Facilities (GB 50863-2013), Code for Construction and Acceptance of Tailings Facilities (GB-T 50864-2013), Geotechnical Engineering Survey Code (GB50021-2009) and Safety Regulations for Tailings Pond (GB GB39496-2020). Because this evidence was compiled by the most experienced experts in the industry for different types of tailings ponds, and after a long period of application and multiple rounds of corrections, essentially cover all the hazards that may occur during the reclamation stage of the tailings ponds. At the same time, this paper also supplements the hazards with literatures, accident cases and other evidence related to the tailings reclamation. After integration, a complete list of hazards can be obtained. Compared with the hazard identification methods used in the reclamation stage of tailings ponds in the past [19,20], the hazards identified by TDIFAHE are more complete and objective.

By using nodes to represent hazards and edges to represent the relationship between hazards, this paper constructs the ENRRTP to characterize the propagation process of reclamation risk for tailings ponds firstly. Using complex network theory to analyze the statistical characteristics of this network, we can find that: (a) the cumulative degree distribution of the ENRRTP presents a power-law distribution, indicating that the ENRRTP is scale-free; (b) the ENRRTP has a larger clustering coefficient and a smaller characteristic path length, which indicates that the network has a small-world characteristic; (c) in the ENRRTP, the distribution of node's betweenness centrality presents a strong heterogeneity, and a small number of nodes have a large betweenness centrality. The above results indicate that reclamation risks have many coupling factors and short transmission paths. By applying the above method to the Gaoxi tailings pond, we can find that if nodes with larger degree value and betweenness centrality are prioritized, the propagation efficiency of the accident risk can be greatly reduced, indicating that the hazards with a larger degree value and betweenness centrality plays a dominant role in the spread of reclamation risk.

In order to better analyze the propagation laws of reclamation risk, the paper uses four indicators of degree, path length, clustering coefficient, and betweenness centrality to analyze the statistical characteristics of ENRRTP. These four indicators are relative indicators to measure the reclamation risk, so there is no indicator threshold for dividing high-risk and low-risk. For example, in ENRRTP, although the value range of the degree is 1 to 157, only a few hazards have a larger degree value due to the scale-free nature of the network. From the perspective of the relationship number between hazards, a hazard with a large degree value may have a relatively higher risk than a hazard with a small degree value, but we cannot accurately give a threshold value (when the degree value of a hazard is greater than a certain value, the hazard must enter a high-risk state), because the degree value can only represent one aspect of the strength of the hazard.

The paper successfully uses the complex network to characterize the propagation process the reclamation risk of tailings ponds, and find some important propagation laws and key hazards, but there are still some shortcomings. Due to the lack of real-time monitoring data, the paper simplifies the weights between the continuously changing reclamation hazards to the equal weight during analyzing the network characteristics, which hinders the real-time representation of reclamation risks by the network proposed in the paper. In order to better solve the above problems, the author of the paper next plans to apply the above methods to specific reclamation cases that have secured real-time monitoring equipment, and determine the weights between hazards based on monitoring data, work logs, safety evaluations and other imformation, so as to improve the practicality of the above methods [23]. Therefore, the authors of the paper will look for a tailing pond with a real-time monitoring system in the reclamation stage for further research.

6. Conclusions

To more systematically and objectively identify the accident hazards in the reclamation stage of tailings ponds, this paper proposes the TDIFAHE method. Based on the method and the integration of the same hazards, this paper obtains a hazard list containing 191 types of reclamation hazards and 1207 relationships between hazards.

This paper uses hazards to represent the nodes of a complex network, and the relationships between hazards represent the edges of the network, and constructs an ENRRTP model that characterizes the propagation of reclamation risks. Through analysis of its statistical characteristics, it is found that the propagation of reclamation risk is small world and scale-free.

Combining the Gaoxi tailings pond, this paper finds that compared with the random hazard remediation, the hazards with larger degree value and betweenness centrality remedied preferentially can reduce the propagation efficiency of reclamation risks more quickly. The above findings are of great significance for preventing accident risks during the reclamation of tailings ponds.

Author Contributions: Conceptualization, Z.Z. and Y.Z.; methodology, Z.Z.; software, Z.Z.; validation, Z.Z. and M.H.; resources, Y.Z.; writing—original draft preparation, Z.Z.; writing—review and editing, M.H.; visualization, Z.Z. All authors have read and agreed to the published version of the manuscript.

Funding: This work is supported by The National Key Research and Development Program of China (Grant No. 2017YFC0804605).

Institutional Review Board Statement: Not applicable.

Informed Consent Statement: Not applicable.

Data Availability Statement: The data that supports the findings of this study are available within the paper.

Conflicts of Interest: The authors declare that they have no known competing financial interests or personal relationships that could have appeared to influence the work reported in this paper.

Appendix A

Table A1. List of reclamation hazards in tailings ponds.

Tailings Pond Subsystem	Elements of the System	Number (v)	Hazard Name	Number of Hazards Caused
Pond area system	Pond environment	1	Environmental pollution	338, 358, 359
		2	Flood	1, 7, 9, 19, 60, 62, 64–67, 69, 150, 156, 158, 167, 190–193, 195, 273, 325, 358
		3	Ice and snow	19, 67, 195
		4	Strong wind	1, 7, 19, 66, 142, 325
		5	Heavy rainfall	9, 19, 67, 69, 150, 193, 195
		6	Extreme temperature changes	19, 62, 65–67, 191, 267, 325, 358
		7	Surge	62, 65–67, 69, 150, 190, 193, 358
		8	Beyond standard earthquake	19, 60, 62, 64–66, 70, 136, 150, 191–192, 267, 273, 325, 358
		9	Mudslide	1, 39, 358
		10	Gravel foundation	157
		11	Liquefied soil, soft clay and collapsible loess foundation	68, 70, 135–136, 157
		12	Water burst in the tailings pond	158
		13	Karst cave or existence of mine shafts	68, 135–136, 158
		17	Inadequate research on adverse geological problems and improper handling measures	19
		19	Landslides in the tailings pond	1, 7, 9, 39, 195, 358
		20	The overburden of the bank slope connected to both ends of the tailings dam is thin	158
		21	The rock on the bank slope is broken, joints are developed, or faults pass through	19, 158
		22	Animals burrow, camp, and graze illegally	19, 64, 66, 142, 150, 158
		23	Private digging in the tailings impoundment	19, 64, 66, 142, 150
		24	Illegal soil borrowing behind the dam	64, 66, 142, 157
		25	There are mining activities near the site	19, 62, 64, 66, 142

Table A1. *Cont.*

Tailings Pond Subsystem	Elements of the System	Number (v)	Hazard Name	Number of Hazards Caused
	Selection of pond location	27	Located on the upstream of important facilities and residential areas	338, 358
		28	Located on the upwind side of the dominant wind in a concentrated residential area	338, 358
		29	Occupy a lot of arable land and move a large number of residents	338
		30	Located on a valuable deposit	338
		32	Insufficient impoundment length (upstream wet tailings impoundment)	39
		34	Large catchment area	195
		39	Insufficient storage capacity of tailings pond	190
		40	The grade of tailings pond does not match the grade of the structures	338
		42	No antifreeze measures have been taken for tailings facilities	66, 191
		43	Anti-freezing measures have not been finished before freezing	66, 191, 338
		44	Blasting construction does not meet the technical specifications	19, 62, 64–66, 191
Material system		45	Tailings particle size/gradation does not meet the requirements	47, 51, 66, 68, 70, 61, 142, 234
		47	Excessive tailings unit weight	51–52, 68, 61
		49	Strongly corrosive tailings	238
		51	Unqualified dry beach covering materials	53, 142, 157–158, 195
		52	Unqualified filling materials	64–68, 70, 73, 135–136, 157
		53	Erodible tailings exposure	157
		54	Mismatch between tailings pond type and stockpiled waste	1
	Dam body	60	Dam break	1, 9, 338, 358, 359
		62	Local landslide and collapse of the dam	1, 9, 60, 63, 338
		63	Decrease of dam elevation	39, 190, 194
		64	Dam instability	60, 62
		65	Dam deformation	62, 64, 157, 267, 273
		66	Dam crack	62, 64, 73, 158
		67	Dam surface water saturation	62, 64–66, 70, 73, 157
		68	Uneven settlement of the dam	62–66, 191–192, 267, 273
		69	Scour the dam	1, 62, 64–66
		70	Tailings liquefaction	62, 64, 68, 136, 156–158
		73	Poor stability of tailings dam slope	62, 64, 70

Table A1. *Cont.*

Tailings Pond Subsystem	Elements of the System	Number (v)	Hazard Name	Number of Hazards Caused
		77	The tailings dam slope ratio is unreasonable	62, 64–65, 70, 73, 157
		78	Unreasonable width of dam crest	62, 64–65, 157
		79	Improper dam type selection for the initial dam	39, 64, 157
		80	The height of initial dam is unreasonable	39, 64–65, 73, 81, 194, 228, 338
		81	The ratio of the initial dam height to the total dam height of the upstream tailings dam is unreasonable	64–65, 73
		61	Poor control of tailings deposits	64–65, 68, 77, 142, 152, 157
		85	The accumulation dam is too high	62, 64–65, 81, 338
		86	The height of the accumulation dam is lower than the height of tailings accumulation	39, 65, 190, 194
		89	Use the upstream method to build dams on the seismic zone	60, 62, 70
		90	Fine-grained tailings dams using direct method	64–65, 61
		93	No filtration water and sediment storage dams are built in the centerline and downstream tailings dams	64
		94	Unreasonable height of the filtration water and sediment storage dams	39, 64–65
		110	Improper paving	64–66, 68
		113	Improper unloading method	65–66
		114	Filling and slope adjustment are not carried out at the same time	64, 66, 77
		115	Construction machinery and personnel crossing the dam surface in violation of regulations	65–66
		116	Resuming work in violation of regulations	119
		117	The maintenance platform in downstream dam slope is defective	65–66, 73, 122
		119	Construction under environmental indicators exceeding limits	64, 77–78
		120	The subsidence allowance of the dam filling is unreasonable	358
		122	There is a horizontal weld on the slope	64, 66, 73
		123	Improper selection and care of slope protection turf	73
		124	Slope cutting did not follow the design requirements	19, 64–65
		125	Slope protection was not carried out in time	19, 62, 64–65, 73, 122

Table A1. *Cont.*

Tailings Pond Subsystem	Elements of the System	Number (v)	Hazard Name	Number of Hazards Caused
		126	Unreasonable design of cast-in-place protective surface	19, 62, 64–66, 73, 77, 122, 157
	Dam foundation	132	No effective filter layer is set on the dam foundation	157, 165
		135	Uneven foundation subsidence	63–66, 68, 73, 136, 191, 267, 273
		136	Dam foundation instability	64–66, 68, 73
	Dry beach	142	Raise dust in the pond area	1, 338, 358–359
		143	No watering to reduce dust	142
		144	Insufficient dust prevention measures in the tailings accumulation area(dry)	142
		145	No coverage measures in the pond area	53, 142, 157
		146	The main dam has not been reclaimed and greened in time	53, 142
		147	Insufficient soil cover or greening on the dam slope(dry)	53, 142
		148	Weakness of paving has not been reinforced	158, 142
		130	Poor construction quality of horizontal paving	157, 142
		149	The length or thickness of the horizontal paving in front of the dam is insufficient	157, 142
		150	Natural paving (covering) is destroyed	158, 142
		152	Poor deposition control for dry beach face	142, 157
	Seepage system	156	Seepage damage	1, 60, 62, 64, 338, 359
		157	Filter failure	64, 67, 136, 156, 167, 195
		158	Leakage channel	64, 68, 135–136, 156
		163	The substandard seepage water has not been collected, recovered and treated	1
		164	The dam foundation area between the initial dam and sediment storage dam is not equipped with drainage facilities	157
		165	Defects of dam foundation drainage facilities	157
		167	Seepage line is higher than control seepage line	65–67, 70, 156
		168	Improper measures to reduce the seepage line	167
		170	Insufficient protection measures for seepage prevention facilities	158, 165, 183
		174	Unqualified geomembrane	157, 165

Table A1. *Cont.*

Tailings Pond Subsystem	Elements of the System	Number (v)	Hazard Name	Number of Hazards Caused
		175	No drainage measures under the geomembrane protective layer	157, 165
		176	Poor drainage of composite geotechnical drainage network	157, 165
		182	Unqualified filter material	183
		183	Filter failure	65, 157, 165
		187	Geotextile clogged	165, 183
Drainage system	Drainage plan	190	Overtopping	1, 60, 62, 64, 69, 338, 358–359
		191	Fracture of drainage structure	66, 69, 158, 192, 200
		192	Leaking drainage structure	1, 66–67, 69, 150, 158, 195, 200
		193	Scour or cavitation drainage structures	191–192
		194	Insufficient regulating water storage	39
		195	Rapid rise of pond water level	39, 65, 67, 152, 167, 190, 194
		196	No drainage facilities	195, 200
		197	The foundation pit at the higher groundwater level has no drainage facilities	195, 200
		200	Insufficient flood discharge capacity	193, 195
		201	Blocking defects of flood drainage facilities	192–193, 195, 200
		202	Unreasonable temporary flood control plan during construction period	195, 200
		203	Improper diversion measures	195, 200
		205	The installation location and elevation of drainage facilities do not meet the design requirements	193, 195, 200
		206	Insufficient elevation of drainage holes in front of the dam	200
		207	Flood drainage structures are directly located on the tailings sediment beach	191
		208	The foundation of the flood drainage structure is set in the area with poor engineering geology	191
		209	Insufficient foundation bearing capacity of underground flood drainage structures	191
		211	The dry tailings pond of third–class and above adopts flood interception ditch for flood discharge	200
		212	Drainage facilities are not located in front of the blocking dam	193
		213	Use mechanical flood drainage	200

Table A1. *Cont.*

Tailings Pond Subsystem	Elements of the System	Number (v)	Hazard Name	Number of Hazards Caused
		214	The on–site line setting is inconsistent with the construction drawing	77–78, 205, 273
		218	Improper installation of energy dissipation facilities	191, 193
		219	No energy dissipation measures have been taken in the tailings facility	191, 193
	Backwater plan	221	The clarified water of the tailings pond is not used for backwater utilization	195, 338
		222	One–sided pursuit of backwater quality	195
		223	Excessive pursuit of backwater in tailings pond	228
		224	Low tailings water recovery rate	195, 338
		225	The backwater design of the tailings pond does not utilize the potential energy	338
		227	Insufficient volume of backwater pond	224
		228	Unqualified backwater quality	1
		229	Turbid tailings water	1
		230	Unexpected discharge or misdischarge water that does not meet the discharge standard	1, 359
Conveyor system	Conveying facilities	234	Blockage or siltation	165, 176, 191, 195, 200
		236	No flow and pressure detection instrument	191, 234
		238	Serious corrosion of equipment	191, 228, 231, 325, 338
		240	No anti–corrosion treatment in tailings facilities	238
		241	Unqualified anti–corrosion materials	193, 238
		260	Improper handling of local hydraulic phenomena	234, 238, 267
	Other transportation facilities	267	Pipes and grooves deformation	191, 193, 234,
		268	Defects of the interception ring in pipe body	69, 192–193
		269	The pipe body is in direct contact with the big rocks	191, 267
		270	The outer wall of the pipe is not protected	191, 267
		271	The dimensions of pipes, grooves, tunnels, etc. do not meet the requirements	191, 193, 234, 267
		272	Pipes and grooves material unqualified	191, 193, 267

Table A1. *Cont.*

Tailings Pond Subsystem	Elements of the System	Number (v)	Hazard Name	Number of Hazards Caused
		273	Subsidence or deformation of supporting facilities such as pipes, trenches and tunnels	191, 267
		275	Excessive slope deviation for laying pipes, trenches, tunnels, etc.	191, 193, 234, 267
		281	Poor quality of fill around the pipeline	191, 267
		285	No settlement joints between pipe and well	191, 267
		286	The joint length of the drain pipe is unreasonable	191–192, 267
		287	Deformation joints are not provided at both ends of the drain pipe according to the design requirements	191, 267
		289	Improper excavation (pipes, trenches, tunnels, etc.)	65–66, 117, 120, 191
	Pump	291	Poor construction ventilation	358
		296	Poor pump quality	192–193, 234, 307
		307	Pump failure	61, 192, 195, 200, 228, 338
		373	No liquid can be discharged from the sand pump	307
		374	Insufficient liquid output from sand pump	307
		375	Pump consumes too much power	307
		376	Pump bearing heat	307
		377	Deformed or broken pump shaft	307
		310	The surrounding environment improvement does not meet the requirements	1, 9, 19, 142
		312	Dam body renovation does not meet the requirements	62–70, 73, 135–136, 142, 148, 157–158, 167, 183
		313	The improvement of flood discharge system does not meet the requirements	191–192, 195, 228, 234, 267, 273, 307
		315	Unreasonable reclamation plan	1, 19, 310, 312–313
Monitoring System		324	Improper selection of monitoring instruments and equipment	327, 338, 343
		325	Monitoring instrument failure, work interruption	327, 338, 343
		326	The third–class and above tailings ponds are not equipped with monitoring facilities that combine manual and automatic monitoring	1, 19, 22–24, 142, 327
		327	Safety monitoring facilities cannot fully reflect the operating status of the tailings pond	1, 7, 9, 19, 22–24, 37, 45, 47, 49, 54, 65–69, 135–136, 163, 191–192, 200, 228–230, 267, 343

Table A1. *Cont.*

Tailings Pond Subsystem	Elements of the System	Number (v)	Hazard Name	Number of Hazards Caused
		328	No monitoring points are arranged outside the dam toe	9, 19, 22–23, 142, 144, 310, 327
		329	No additional monitoring facilities are installed at the dam abutment, bedrock faults, and buried pipes in the dam	65–69, 135–136, 191–192, 267, 327
		331	No external drainage and composition monitoring	1, 163, 200, 228–230, 327
		332	No monitoring of groundwater and surrounding water bodies	1, 230, 327
		334	The number of water quality monitoring wells around the tailings pond is insufficient	1, 230, 327
Management system		338	Economic losses	359
		339	Insufficient capital investment	17, 24, 42, 52, 54, 89–90, 93, 132, 145–148, 163–164, 196–197, 221, 230, 236, 240, 268, 291, 310, 312, 324, 326, 328, 329, 331–332, 334, 351
		340	Insufficient safety supervision	7, 9, 17, 19, 22–24, 40, 42, 43–45, 47, 49, 51–52, 54, 60, 62–70, 73, 77–81, 85–86, 89–90, 93–94, 110,113–117, 119–120, 122–126, 130, 132, 135–136, 142–150, 152, 156–158, 163–168, 170, 174–176, 182–183, 187, 190–197, 200–203, 205–209, 211–214, 218–219, 221–225, 227–230, 234, 236, 238, 240–241, 260, 267–273, 275, 281, 285–287, 289, 291, 296, 307, 310, 312–313, 315, 324–329, 331–332, 334, 346, 348, 351–352, 354–355, 358
		343	Inadequate safety evaluation	1, 19, 60, 142, 156, 190, 200, 224, 228, 327, 358–359
		344	Outdated specifications and standards for survey, design, construction, and acceptance	17, 23–24, 27–30, 32, 40, 42–43, 61, 73, 77–81, 85–86, 89–90, 93–94, 110, 115–116, 120, 123–124, 130, 145, 157–158, 164, 168, 170, 174, 191–192, 194, 196–197, 200, 207–208, 211–212, 214, 221, 225, 227, 238, 240, 268–271, 275, 281, 285–287, 289, 296, 310, 312–313, 315, 324, 326–329, 331–332, 334, 343, 346, 351, 354, 358
		345	Defects in safety production rules and regulations and operating procedures	23–24, 39, 42–43, 54, 80, 85–86, 90, 93–94, 110, 113–115, 119–120, 122–125, 130, 132, 142–148, 150, 157–158, 163–165, 167–168, 170, 174–175, 183, 191–193, 201, 203, 205–208, 213–214, 218–219, 221–224, 229–230, 234,236, 238, 240, 260, 267–268, 270, 275, 281, 285–289, 291, 307, 310, 312–313, 325, 343, 352, 358

Table A1. *Cont.*

Tailings Pond Subsystem	Elements of the System	Number (v)	Hazard Name	Number of Hazards Caused
		346	Improper data management	17, 42–43, 79, 197, 205, 207–208, 225, 315, 324, 327, 343, 352
		347	Insufficient or wrong hydrological and geological data	17, 42–43, 79, 197, 205, 207–208, 225, 309, 315, 324, 327, 343, 352
		348	Improper quality acceptance	19, 60, 62–70, 135–136, 142, 156–158, 167, 183, 190–193, 200, 234, 238, 267, 307, 310, 312–313
		351	Improper maintenance	60, 62, 64–70, 142, 156–158, 167, 183, 190–193, 234, 238–239, 254, 267, 307, 325
		352	Design defects of emergency plan	1, 19, 60, 62, 142, 156, 190–191, 195, 338, 358
		354	Insufficient emergency plan drills	1, 19, 60, 62, 142, 156, 190–191, 195, 338, 358
Personnel system		355	Insufficient experience in personnel or organization qualification problems	17, 23–24, 44, 54, 61, 79, 110, 113–116, 119, 123–126, 130, 132, 143–149, 163–164, 168, 170, 174–176, 187, 196–197, 201–203, 205–209, 211–214, 218–219, 221–225, 230, 240, 260, 268–272, 275, 281, 285–289, 291, 310, 312–315, 324, 326, 328–332, 334, 343, 346–348, 351–352, 354
		358	Personal injury	338, 359
		359	Loss of reputation	338

References

1. Chen, Y. Current status and existing problems of comprehensive utilization of tailings. *Des. Res. Nonferrous Metall.* **2018**, *39*, 129–131.
2. Scotti, A.; Milia, S.; Silvani, V.; Cappai, G.; Guglietta, D.; Trapasso, F.; Tempesta, E.; Passeri, D.; Godeas, A.; Gómez, M.; et al. Sustainable Recovery of Secondary and Critical Raw Materials from Classified Mining Residues Using Mycorrhizal-Assisted Phytoextraction. *Metals* **2021**, *11*, 1163. [CrossRef]
3. *ISO: 31000: Risk Management Guide*; ISO: Geneva, Switzerland, 2018.
4. Liu, Q.; Peng, Y.; Li, Z.; Zhao, P.; Qiu, Z. Hazard identification methodology for underground coal mine risk management—Root-State Hazard Identification. *Resour. Policy* **2021**, *72*, 102052. [CrossRef]
5. Divine, C.; Griffin, A.; Christoffersen, L.; Gillow, J.; Preston, K.; DeDycker, P. Tracer Testing to Evaluate Tailing Pile Hydrogeologic Characteristics to Support Closure and Reclamation. In Proceedings of the International Mine Water Association Annual Conference on Reliable Mine Water Technology, Golden, CO, USA, 6–9 August 2013; pp. 967–972.
6. Zhang, Z.; Wang, J.; Li, B. Determining the influence factors of soil organic carbon stock in opencast coal-mine dumps based on complex network theory. *Catena* **2019**, *173*, 433–444. [CrossRef]
7. Lam, C.Y.; Tai, K. Network topological approach to modeling accident causations and characteristics: Analysis of railway incidents in Japan. *Reliab. Eng. Syst. Saf.* **2020**, *193*, 106626. [CrossRef]
8. Xu, W.; Wa Ng, J.; Zhang, M.; Li, S. Construction of landscape ecological network based on landscape ecological risk assessment in a large-scale opencast coal mine area. *J. Clean. Prod.* **2021**, *286*, 125523. [CrossRef]
9. Qin, X.; Li, Z.; Zhao, Y. Complex network model of tailing pond risk evolution and analysis of key hazards. *Syst. Eng. Theory Pract.* **2017**, *37*, 1648–1653.
10. Zhen, Z.; Li, J.; Li, Z. Exploration and research on the propagation law of failure risk of tailings dam based on complex networks. *Phys. A Stat. Mech. Its Appl.* **2022**, in press.
11. Zhou, J.; Xu, W.; Guo, X.; Ding, J. A method for modeling and analysis of directed weighted accident causation network (DWACN). *Phys. A Stat. Mech. Its Appl.* **2015**, *437*, 263–277. [CrossRef]
12. Guo, S.; Zhou, X.; Tang, B.; Gong, P. Exploring the behavioral risk chains of accidents using complex network theory in the construction industry. *Phys. A Stat. Mech. Its Appl.* **2020**, *560*, 125012. [CrossRef]

13. Chen, W.; Zhang, H.; Comfort, L.K.; Tao, Z. Exploring complex adaptive networks in the aftermath of the 2008 Wenchuan earthquake in China. *Saf. Sci.* **2020**, *125*, 104607. [CrossRef]
14. Zhou, C.; Ding, L.; Skibniewski, M.J.; Luo, H.; Jiang, S. Characterizing time series of near-miss accidents in metro construction via complex network theory. *Saf. Sci.* **2017**, *98*, 145–158. [CrossRef]
15. Watts, D.J.; Strogatz, S.H. Collective dynamics of 'Small World' networks. *Nature* **1998**, *393*, 440–442. [CrossRef] [PubMed]
16. Latora, V.; Marchiori, M. Efficient behavior of small-world networks. *Phys. Rev. Lett.* **2001**, *87*, 198701. [CrossRef]
17. Latora, V.; Marchiori, M. Economic small-world behavior in weighted networks. *Eur. Phys. J. B-Condens. Matter Complex Syst.* **2003**, *32*, 249–263. [CrossRef]
18. Gao, S.; Zhen, Z.; Li, Z.; Zhao, Y.; Qin, Z. Complex Network Model for Characterizing Hazards and Risks Associated with Mine-tailings Facility. *Resour. Environ. Eng.* **2017**, *2*, 101–107. [CrossRef]
19. Wang, S. Theory and Application for the Safety Integrity of Tailings Ponds. Master's Thesis, University of Science and Technology Beijing, Beijing, China, 2018.
20. Zhao, Y.; Tang, L.; Li, Z.; Qin, J.; Lin, L. Identification of accident hazards of tailing ponds based on process-cause grid method. *China Work. Saf. Sci. Technol.* **2013**, *9*, 91–98.
21. Ma, B.; Zhen, Z.; Chen, C.; Li, Z. Failure Analysis and Prevention of Tailings Industrial System Containing Heavy Metals. *Fail. Anal. Preven.* **2021**. [CrossRef]
22. Li, J. Characterization and Visualization of Accident Hazards With Molten Iron and Steel Workplace. Master's Thesis, University of Science and Technology Beijing, Beijing, China, 2020.
23. Li, S.; Yuan, L.; Yang, H.; An, H.; Wang, G. Tailings dam safety monitoring and early warning based on spatial evolution process of mud-sand flow. *Saf. Sci.* **2020**, *124*, 104579. [CrossRef]
24. Zhao, Y.; Qin, X.; Li, Z.; Tang, L. System dynamics simulation and simulation of hazards and risk evolution of tailings ponds. *J. Univ. Sci. Technol. Beijing* **2014**, *36*, 1158–1165.
25. Zhen, Z.; Ma, B.; Zhao, H. Propagation Network of Tailings Dam Failure Risk: An Empirical Research and The Identification of Key Hazard Node. *Sci. Rep.* **2021**. Available online: https://www.researchsquare.com/article/rs-778838/v1 (accessed on 9 August 2021). [CrossRef]
26. Zhao, Z.; Zhou, Z.; Liang, D.; Bai, Z. Pollution prevention and ecological reclamation design in metal mine reclamation: Taking the Gaoxi lead-zinc mine in Yunan County, Guangdong as an example. *Environ. Sci. Technol.* **2010**, *36*, 106–110.

metals

MDPI

Article

Study and Analysis on the Influence Degree of Particle Settlement Factors in Pipe Transportation of Backfill Slurry

Chonghao Wang * and Deqing Gan

School of Mining Engineering, North China University of Science and Technology, Tangshan 063210, China; jtgdq@ncst.edu.cn
* Correspondence: wangchonghao@ncst.edu.cn

Abstract: In this study, we developed a pipeline transport model to investigate the influence of particle sedimentation factors on slurry transportation through pipelines. The particle tracking module of the software was used to simulate the transport process, and the influences on the sedimentation rate were analyzed considering the slurry concentration, particle size, and flow velocity. The established model exhibited small calculation errors. In addition, the results revealed that the proposed model is reliable for calculating the degree of influence of various factors on particle sedimentation. The effect of the particle sedimentation rate on the pipeline slurry was explored considering the particle size, slurry concentration, and flow velocity. The sedimentation rate was positively related to particle size and adversely related to the slurry concentration and flow velocity. Indeed, study on the sedimentation rate requires considering a reasonable range of particle sizes, preparing a slurry with an appropriate concentration, and adjusting an appropriate flow velocity. Numerical simulations were performed using the filling data as the background for a sample mining area. The experimental results showed optimal slurry concentration and particle size of 60% and 25.25 μm, respectively.

Keywords: particle sedimentation; filling mining; degree of influence; pipeline transportation; solid waste utilization

Citation: Wang, C.; Gan, D. Study and Analysis on the Influence Degree of Particle Settlement Factors in Pipe Transportation of Backfill Slurry. *Metals* **2021**, *11*, 1780. https://doi.org/10.3390/met11111780

Academic Editor: Jürgen Eckert

Received: 7 October 2021
Accepted: 3 November 2021
Published: 5 November 2021

1. Introduction

In recent years, pipeline filling technology has been widely used while the mining depth has been gradually increased [1–3]. Therefore, both the rules of sedimentation and influence on the pipeline transportation must be determined [4–6]. The production capacity of a mine can be increased by selecting an appropriate technology of particle sedimentation, and avoiding pipe blocking to reduce the failure rates in mine pipelines [7–9].

There are many methods to study particle sedimentation: the bottom suction pipe method based on the Oden theory [10,11], repetition depth pipette method based on the McLaughlin formula [12,13], and sand sink method to study the effect of flow turbulence on the particle settling rate [14–16]. Some studies were conducted to perform sedimentation tests on a sediment suspension in a sedimentation cylinder, and formulae for the sedimentation rate were obtained [17,18]. Other studies analyzed experimental data to design a conical settlement cylinder and establish a settlement rate formula for nonuniform sand groups [19]. Some study results have shown that both sediment concentration and particle size are important factors affecting the sedimentation rate [20].

Foreign research is more advanced. Debadutta Das et al. studied the surface activity of the natural surfactant A. auriculiformis isolated by a chemical and aqueous extraction method by calculating its critical micellar concentration (CMC) on the basis of surface tension measurement. The stability of fly ash slurry was determined from its rheological parameters, dispersant concentration, and stabilization mechanism. The stabilization of high-concentration fly ash slurry has been studied through its rheological behavior by variation of temperature and dispersant and ash concentration [21,22]. Anupama

Routray et al. researched the flow behavior of coal water slurry (CWS) characterized with a mixture of non-ionic surfactant saponin from shikakai and a cationic surfactant di-docyl ammonium bromide (DDAB) at different coal loadings, pH, and dispersant dosages. A suitable mechanism of interaction between the mixture of saponin and DDAB on coal surfaces was given on the basis of zeta potential measurement [23]. Subrata Narayan Das et al. reported the stabilization of various types of slurries (particularly iron ore slurry) with their rheological behaviors and CFD analysis for economic pipeline transportation [24].

In this study, we established a filling pipeline model using COMSOL software to simulate the flow regime of slurry in pipelines and to determine the degree of influence of various factors on particle sedimentation [25–27]. This will provide a theoretical basis for improving the stability of mine filling pipeline transportation.

Therefore, we analyzed the factors that affect particle settlement in the slurry and obtained the influence degree of each factor, which provides a reference for the calculation of pipeline transportation parameters. COMSOL software has noticeable advantages in the simulation of particle settlement experiments, simple calculations, and a high visualization degree [28–30].

2. Materials and Methods

This experiment refers to the "Technical Specification for the Total Tailings Paste Backfill" standard (GB/T 39489-2020).

2.1. Test Material

The tailings of the test material were chosen from the Iron Mine of Gao Guan Ying. It was ground by the ore powder grinding machine of our laboratory (YGM130). The density of tailings was determined to be 2620 kg/m^3. The contents of TFe, SiO_2, CaO, MgO, and Al_2O_3 in the tailings were 6.00, 67.58, 4.04, 5.60, and 7.30%, respectively. A NKT6100-D (Haixinrui Technology, Beijing, China) laser particle size analyzer was used to detect the particle size composition. The volume proportion and cumulative proportion of particle size are shown in Table 1 and Figure 1a,b.

Table 1. Particle size, volume proportion, and cumulative proportion.

No.	Size μm	Volume %	Accumulation %	No.	Size μm	Volume %	Accumulation %	No.	Size μm	Volume %	Accumulation %
1	0.36	0.01	0.01	13	2.53	2.72	14.67	25	17.84	6.43	65.84
2	0.42	0.02	0.03	14	2.98	2.75	17.42	26	20.99	6.67	72.51
3	0.5	0.03	0.06	15	3.50	2.81	20.23	27	24.70	6.60	79.11
4	0.59	0.09	0.15	16	4.12	2.93	23.16	28	29.06	6.15	85.26
5	0.69	0.19	0.34	17	4.85	3.14	26.30	29	34.20	5.34	90.60
6	0.81	0.42	0.76	18	5.71	3.53	29.83	30	40.24	4.17	94.77
7	0.95	0.76	1.52	19	6.72	4.01	33.84	31	47.36	2.87	97.64
8	1.12	1.25	2.77	20	7.91	4.42	38.26	32	55.73	1.65	99.29
9	1.32	1.78	4.55	21	9.30	4.72	42.98	33	65.58	0.66	99.95
10	1.55	2.22	6.77	22	10.95	5.02	48.00	34	77.17	0.05	100.00
11	1.83	2.51	9.28	23	12.88	5.44	53.44				
12	2.15	2.67	11.95	24	15.16	5.97	59.41				

According to Table 1, weighted average particle size, $d_{av} = \sum_{i=1}^{n} d_i G_i / 100 = 15.05$ μm; uniformity coefficient, $K_0 = \frac{d_{60}}{d_{10}} = 7.16$; and curvature coefficient, $C_c = \frac{d_{30}^2}{d_{60} \times d_{10}} = 1.76$, where d_i represents the particle size of tailings, G_i represents the weight of tailings, d_{60} is the particle size less than 60%, and d_{30} and d_{10} are 30 and 10%, respectively.

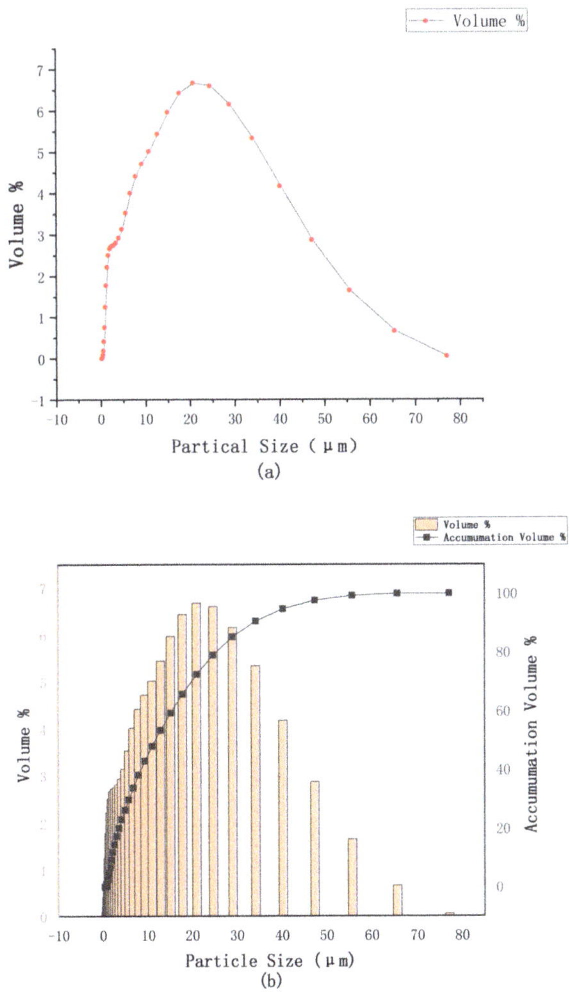

Figure 1. (**a**) Volume of the particle size; (**b**) cumulative proportions of the particle size.

According to the development status of domestic mining science, in the process of tailings cementing, filling and the pipe gravity transport filling times line is typically not more than 5–6; thus, the concentration of filling slurry should be reduced or a method of increasing the pressure should be adopted when the filling timeline is extremely large. The filling timeline of the test was 2.5 and the diameter of the pipe was 200 mm. Viscosity, slurry stability, and pressure are the most important parameters to explain the flow behaviors. Viscosity is a physical and chemical property of a substance. Due to the action of viscosity, the object is subject to friction resistance and differential pressure resistance when moving in the fluid, resulting in the loss of mechanical energy. The stability of slurry is affected by many factors, the most important of which are its concentration and flow velocity. Appropriate concentration and velocity will make the slurry flow steadily and the flow state will not change greatly. In a broad sense, pressure pipes refer to all pipes subjected to internal or external pressure, regardless of the medium in the pipe. The pressure pipeline is a part of the pipeline; the pipeline is used for transportation, distribution, mixing, separation, discharge, metering, control, and stopping the flow of fluid by pipes, pipe fittings, flanges, bolt connections, gaskets, valves, other components, or compression parts and supporting parts of the assembly.

Pipeline transportation is a complex process; therefore, the results of the test can be affected by the pipe diameter, pipe wall, flow velocity, particle size, concentration, temperature, height, friction, and other factors. The optimal value of each factor was calculated by the resistance loss experiment, and the concentration range was 50–60%; the particle size range was 20–60 μm, the flow rate was not more than 2.5 m/s, and the number of experiments in each group was less than 10.

To obtain the test data more scientifically and reduce redundant work, the degree of influence of each factor on sedimentation was explored considering the concentration, particle size, and velocity (Table 2). Orthogonal experimental design is an experimental design method of multiple factors and multiple levels. According to the orthogonality, some representative points are selected from the comprehensive test to carry out the test. These representative points have the characteristics of uniform dispersion and uniform comparability. Orthogonal experimental design is the main method of fractional factorial design. Orthogonal assistant software was used to obtain the experimental scheme and plan. The details are presented in Tables 3 and 4.

Table 2. Simulation test plan parameters.

Number	Concentration %	Particle Size (μm)	Flow Velocity (m/s)
1	60	15.05	2.0
2	55	30.25	2.2
3	50	50.95	2.4

Table 3. Factor and level chart.

Number	1	2	3
Name	Concentration %	Particle Size (μm)	Flow Velocity (m/s)
Level 1	60	15.05	2.0
Level 2	55	30.25	2.2
Level 3	50	50.95	2.4

Table 4. Experimental plan.

Factor	Concentration %	Particle Size (μm)	Flow Velocity (m/s)
Test 1	60	15.05	2.0
Test 2	60	30.25	2.2
Test 3	60	50.95	2.4
Test 4	55	15.05	2.2
Test 5	55	30.25	2.4
Test 6	55	50.95	2.0
Test 7	50	15.05	2.4
Test 8	50	30.25	2.0
Test 9	50	50.95	2.2

After drying the material, the particle sizes were separated, the attrition treatment was performed, and the particle size was measured. Three kinds of tailings with average weighted particle sizes of 15.05, 30.25, and 50.95 μm were selected, whereas the cement–sand ratio of the test samples was 1:6. The concentrations of 60, 55, and 50% were prepared for the three types of tailings, respectively, as shown in Table 5.

Table 5. Tailings test scheme parameters.

Concentration	Cement–Sand Ratio	Quality1 (g)	Quality2 (g)	Quality3 (g)
60%	1:6	25.71	154.29	300
55%	1:6	23.57	141.43	300
50%	1:6	21.43	128.57	300

Mixing the cement with tailings: The tailings slurry was prepared according to the ratios listed in Table 5. Slag Portland cement was used in this test. The prepared slurry of each group was placed in the rheometer. The rheological characteristic curve of the slurry was obtained using a HAAKES series rheometer manufactured by Thermo Fisher Scientific Company (Shanghai, China), USA. A four-blade pulp rotor (FL16SK01140438, Thermo Fisher Scientific Company, Shanghai, China) was used in the test. The type of sleeve used for filling the slurry was L13092 (Thermo Fisher Scientific Company, Shanghai, China), with an inner sleeve diameter of 26.20 mm and an outer diameter of 30.00 mm. Filling slurries were prepared with particle sizes of 15.05, 30.25, and 50.95 μm, and mass concentrations of 60, 55, and 50%. The rheological characteristic curves of filling slurries with different particle sizes and mass concentrations were obtained by controlling the shear rate (the CR method). Compared with the test method of controlling shear stress (the CS method), the rheological characteristic curve of the slurry tested using the CR method is more comprehensive, which can reflect the rheological characteristics of the slurry during the entire process from the low-speed to the high-speed state. The test was carried out at room temperature (22 °C) to avoid the influence of temperature changes on the test results.

The test parameters are presented in Table 6. The fluid flow can be characterized by many physical quantities, among which the Reynolds number has been used frequently [31–33]. The linear dimension of pipe diameter was used in determining the Reynolds number. The Reynolds numbers of the materials used in this experiment were all less than 2300 and belonged to the laminar flow state; they were calculated using $R_e = \rho v d / \mu$.

Table 6. Rheological characteristics of the tailings.

Concentration	Cement–Sand Ratio	Particle Size (μm)	Yield Stress (Pa)	Plastic Viscosity (Pa·s)
60%	1:6	15.05	50.05	0.513
60%	1:6	30.25	51.23	0.522
60%	1:6	50.95	52.34	0.568
55%	1:6	15.05	20.58	0.339
55%	1:6	30.25	21.36	0.348
55%	1:6	50.95	22.59	0.362
50%	1:6	15.05	15.33	0.248
50%	1:6	30.25	16.47	0.269
50%	1:6	50.95	17.26	0.297

2.2. Geometric Modeling and Meshing

First, we made the following assumptions: The size of slurry particles is uniform. The fluid in the pipeline is a single-phase laminar homogeneous body, and the particles rebound when the fluid flows to the pipeline wall. There is no heat exchange phenomenon inside the pipeline; thus, the influence of water viscosity on particle and interference sedimentations on the particle sedimentation velocity are ignored.

In this test, COMSOL software was used to conduct a simulation on the filling pipeline, which was used to build a model of the filling elbow pipe, as shown in Figure 2. As the pipeline had an axisymmetric model, the physical structure of the pipeline could be sufficiently reflected by a two-dimensional model. Therefore, an "L"-shaped geometric model was drawn, and the filling timeline of the test was 2.5. The height of the standpipe part, outer diameter of the elbow pipe, inner diameter of the elbow pipe, and length of the horizontal straight pipe were 1.0, 1.0, 0.8, and 4.0 m, respectively. Particles follow the flow, with the inlet and outlet at the top and bottom of the pipe, respectively. COMSOL software (Multiphysics 5.5., Shanghai, China.)includes a mesh division tool, which can edit the modularized physical model. The mesh division was based on the laminar flow and FPT modules, as shown in Figure 3. A contour of the flow pattern of particle deposition and the pressure result are shown in Figures 4 and 5.

Laminar flow module conditions:

(1) Fluid properties: density 1680 (kg/m^3), dynamic viscosity 0.513 (Pa·s).
(2) Initial values: both are 0.
(3) Wall: Wall condition—no slip.
(4) Gravity: x axis is "0", y axis is "g_const".
(5) Inlet: Normal speed is 2.0 m/s, 2.2 m/s, 2.4 m/s.
(6) Outlet: Pressure condition is "0" Pa.

Fluid particle tracking module:

(1) Particle properties: Particle density is 2620 (kg/m^3). Particle diameter is 15.05 μm, 30.25 μm, 50.95 μm
(2) Entrance: Release time—range (0, 0.05, 10), number of particles released each time $n = 20$.

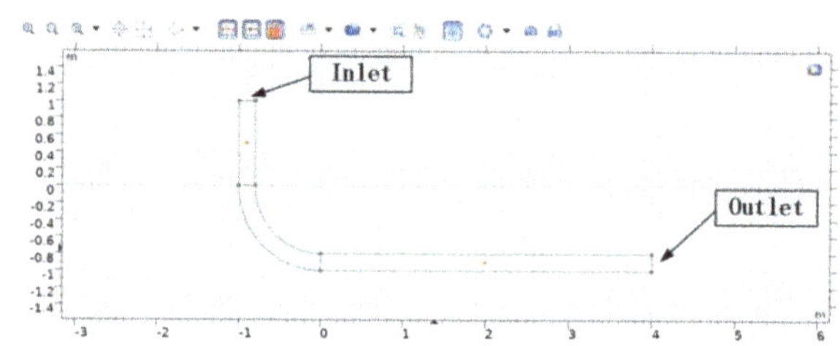

Figure 2. Geometric simulation.

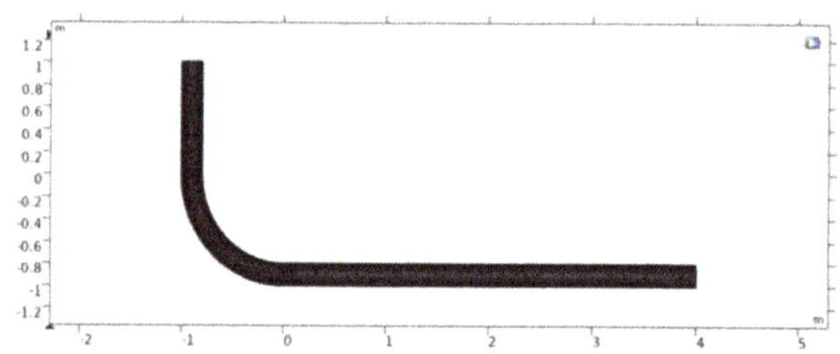

Figure 3. Meshing pattern.

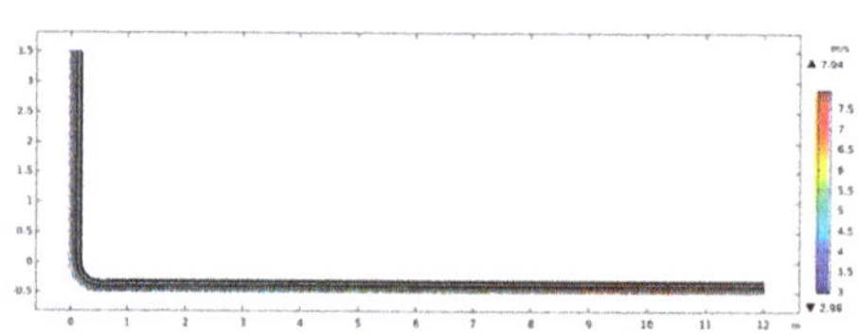

Figure 4. A contour of the flow pattern of particle deposition.

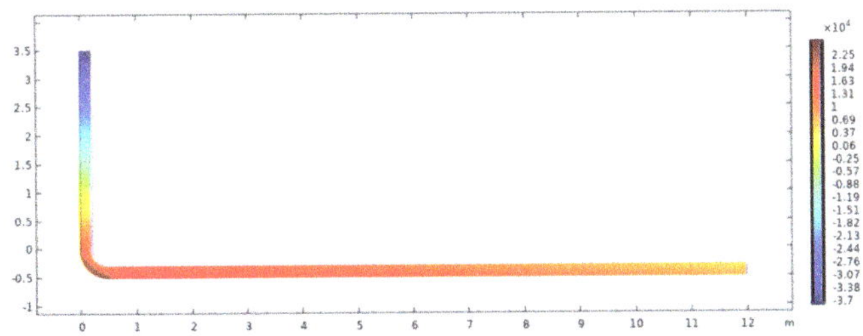

Figure 5. Pressure result.

The unit size was selected to be ultrafine, and the sequence type was the physical field control network. Finally, the partition results contained 5194 domain units and 566 boundary elements. The minimum cell mass of the mesh was 0.4462, average cell mass of the mesh was 0.8112, unit area ratio was 0.2884, and the time step of the fluid flow was in the range (0, 0.05, 10). This meant release every 0.05 s from 0 to 10 s.

3. Results

The resistance of particles in the pipeline can be expressed as follows:

$$F_D = C_D \frac{\rho u_0^2}{2} A_D = C_D \frac{\rho u_0^2}{2} \frac{\pi d^2}{4} \tag{1}$$

where ρ is the slurry density and C_D is the coefficient of resistance, which varies with Reynolds number R_e and ranges from 0.01 to 10^5. The coefficient of resistance can be expressed as

$$C_D = 2\left(1.84R_e^{-0.31} + 0.293R_e^{-0.06}\right)^{3.45} \tag{2}$$

According to Newton second law, the forces of the particles in the pipeline are gravity, buoyancy, and drag, which can be expressed as follows:

$$F_g - F_b - F_D = m\frac{d_u}{d_t}, \tag{3}$$

where $F_g = mg$, $F_b = \frac{m}{\rho_p}\rho g$, and d is the diameter. F_b is lift force, which is not significant in this test.

The differential equation of motion of particles can be expressed as

$$\frac{d_u}{d_t} = \left(\frac{\rho_p - \rho}{\rho_p}\right)g - \frac{3C_D}{4d}\frac{\rho}{\rho_p}u^2 \tag{4}$$

where ρ_p is the solid density and $d_u/d_t = 0$. The relative rate of particle-free sedimentation can be written as follows:

$$u = \sqrt{\frac{4(\rho_p - \rho)gd}{3\rho C_D}} \tag{5}$$

According to the corresponding relationship between the coefficient of resistance and Reynolds number, the free sedimentation velocity in a laminar flow can be expressed as

$$u = \frac{(\rho_p - \rho)gd^2}{18\mu}, \tag{6}$$

where μ is the coefficient of viscosity, ρ_p is the particle density, ρ is the slurry density, and d is the particle size.

The density of the particles is high and some forces need to be taken into account such as added mass. In actual transportation, particles are moving in groups. Collision and friction will inevitably occur between particles and the pipe wall. At this time, if you want to accelerate the movement, this will cause the surrounding fluid to accelerate the movement. Because the fluid has inertia, it acts as a reaction force on the particles. At this point, the force pushing the particle will be greater than the inertial force of the particle itself, as if the particle mass had increased. The force greater than the inertial force of the particle itself is called the additional mass force.

$$F = (m + \delta)a \tag{7}$$

where δ is mass force; a is accelerated velocity.

The sedimentation velocity of the particles cannot be affected when the slurry flows horizontally. The solid particle sizes are in a wide range in the slurry transported by the mine pipelines. The effect of water viscosity on sedimentation is ignored. In addition, when $d_i \leq 0.3a$, the sedimentation velocity of particles in a laminar flow can be calculated using a simplified Stokes formula as follows [34]:

$$\left\{ \begin{array}{c} a = \sqrt[3]{0.0001 + (\rho_g - 1)} \\ V_s = 5450 d_i^2 (\rho_g - 1) \end{array} \right\} \tag{8}$$

The calculation results using Equation (8) were compared with the numerical simulation results to verify the reliability of COMSOL software in the numerical simulation of slurry particle settlement. The details are presented in Table 7.

Table 7. Results of experimental simulation and actual calculation.

No.	Concentration %	Particle Size μm	Flow Velocity m/s	Simulation Speed cm/s	Calculation Speed cm/s	Relative Error %
1	50%	15.05	2.0	0.0182	0.0199	8.54
2	50%	30.25	2.2	0.0759	0.0807	5.95
3	50%	50.95	2.4	0.2193	0.2291	4.28
4	55%	15.05	2.2	0.0175	0.0199	12.06
5	55%	30.25	2.4	0.0724	0.0807	10.29
6	55%	50.95	2.0	0.2135	0.2291	6.81
7	60%	15.05	2.4	0.0166	0.0199	16.58
8	60%	30.25	2.0	0.0701	0.0807	13.14
9	60%	50.95	2.2	0.2099	0.2291	8.38

SPSS software (Beijing Netnumber Times Technology, Beijing, China) was used to analyze the correlation between the simulation and calculation results (Table 8). As shown in Table 8, the value of Sig. (double tail) was less than 0.01, confirming a significant correlation between the simulation and calculation results. Additionally, the correlation coefficient between the simulation and calculation results is 0.999437, which is close to 1, confirming an extreme correlation [35–37].

Table 8. Correlation between the simulation and actual calculation results.

Name	Simulation Speed (cm/s)			Calculation Speed (cm/s)		
	Pearson Correlation	Sig. (Double Tail)	Case Number	Pearson Correlation	Sig. (Double Tail)	Case Number
Simulation speed (cm/s)	1	0	9	0.999437	$1.3948388027 \times 10^{-11}$	9
Calculation speed (cm/s)	0.999437	$1.3948388027 \times 10^{-11}$	9	1	0	9

As shown in Table 7, the relative error between the calculated results of Equation (1) and those of the numerical simulation is less than 20%, indicating that the numerical model of tailing particle sedimentation established in this experiment is relatively reliable.

The range analysis method was adopted to examine the influence degree of slurry concentration, particle size, and flow velocity on the particle sedimentation velocity. Table 9 shows the degree of influence of each factor on the sedimentation velocity. Software was used to establish the regression equation of sedimentation velocity with respect to the three factors. The regression equation can be expressed as follows:

$$Y = 2.811 - 4.654x_1 + 7.153 \times 10^{-3}x_2 - 1.482x_3 - 1.752 \times 10^{-2}x_1x_2 + 2.407x_1x_3 + 3.943 \times 10^{-3}x_2x_3,$$

where Y is the sedimentation velocity, x_1 is the concentration, x_2 is the particle size, and x_3 is the flow velocity. The correlation coefficients of this regression equation before and after adjustment were R = 0.99365 and Ra = 0.97433, respectively. Therefore, the regression equation is significant and its reliability is high.

Table 9. Degree of influence of each factor on the sedimentation velocity.

Range	Test Factor		
	Concentration %	Particle Size (μm)	Flow Velocity (m/s)
K_1	0.3134	0.0523	0.3018
K_2	0.3034	0.2184	0.3033
K_3	0.2966	0.6427	0.3083
$\overline{K_1}$	0.1045	0.0174	0.1006
$\overline{K_2}$	0.1011	0.0728	0.1011
$\overline{K_3}$	0.0989	0.2142	0.1028
R	0.0056	0.1968	0.0022

In Table 9, K_1, K_2, and K_3 are the sums of ranges for each factor; $\overline{K_1}$, $\overline{K_2}$, $\overline{K_3}$ are the mean ranges of each factor at each level.

The pipeline circulation monitoring system designed and built by our research group in the laboratory was used to verify the above model. The experimental platform is shown as Figure 6. The platform is composed of transparent pipe, ITS detector, mixing tank, pressure pump, etc. Among them, the ITS mainly collects the data of the front and rear sections and transmits them to the computer. Finally, the results were analyzed, as shown in the table below. According to the data in Table 10, the relative errors are within 20% and the model established by the software has certain reliability.

Figure 6. Experimental platform.

Table 10. Results of simulation and experiment.

	Concentration %	Particle Size µm	Flow Velocity m/s	Simulation Speed cm/s	Experimental Speed cm/s	Relative Error %
1	50%	15.05	2.0	0.0182	0.0198	8.08
2	50%	30.25	2.2	0.0759	0.0805	5.71
3	50%	50.95	2.4	0.2193	0.2293	4.36
4	55%	15.05	2.2	0.0175	0.0195	10.26
5	55%	30.25	2.4	0.0724	0.0809	10.51
6	55%	50.95	2.0	0.2135	0.2289	6.73
7	60%	15.05	2.4	0.0166	0.0194	14.43
8	60%	30.25	2.0	0.0701	0.0802	12.59
9	60%	50.95	2.2	0.2099	0.2288	8.26

4. Discussion

4.1. Analysis of the Influence Degree

The obtained results confirm that the particle size has a significant influence on the slurry particle sedimentation process.

In addition, the particle density affected the sedimentation velocity, which had a certain influence on the degree of settlement [38]. Particles in the slurry with a larger apparent density presented a higher difference between the density of the particles and filling slurry, resulting in a greater sedimentation tendency.

In this study, ultrafine unclassified tailings were used as the experimental material, whereas the weighted mean particle size was 15.05 µm. Both the apparent and bulk densities are approximately the same [39–42]. Experimental materials were treated with extremely fine and approximately uniform shapes. The densities of each sample group were approximately equal, ranging from 2620.2 to 2620.5 kg/m^3. In the future, we will conduct a thorough study on the factors influencing the particle density.

The grading of tailings in the filling pipeline is crucial. Filling slurries prepared with coarse, fine, or ultrafine tailings have different influences on the entire filling process. Finer tailings are easier to prepare from approximately homogeneous fluids, which makes the flow smoother. When the particle size remains the same, slurry of different concentrations affects the sedimentation velocity of particles and the flow pattern. When the concentration is low, the slurry flows and settles at the same time, and becomes thin when it reaches the stope, which influences the filling effect. However, when the concentration is high, the viscosity of the slurry increases so that the filling speed slows down, the pumping pressure increases, and even excessive sedimentation leads to silting. In serious cases, pipes burst, resulting in serious accidents. When only the influence of velocity is considered, as the particle is affected by the drag force in the fluid, the faster velocity produces a larger drag force, pulling the particle forward; hence, it is not easy to settle.

In conclusion, particle size, slurry concentration, and flow velocity have a certain influence on the sedimentation of the filling slurry. Therefore, it is crucial to study the degree of influence of the three factors on particle sedimentation for the entire filling system. A three-dimensional direct view is shown in Figure 7.

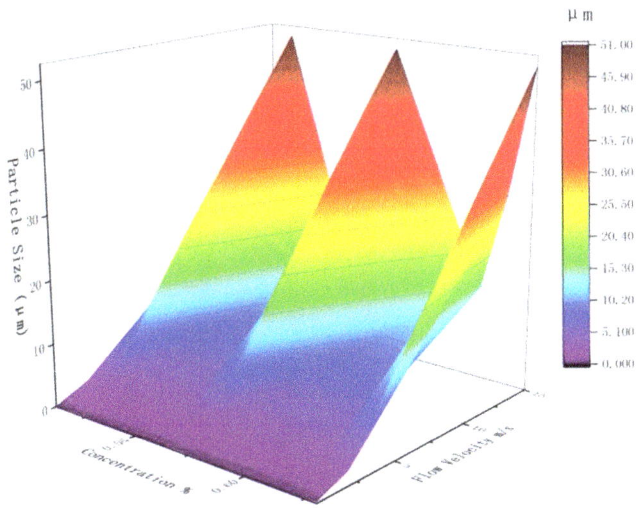

Figure 7. Three-dimensional direct view.

4.2. *Effect of Particle Size on the Sedimentation Velocity*

When the slurry is transported by a mine filling pipeline, the size of the solid particles is distributed over a wide range. In the filling slurry of fine tailings, the particle size of one stage can be selected as the basis for the sedimentation velocity calculations. Because particle size and shape are the key factors affecting particle sedimentation, we assumed that the shape of tailings was approximately the same after refining treatment; hence, the correction coefficient of the sedimentation velocity of non-spherical particles was not considered. In addition, an increase in the particle size generates a certain head effect. The influence of particle size on sedimentation velocity is shown in Figure 8. As shown, with an increase in particle size, the sedimentation velocity gradually increases. Under the conditions of the same particle size and different concentrations, the sedimentation law is roughly the same; however, the sedimentation velocity gradually decreases with an increase in the concentration. The sedimentation of a single particle causes the movement of the surrounding water in the fluid, which considerably affects the sedimentation velocity. However, solid particles in the filling pipeline are distributed in clusters, the resistance of the nearby water increases when a single particle settles down, and the viscosity reduces its sedimentation velocity. When a group of particles settles down, the surrounding water rises, and the sedimentation velocity increases.

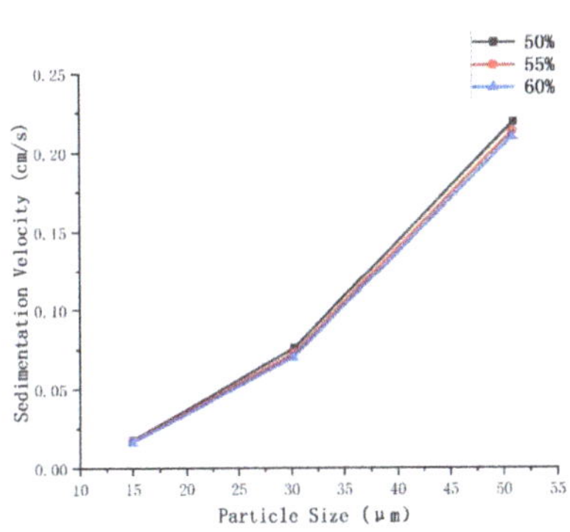

Figure 8. Relationship between the particle size and sedimentation velocity.

4.3. Effect of Concentration on the Sedimentation Velocity

Concentration is not only an important parameter affecting the hydraulic gradient but is also a key factor in the reasonable operation of the entire filling system. An appropriate slurry concentration can not only guarantee stable transportation of the filling pipeline but also avoid pipe blocking accidents caused by the concentration problem. A low concentration of slurry causes the center point of the slurry to move up; thus, the speed of all sides is larger at the top and smaller at the bottom; therefore, the bottom of the pipeline can readily produce siltation, or even pipe blocking, causing serious accidents that affect production. The influence of concentration on the sedimentation velocity is shown in Figure 9. As shown, there is an adverse relationship between the concentration and sedimentation velocity. When different types of tailings are selected to prepare the slurry with the same concentration, the larger the content of large particles, the easier is the settlement, and the faster is the sedimentation velocity. When concentration increases, the sedimentation velocity of the slurry with a small particle size remains unchanged. When the pipe diameter is constant and the concentration is extremely high, the conveying resistance increases accordingly so that the horizontal direction speed in the pipe decreases, and the free sedimentation velocity of the particles in the vertical direction increases. When the concentration was minimal, the particles in the slurry appeared to be suspended. The resistance loss in the horizontal direction was slight; thus, the slurry flow in the pipeline was smooth, and the vertical direction of the sedimentation velocity was reduced, indicating a state of stable flow.

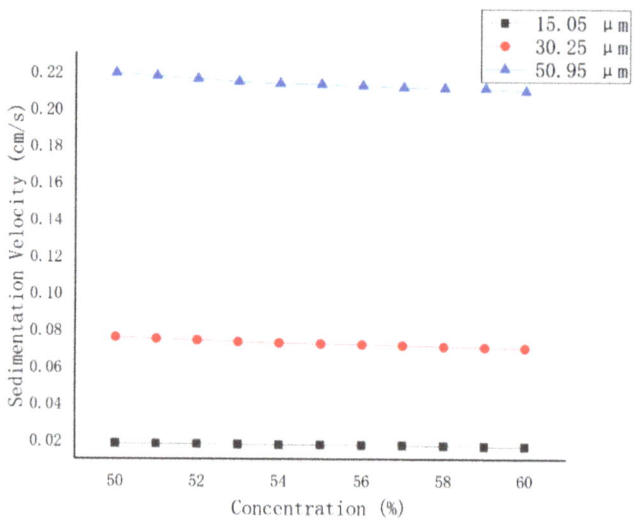

Figure 9. Relationship between the concentration and sedimentation velocity.

4.4. Effect of Flow Velocity on the Sedimentation Velocity

The calculation of the hydraulic gradient is significant in the hydraulic transport of solid materials. Many factors affect the hydraulic gradient, whereas the flow velocity has the highest influence. Solid–liquid mixtures are usually divided into structural, Newtonian, homogeneous, and heterogeneous flows. In general, cream-body fill belongs to the category of structural flow. In practical production, however, Newtonian flow is frequently used for gravity flow conveying. A heterogeneous fluid is affected by the drag force of the fluid to produce a forward force when it flows in a pipeline. The particles in the fluid do not easily undergo sedimentation if the flow velocity is considerably fast, and the drag force is significantly large. Otherwise, the flow velocity is moderate, and the sedimentation velocity is swift. The influence of the concentration on the sedimentation velocity is shown in Figure 10. With an increase in the slurry velocity, the sedimentation velocity exhibits a slight downward trend. The faster the flow velocity of the slurry in the pipeline, the slower the sedimentation velocity. The particles are subjected to a drag force in the horizontal direction, and the vertical direction of the sedimentation velocity decreases; thus, an adverse relationship exists. As shown in Figure 7, under the same flow velocity, the larger the particle size, the faster the sedimentation velocity. When the particle size varies from 30.25 to 50.95 μm, the sedimentation velocity presents a step change. We cannot set the flow velocity more widely during the process of actual transportation. A film was formed between the precipitated particles and the pipe wall. If the flow velocity is extremely high, the friction between the slurry and the pipe wall is aggravated. It is easy to wear the pipe wall, which causes the slurry to settle and hoard. In long-term operations, it may also lead to serious leakage and plugging accidents, causing vast economic losses.

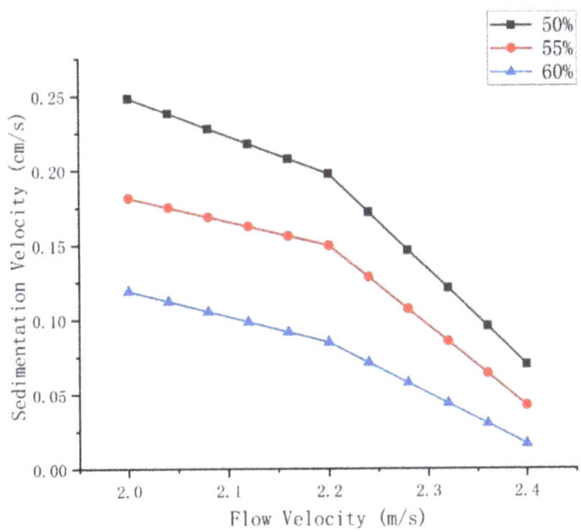

Figure 10. Relationship between the flow velocity and sedimentation velocity.

4.5. Case Study and Analysis

The mining area is an underground mine, and waste tailings need to enter the filling station for secondary utilization. They are conveyed to the underground goaf by a pipeline to maintain the stress balance of the stope. The system adopts the entire tailings cemented filling technology. The ground station is equipped with a slurry at a concentration range of 55–60%, and a particle size range of 18.25 to 32.25 μm. A large flow rate in a pipeline with an inner diameter of 200 mm was selected, in which the straight pipe part, horizontal, vertical, slope angle, and the flow rate per unit time were approximately 2000 m, 1500 m, 500 m, 45°, and 120–160 m^3/h, respectively. The filling system is shown in Figure 11.

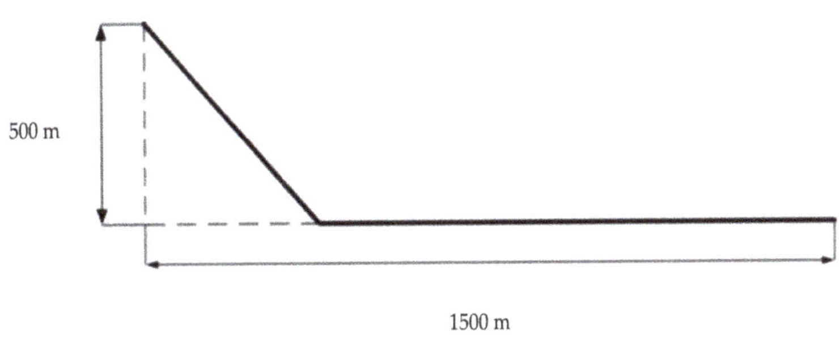

Figure 11. Diagram of the mine filling system.

To restore the scene, the sedimentation velocity of the slurry in the pipe during the conveying process of the filling system was calculated, and a geometric model with a model size of 1:1 was established. A slurry with a concentration of 55% was selected according to the actual filling conditions. Simultaneously, based on the analysis of the three influencing factors, the optimal parameters were selected for verification, and the tailings of particle sizes of 18.25, 25.25, and 32.25 μm were screened for comparison with those of the concentration of 60%, as shown in Figure 12.

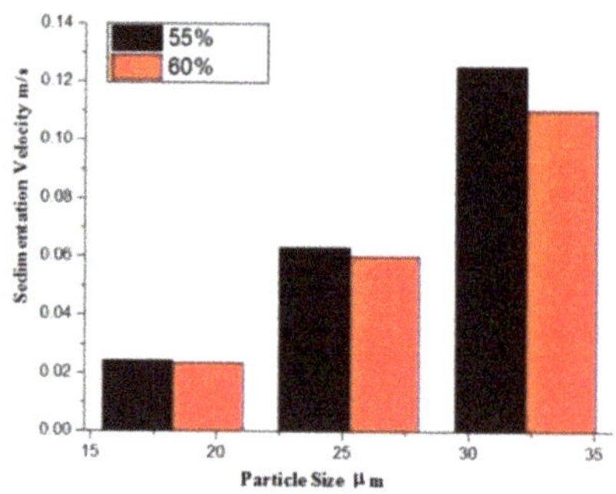

Figure 12. Results of simulation experiment.

The figure shows that when filling is carried out in accordance with an engineering practice, the particle sedimentation velocity in the pipeline is faster. The sedimentation velocity of the particles decreases with an increase in the concentration. To avoid the risk of pipe plugging caused by sedimentation, a larger concentration range can be selected; however, it is necessary to balance the resistance loss caused by high concentrations. This is shown in Table 11 below. The larger the particle size, the greater the resistance loss will be. In addition, the larger the particle size, the faster the sedimentation rate. This will lead to the formation of a slip layer or even a settlement area in the pipe, which will affect the transportation of the whole pipe and lead to greater resistance loss. Although a small particle size is helpful for pipeline transportation, too small a particle size will affect the formation of the early strength of the backfill. The transport resistance of the filling pipeline increases rapidly with the increase in slurry concentration. When the slurry concentration is less than 60%, the energy loss in the slurry pipeline transportation can be greatly reduced, and the risk of pipe blocking and pipe explosion can also be reduced. Considering the strength demand of the filling structure in the mine, in order to effectively improve the mine production efficiency and reduce the water seepage, the concentration of the filling body should not be lower than 60%.

Table 11. Case analysis data.

	Concentration %	Particle Size μm	Flow Velocity m/s	Sedimentation Velocity cm/s	Resistance Loss Pa/m
1	55%	18.25	2.0	0.0256	4532.6
2	55%	25.25	2.0	0.0618	4705.7
3	55%	32.25	2.0	0.1393	4932.4
4	60%	18.25	2.0	0.0248	5087.5
5	60%	25.25	2.0	0.0601	5365.2
6	60%	32.25	2.0	0.1154	5736.6

Therefore, the optimal concentration for this experiment was 60%. As the particle size of the tailings is variable, screening procedures are added to classify the tailings. Therefore, the larger the particle size, the greater the sedimentation velocity; thus, the selection of tailings slurry with small particle size is the best choice. Nevertheless, an extremely small particle size is not conducive to increase the viscosity of the slurry. Thus, a moderate particle size of 25.25 μm can be selected. Consequently, the optimal parameters of this project were a concentration of 60% and particle size of 25.25 μm. In addition, when the pipe diameter is less than or equal to 200 mm, the optimal flow rate can be calculated using the Durard formula ($v = F_l\sqrt{2gD\frac{\rho_g-\rho_1}{\rho_1}}$), where F_l is the correlation coefficient, g is gravitational acceleration, D is pipe diameter, ρ_g is the density of the mixture, and ρ_1 is the density of the slurry. In our experiment, the optimal flow rate was 1.44 m/s.

5. Conclusions

In this study, we established a pipeline transport model to investigate the influence of particle sedimentation factors on slurry transportation. The following conclusions can be drawn:

1. The simulation transport process model of the filling slurry pipeline was established and verified using COMSOL software. Comparing the simulation results with those obtained using a simplified Stokes formula and ignoring the influence of medium viscosity on sedimentation velocity yielded a relative error in the range 4–17%, which proved the reliability of the proposed model.
2. Under different conditions, the model was used to calculate the variation characteristics of the sedimentation velocity. The sedimentation velocity was positively related to the particle size and adversely related to the concentration and flow velocity. Setting a reasonable range of particle sizes, preparing a slurry with a reasonable concentration, and adjusting an appropriate flow velocity are key factors in examining the sedimentation velocity. Therefore, this study provides a theoretical basis for investigating the sedimentation law.
3. Using the range analysis method showed that the degree of influence on the sedimentation velocity is as follows: particle size > concentration > flow velocity. In a sample mining area, the optimal slurry concentration and particle size were 60% and 25.25 μm, respectively. Consequently, finding optimal parameters is significantly important in reducing the sedimentation velocity of particles and improving the efficiency of pipeline transportation.

Author Contributions: Conceptualization, C.W. and D.G.; methodology, C.W.; software, C.W.; validation, C.W. and D.G.; formal analysis, C.W.; investigation, C.W.; resources, C.W.; data curation, C.W.; writing—original draft preparation, C.W.; writing—review and editing, C.W.; visualization, D.G.; supervision, D.G.; project administration, D.G.; funding acquisition, D.G. All authors have read and agreed to the published version of the manuscript.

Funding: This research was funded by the National Natural Science Foundation of China, grant number 51774137.

Institutional Review Board Statement: Not applicable.

Informed Consent Statement: Not applicable.

Data Availability Statement: The raw data supporting the conclusions of this article will be made available by the authors, without undue reservation.

Conflicts of Interest: The authors declare no conflict of interest.

References

1. Zhang, C.; Tan, Y.; Zhang, K.; Zhang, C.; Song, W. Rheological Parameters and Transport Characteristics of Fresh Cement Tailings Backfill Slurry in an Underground Iron Mine. *Adv. Civ. Eng.* **2021**, *2021*, 1–16. [CrossRef]
2. Li, J.; Yilmaz, E.; Cao, S. Influence of Solid Content, Cement/Tailings Ratio, and Curing Time on Rheology and Strength of Cemented Tailings Backfill. *Minerals* **2020**, *10*, 922. [CrossRef]
3. Duan, X.; Huang, Y.; Zheng, W. Study on the Influence of Particle Size and Grade of Total Tailings on the Fluidity of High-Concentration Filling Slurry considering Strength. *Geofluids* **2021**, *2021*, 1–9. [CrossRef]

4. Cunliffe, C.J.; Dodds, J.M.; Dennis, D. Flow correlations and transport behaviour of turbulent slurries in partially filled pipes. *Chem. Eng. Sci.* **2021**, *235*, 116465. [CrossRef]
5. Ozcan, N.T.; Ulusay, R.; Isik, N.S. A study on geotechnical characterization and stability of downstream slope of a tailings dam to improve its storage capacity (Turkey). *Environ. Earth Sci.* **2013**, *69*, 1871–1890. [CrossRef]
6. Zhang, R.J.; Zheng, Y.L.; Dong, C.Q.; Zheng, J.J. Strength behavior of dredged mud slurry treated jointly by cement, flocculant and vacuum preloading. *Acta Geotech.* **2021**, 1–16. [CrossRef]
7. Wei, H.; Xiao, B.; Gao, Q. Flow Properties Analysis and Identification of a Fly Ash-Waste Rock Mixed Backfilling Slurry. *Minerals* **2021**, *11*, 576. [CrossRef]
8. Leong, Y.K. Controlling the rheology of iron ore slurries and tailings with surface chemistry for enhanced beneficiation performance and output, reduced pumping cost and safer tailings storage in dam-ScienceDirect. *Miner. Eng.* **2021**, *166*, 106874. [CrossRef]
9. Xing, K.; Hou, H.; Min, Z.; Yan, W.; Xian, Z. Effect of particle gradation on properties of fresh and hardened cemented paste backfill. *Constr. Build. Mater.* **2015**, *96*, 378–382.
10. Wang, Y.; Song, Y.; Yu, J.; Chen, F. Effect of the injection orientation and position on the leakage flow in a honeycomb-tip turbine cascade. *Int. J. Heat Mass Tranf.* **2019**, *144*, 118633. [CrossRef]
11. Yong, T.; Xiang, J.; Wan, Z.; Wei, Z.; Lei, W. A novel miniaturized loop heat pipe. *Appl. Therm. Eng.* **2010**, *30*, 1152–1158.
12. Hz, A.; Ss, A.; Yy, A.; Jie, Z.; Dong, L. Column studies on the adsorption of As(V) by granular chitosan adsorbent prepared with backwashing iron-containing sludge. *Colloids Surf. A Physicochem. Eng. Asp.* **2021**, *627*, 127247.
13. Zou, X.; Zhang, Z.; Wu, M.; Wan, Y. Slope scale spatial variability of fractal dimension of soil particle size distribution at multiple depths. *Soil Sci. Soc. Am. J.* **2021**, *85*, 117–131. [CrossRef]
14. Yin, J.; Zhang, S.; Wu, Z. Provenance Analysis of the Paleogene Strata in the Northern Qaidam Basin, China: Evidences from Sediment Distribution, Heavy Mineral Assemblages and Detrital Zircon U-Pb Geochronology. *Minerals* **2020**, *10*, 854. [CrossRef]
15. Huang, X.; Song, J.; Yue, W.; Wang, Z.; Mei, X.; Li, Y.; Li, F.; Lian, E.; Yang, S. Detrital Zircon U-Pb Ages in the East China Seas: Implications for Provenance Analysis and Sediment Budgeting. *Minerals* **2020**, *10*, 398. [CrossRef]
16. Kumar, D.; Iveson, S.M.; Galvin, K.P. Novel jamming mechanism for dry separation of particles by density. *Miner. Eng.* **2020**, *148*, 106185. [CrossRef]
17. Rouchalova, D.; Rouchalova, K.; Janakova, I.; Cablik, V.; Janstova, S. Bioleaching of Iron, Copper, Lead, and Zinc from the Sludge Mining Sediment at Different Particle Sizes, pH, and Pulp Density Using Acidithiobacillus ferrooxidans. *Minerals* **2020**, *10*, 1013. [CrossRef]
18. Singh, R.; Akella, V.S.; Sethy, S.N. Exploration for strategic placer mineral deposits in a fluctuating shoreline: Depositional environment and mineralogical characterization of the NE Odisha coast placers, India. *Ore Geol. Rev.* **2020**, *127*, 103850. [CrossRef]
19. Peng, H.F.; Kong, G.Q.; Liu, H.L.; Abuel-Naga, H.; Hao, Y.H. Thermo-mechanical behaviour of floating energy pile groups in sand. *J. Zhejiang Univ. Ser. A* **2018**, *19*, 638–649. [CrossRef]
20. Peng, Y.; Xiao, J.; Deng, B.; Wang, Z.; Wu, Q. Study on separation of fine-particle ilmenite and mechanism using flocculation flotation with sodium oleate and polyacrylamide. *Physicochem. Probl. Miner. Process.* **2021**, *1*, 162–173.
21. Dd, A.; Rkm, B.; Hb, C.; Ar, D.; Pkp, E.; Eh, C. Combined effect of natural dispersant and a stabilizer in formulation of high concentration coal water slurry: Experimental and rheological modeling. *J. Mol. Liq.* **2020**, *320*, 114441.
22. Das, D.; Pattanaik, S.; Parhi, P.K.; Mohapatra, R.K.; Jyothi, R.K.; Lee, J.; Kim, H.I. Stabilization and Rheological Behavior of Fly Ash–Water Slurry Using a Natural Dispersant in Pipeline Transportation. *ACS Omega* **2019**, *4*, 21604–21611. [CrossRef] [PubMed]
23. Routray, A.; Senapati, P.K.; Padhy, M.; Das, D.; Mohapatra, R.K. Effect of mixture of a non-ionic and a cationic surfactant for preparation of stabilized high concentration coal water slurry. *Int. J. Coal Prep. Util.* **2019**, 1–16. [CrossRef]
24. Das, S.N.; Biswal, S.K.; Mohapatra, R.K. Recent advances on stabilization and rheological behaviour of iron ore slurry for economic pipeline transportation. *Mater. Today Proc.* **2020**, *33*, 5093–5097. [CrossRef]
25. Narkuniene, A.; Poskas, P.; Justinavicius, D. The Modeling of Laboratory Experiments with COMSOL Multiphysics Using Simplified Hydromechanical Model. *Minerals* **2021**, *11*, 754. [CrossRef]
26. Crha, J.; Basařová, P.; Ruzicka, M.C.; Kašpar, O.; Zednikova, M. Comparison of Two Solvers for Simulation of Single Bubble Rising Dynamics: COMSOL vs. Fluent. *Minerals* **2021**, *11*, 452. [CrossRef]
27. Wang, L.M.; Yin, S.H.; Wu, A.X. Visualization of flow behavior in ore-segregated packed beds with fine interlayers. *Int. J. Min. Met. Mater.* **2020**, *27*, 900–909. [CrossRef]
28. Zeng; Tong; Yi; Chen. Selective Capture of Magnetic Wires to Particles in High Gradient Magnetic Separation. *Minerals* **2019**, *9*, 509. [CrossRef]
29. Ying, H.W.; Zhu, C.W.; Gong, X.N.; Wang, X. Analytical solutions for the steady-state seepage field in a finite seabed with a lined tunnel. *Mar. Georesour. Geotechnol.* **2019**, *37*, 972–978. [CrossRef]
30. Rezaei Niya, S.M.; Selvadurai, A.P.S. Correlation of joint roughness coefficient and permeability of a fracture. *Int. J. Rock Mech. Min.* **2018**, *113*, 150–162. [CrossRef]
31. Larsson, S.; Rodríguez Prieto, J.M.; Heiskari, H.; Jonsén, P. A Novel Particle-Based Approach for Modeling a Wet Vertical Stirred Media Mill. *Minerals* **2021**, *11*, 55. [CrossRef]
32. Islam, M.T.; Nguyen, A.V. Bed expansion and gas holdup characteristics of bubble–assisted fluidization of liquid–particle suspensions in a HydroFloat cell. *Miner Eng.* **2021**, *160*, 106678. [CrossRef]

33. Wang, S.; Guo, J.; Tang, L.; He, H.; Tao, X. Investigation of the induction time of low-rank coal particles on rising bubble surfaces. *Int. J. Coal Prep. Util.* **2017**, *40*, 640–654. [CrossRef]
34. Zhang, Q.L.; Xie, S.Q.; Zheng, J.J.; Wang, X.M. Sedimentation law research and transportation feasibility study of backfilling slurry. *J. Chongqing Univ.* **2011**, 108–112. [CrossRef]
35. Zhang, J.; Lin, X.; Zhang, S. Correlation Analysis of Sprint Performance and Reaction Time Based on Double Logarithm Model. *Complexity* **2021**, *2021*, 1–11. [CrossRef]
36. Lukács, A. The impact of physical activity on psychological well-being and perceived health status during coronavirus pandemic in university students. *J. King Saud Univ. Sci.* **2021**, *33*, 101531. [CrossRef]
37. Hasan, H.H.; Razali, S.F.M.; Razali, N.H.M. Does the Household Save Water? Evidence from Behavioral Analysis. *Sustainability* **2021**, *13*, 641. [CrossRef]
38. Cai, Y.; Liu, Q.; Yu, L.; Meng, Z.; Hu, Z.; Yuan, Q.; Šavija, B. An experimental and numerical investigation of coarse aggregate settlement in fresh concrete under vibration. *Cem. Concr. Compos.* **2021**, *122*, 104153. [CrossRef]
39. Ibáñez-Insa, J.; Woźniak, T.; Oliva, R.; Popescu, C.; Hernández, S.; López-Vidrier, J. Structural and High-Pressure Properties of Rheniite (ReS2) and (Re,Mo)S2. *Minerals* **2021**, *11*, 207. [CrossRef]
40. Costa, F.P.D.; Fernandes, J.V.; Melo, L.R.L.D.; Rodrigues, A.M.; Menezes, R.R.; Neves, G.D.A. The Potential for Natural Stones from Northeastern Brazil to Be Used in Civil Construction. *Minerals* **2021**, *11*, 440. [CrossRef]
41. Aiaa, B. Development of numerical model for simulating resistivity and hydroelectric properties of fractured rock aquifers. *J. Appl. Geophys.* **2021**, *189*, 104319.
42. Luo, J.; Li, X.; Zhang, F.J.; Chen, S.; Ren, D. Sintering of monoclinic SrAl_(2)Si_(2)O_(8) ceramics and their Sr immobilization. *J. Miner. Metall. Mater.* **2021**, *28*, 1057–1062. [CrossRef]

metals

Article

Monitoring and Analysis of Dynamic Response for Open-Pit Mine with Inside Inclined Shafts under Train Loading

Yong Wang [1,2,*], Song-Tao Ni [3], Fa-Wu Yang [4], Zhong-Xin Wang [1], Hong Zhang [1], Ke Ma [1] and Xiao-Jun Li [1]

1 CCTEG Shenyang Engineering Co., Ltd., Shenyang 110000, China; wangzhongxin@zmsyy.com (Z.-X.W.); zhanghong@zmsyy.com (H.Z.); make@zmsyy.com (K.M.); lixiaojun@zmsyy.com (X.-J.L.)
2 College of Civil Engineering and Architecture, Linyi University, Linyi 276000, China
3 College of Civil and Transportation Engineering, Hohai University, Nanjing 210098, China; 191304030005@hhu.edu.cn
4 Laohutai Mine of Fushun Mining Group Co., Ltd., Fushun 113000, China; YFW8519@163.com
* Correspondence: wangyong@zmsyy.com

Abstract: The stability of open-pit mining is a hot issue in geotechnical engineering. A mining railroad is in operation on the slope where the east exhaust inclined shaft and the east sand injection inclined shaft on the Laohutai Mine are located, and it was necessary to determine whether railroad vibration would have an impact on the safety of the inclined shafts. With this project as the background, the dynamic response of the slope with inside two inclined shafts was conducted under train loading. A three-dimensional numerical model by using PLAXIS 3D was established to analyze the stability of the slope. The results show that the dynamic reaction caused by the full-loaded train is significantly greater than the no-load train. The safety factor of the slope under the dynamic load is 1.201, and the maximum displacement of the slope which occurred in the gravel layer directly beneath the train track is about 5 mm. The acceleration responses of the two inclined shafts are almost consistent. The maximum horizontal and vertical acceleration occur at the epidote weak layer. The acceleration directly below the load increases significantly. Therefore, it can be considered that the slopes are stable under the action of train vibration, and the influence on the two inclined shafts is small and negligible.

Keywords: open-pit mine; PLAXIS 3D; dynamic load; safety factor; acceleration

Citation: Wang, Y.; Ni, S.-T.; Yang, F.-W.; Wang, Z.-X.; Zhang, H.; Ma, K.; Li, X.-J. Monitoring and Analysis of Dynamic Response for Open-Pit Mine with Inside Inclined Shafts under Train Loading. *Metals* **2021**, *11*, 1681. https://doi.org/10.3390/met11111681

Academic Editor: Chris Aldrich

Received: 8 September 2021
Accepted: 15 October 2021
Published: 22 October 2021

1. Introduction

Open-pit mining is an important method of mineral resources mining [1,2]. According to the existing data, China's open-pit iron ore production accounts for about 90% of the total iron ore output, non-ferrous metals account for about 52%, chemical minerals account for about 7%, and construction materials are nearly 100%. With good operating conditions, high efficiency, and good safety, the scale and quantity of mines increase year by year. The waste dump is the inevitable product of open-pit mining, because it is generally adjacent to the mining site, and its stability is related to the safety of draining open-air mining age and stripping personnel and equipment [3–5]. Scholars in the open-air mining industry have attached great importance to the research of the waste dump. Its stability analysis is one of the hot issues in the field of open mining and geotechnical engineering [6–10]. With the development of open-pit mining, the stability of open-pit high slope and inclined shaft has also become one of the key technical problems in mine resources development [11]. For a long time, many scholars have explored the deformation and stability of slopes in increasing depth [12–15]. The exploration results show that during the mining process, the geological environment conditions change, especially the blasting operations in mining activities, artificial activities, and geological structure changes such as rock faults and stable structural surfaces with more developed fracture joints have a greater impact on the stability of slopes [16–20]. Therefore, it is important to monitor the stability of slopes and inclined shafts in open-pit mines [6,9].

Qian et al. exceeded at repairing the damaged inclined shaft at the Jintian coal mine in Yulin, China, and the measures taken provide a good reference for controlling similar disasters [21]. Yan et al. applied an automatic load monitoring system and developed an automatic monitoring and data transmission system for inclined shafts on the No. 2 inclined shaft of the Bulianta coal mine in China, which has important engineering significance for evaluating structural safety [22]. Voloshin et al. conducted a feasibility study on the choice of slant shaft construction technology in coal mines in the Tersinsky geological and economic region of Kuzbass [23]. Zevgolis et al. performed geotechnical characterization of the drainage field under optimal ground use and better stability conditions [24]. Liu et al. used the processing function of finite element analysis software MIDAS/GTS to analyze the slope stability under three working conditions of initial excavation, backfill treatment, and subsequent excavation at the mine site, respectively [25]. Huang et al. performed finite element modeling of open-pit slopes to study slope stability under traffic loading [26]. Nie et al. studied the deformation characteristics and mechanism of the landslide in Fushun West Open-Pit Mine and found that the sliding force caused by gravitational stress on the slope was greater than the shear strength of the weak interlayer, causing the slide to slide in line with the layer and continuous deformation [27]. Sakantsev et al. proposed a fast and accurate analysis method for the non-mining pit wall super-flatness problem [28]. Abdellah et al. provided a method for applying the shear strength discounting technique to evaluate the stability of geological features on the slopes of open-pit mines [29]. Abdellah et al. revealed the effectiveness of using numerical simulation analysis in feasibility studies to determine and compare mining costs with safety factors and side slope angles [30]. Empirical and industrial tests of the developed blast design and effective parameters of perimeter blasting were carried out at the Moulong quarry by Zairov et al. The developed effective parameters of perimeter blasting ensured the creation of maximal widest shielding splitting for the designed limit of the disturbance zone thickness with destroyed inter-block connections in the area adjacent to the slope [31].

In this paper, in order to determine whether railroad vibration would have an impact on the safety of the inclined shafts, with the background of topsoil stripping work on the side slope of Fushun LaoHuTai open-pit mine, the vibration of the mine railroad during the upper artificial activity was monitored on site, and a three-dimensional calculation model was established by using geotechnical finite element software PLAXIS 3D to analyze the stability of the side slope where the inclined shaft of open-pit mine is located and the influence of train load on the vibration of the inclined shaft, which provides some guidance for the safety assessment of the side slope and inclined shaft during the train travel.

2. Materials and Methods

2.1. Project Overview

The east exhaust inclined shaft and east sand injection inclined Shaft of Laohutai Mine are located at the southern foot of the East Open Pit in Dongzhou District, Fushun City, near the top of the slope, with an average elevation of 125 m. The east sand injection inclined shaft is mainly used for sand injection in the mine workings and fire-fighting water injection system. The east exhaust inclined shaft is used to exhaust air from the east wing of the mine. Since the ventilation method of the mine is diagonally extracted from both wings, the mine's exhaust shafts are arranged in both wings of the mine, and the coal seam of the Laohutai mine is inclined from south to north, using a combination of inclined and vertical development. From a macro-perspective, the Fushun Coalfield is located in a basin surrounded by the remaining veins of the Changbai Mountains and buried under the alluvial plain of the Hun River. The flat alluvial plain is piled up under the foothills, and the boundary is obvious. The landform type in the area is hilly area with large topographic undulations, and the pit area is a terrace terrain formed by artificial transformation. The geological structure of the Fushun coalfield is relatively complex, with many faults and large folds, as shown in Figure 1a. Topsoil stripping work is currently underway on the slopes where the east exhaust inclined shaft and east sand injection inclined shaft are located. At the same time, two mine roads and a mining railroad are open to traffic above the inclined shaft, which has a certain impact on the stability of the inclined shaft, as shown in Figure 1b.

(a) (b)

Figure 1. Site picture of Laohutai Mine. (**a**) Schematic outline of the mine site; (**b**) road and railroad diagram of the mining area.

The regional geological structure of Fushun coalfield mainly includes the Swire Anshan Group, the Lower Cretaceous Longfengkan Group, the Cenozoic Old Tertiary, and the Fourth Series. According to the survey report, the slope where the inclined shaft is located is divided into three layers: quaternary artificial accumulation layer (Q4ml), with uneven thickness, generally 2–10 m, and an average of 4 m. The uppermost part is artificial accumulation of construction waste, which is mainly composed of broken stones, bricks, and a small amount of clay; quaternary residual slope accumulation layer (Q4dl + el), the thickness of this layer is uneven, generally 4–17 m, average 6 m. It is mainly composed of cohesive soil and gravel (sandstone) with yellowish-brown color and poor sorting and grinding round. The gravel diameter is generally 2–3 cm, and the maximum can reach 8 cm; the lithology of the sandstone formation of the Lizigou Group (E12l), the lithology of this layer is sandstone, with local green curtain stone. Its direction is consistent with the direction of the coal in this layer, with a dip angle of 30–37°, generally 35°, and its tendency is basically consistent with the tendency of the side slope, with a tendency of 33°. That is, the south formation is in the state of cascade, which is not good for the stability of the slope.

2.2. Field Monitoring

The monitoring section is located at the east sand injection shaft of LaoHuTai, and the test site is mainly quaternary artificial accumulation layer. The measuring point arrangement is shown in Figure 2, defining horizontal direction vertical to the railway extending as horizontal (x), horizontal direction parallel to the railway extending as vertical (y) and vertical to the railway as vertical (z). Six measuring points are set along the horizontal direction, with a spacing of 1.5 m, for each measuring point, monitoring both the horizontal and vertical vibration acceleration.

Each measuring point uses a 173A500 acceleration sensor (with a monitored acceleration range: −10 g–+10 g). The CF5920N dynamic signal test analyzer is used for the data acquisition, as shown in Figure 3. CF5920 dynamic signal test and analysis system and SeismoSignal vibration spectrum analysis system were used for data processing. In this test, the monitoring sampling frequency is 1000 Hz.

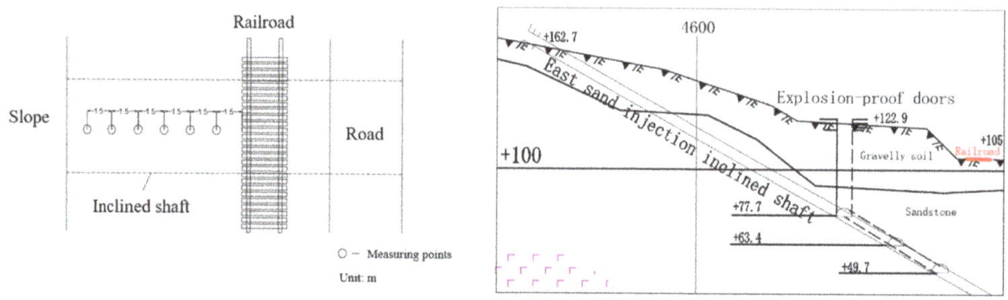

(a) (b)

Figure 2. Monitoring location diagram. (**a**) Schematic diagram of the measuring point layout. (**b**) Cross-sectional view of monitoring location.

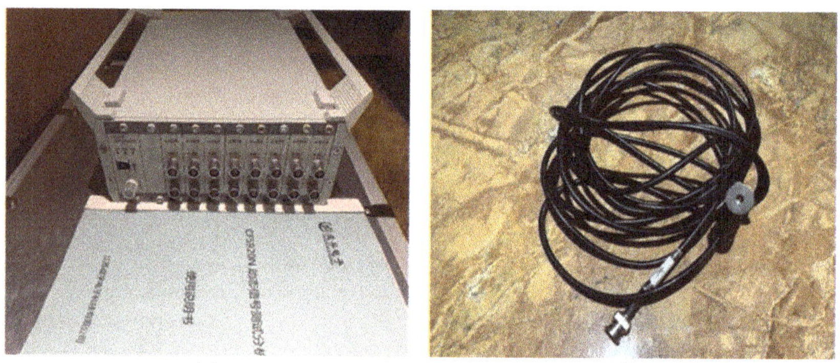

(a) (b)

Figure 3. Monitoring instruments. (**a**) CF5920N dynamic signal test analyzer; (**b**) 173A500 Acceleration Sensor.

The slope location and monitoring site of the railway are shown in Figure 4.

(a) (b)

Figure 4. Field monitoring situation. (**a**) Location of the railway tracks; (**b**) monitoring site.

2.3. Numerical Simulation

2.3.1. Calculation Model and Parameters

PLAXIS 3D program is a 3D finite element program for geotechnical deformation and stability analysis. With the development of PLAXIS, it has been gradually improved into a set of software with solid theoretical foundation, friendly interface, and strong logic applicable to most geotechnical engineering fields. It is successfully applied to the computational work of tedious and time-consuming non-linear finite elements in geotechnical engineering. In order to fully reflect the dynamic response of slope and roadway under train dynamic load, PLAXIS 3D software was used for simulation analysis, and the model size of 200 m × 120 m × 145 m was taken to simulate the area with strong human engineering activities, great influence on slope stability, and possible influence on the safety of underground inclined shaft. As shown in Figure 5, the length of the model is 200 m, and the width is 120 m. The value of the model height depends on the result of geological exploration, considering the effect on the epidote layer and the depth of the inclined shaft—145 m is taken. According to the current diagram of open-pit mine, the cross-sectional diameter of both inclined shafts is 6 m, the length of the inclined shaft is 230 m, and the wall thickness is 0.45 m. Two shafts successively go through the upper complex gravel soil layer, sandstone layer, chlorinated weak layer and bottom sandstone layer. The calculation model is shown in Figure 5, dividing 81,286 units and 123,035 grids.

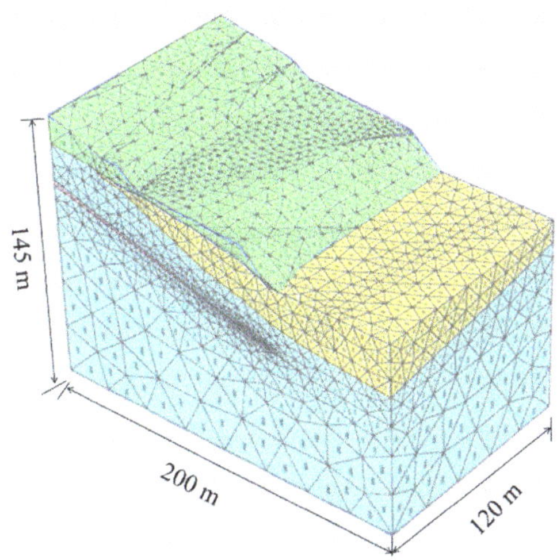

Figure 5. Calculation model diagram.

All the constitutive model of the soil adopts the ideal elastic-plastic model based on the Moore-Coulomb yield criterion and the non-associated flow rule. Two inclined shafts are simulated using plate elements. The bottom surface of the slope calculation area adopts fixed constraints, the model sides are normal fixed constraints, and the others are free boundaries. In practice, the vibration wave propagates all the way along the semi-infinite soil, but after setting the boundary artificially, there is a wave reflection at the boundary, and the reflected-back wave affects the analysis result of vibration response. In order to counteract the effect of reflection of such waves at the boundary on vibration analysis, special measures are taken at the boundary to set up a viscous boundary. Since the theory of viscous boundary conditions is mature and widely used, most of the finite element software provides the boundary condition of viscous boundary; therefore, the viscous boundary condition in PLAXIS 3D is also used in the dynamic calculation after the dynamic load of the train is applied. The displacement of the slope model under the stability analysis of the rock mass. According to the results of the field geological survey, the rock stratification is not obvious, and the rock quality is good; therefore, the suppression of the rock mass layers is not considered in the simulation. This is an elastic-plastic dynamic calculation and does not take into account the dynamic coefficients in the proposed hydrostatic method. The Rayleigh damping coefficients α of 0.1 and β of 0.008 were obtained from the indoor tests. The strata analyzed in this slope are mainly mixed fill, gravel soil, and sandstone with weak layer in the middle section of sandstone, and the physical and mechanical parameters of the stratum are shown in Table 1: where γ_{unsat} is the unsaturated weight, γ_{sat} is the saturated weight, E' is the effective modulus of elasticity, v is the Poisson's ratio, c is the cohesion, and φ is the friction angle.

Table 1. Physical and mechanical parameters of the rock formation.

Category	γ_{unsat} (kN/m^3)	γ_{sat} (kN/m^3)	E' (kN/m^2)	v	c (kN/m^2)	φ (°)
Mixed fill	19	22	10^4	0.3	8	26
Gravelly soil	20	23	4×10^4	0.2	10	23
Sandstone	29	30	2×10^{10}	0.2	38	38
Epidote	20	22	2×10^6	0.3	12	7

The schematic diagram of train dynamic load and two inclined shaft position in the model is shown in Figure 6.

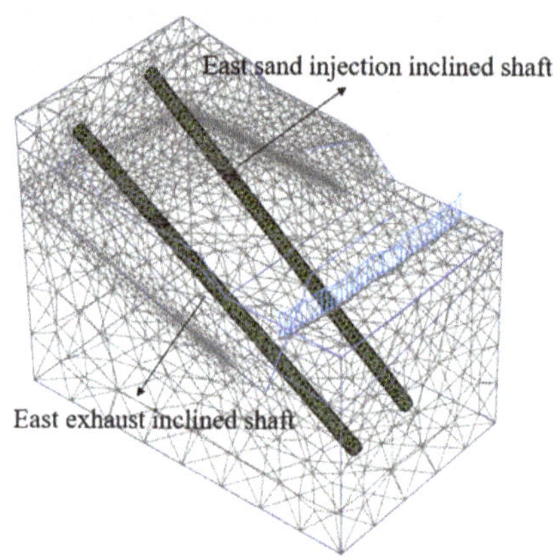

Figure 6. Schematic diagram of dynamic load and inclined shaft position.

2.3.2. Application of the Train Load

Based on the train vibration mechanism, the superposition coefficient and dispersion coefficient are introduced to modify the existing train rail load expression. The revised load expression is:

$$F(t) = k_1 k_2 (A_0 + A_1 \sin \omega_1 t + A_2 \sin \omega_2 t + A_3 \sin \omega_3 t) \quad (1)$$

where A_0 is the static wheel load; A_1, A_2, and A_3 are the peak vibration load corresponding to the circular frequency of rail vibration ω_1, ω_2, and ω_3, respectively; k_1 is the superposition coefficient of wheel-rail action, taking the value range of 1.2–1.7; k_2 is the dispersion coefficient of wheel-rail action, taking the value range of 0.6–0.9.

Let the under-spring mass of the train be m, and then the corresponding vibration load A_i is:

$$A_i = m \alpha_i \omega_i^2 \quad (2)$$

where $i = 1, 2, 3$ corresponds to the three control conditions of smoothness, power surcharge load on the line, and waveform wear. α_i is the typical positive vector, and ω_i is the vibration circle frequency. The formula for ω_i is:

$$\omega_i = \frac{2\pi v}{L_i} \tag{3}$$

where L_i is the wavelength corresponding to α_i, v is the travel speed.

Based on Equations (1) and (2), the time curve of train vibration load at 60 t axle weight and running speed of 15 km/h is calculated and shown in Figure 7.

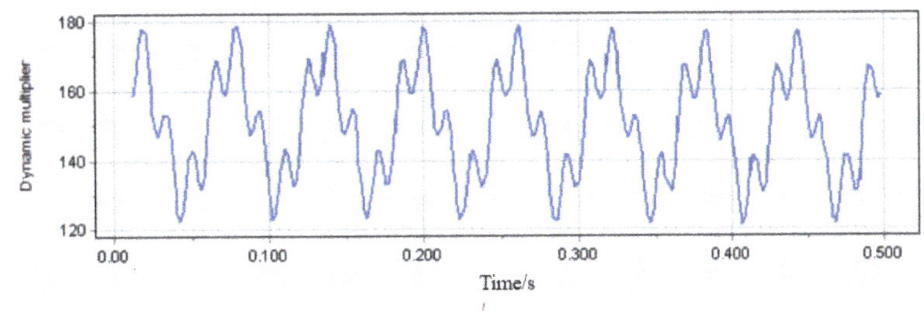

Figure 7. Time range curve of train dynamic load.

2.3.3. Distribution of Inclined Shaft Survey Points

In order to study the safety of east exhaust inclined shaft under train load, the displacement and acceleration of east exhaust inclined shaft and east sand injection inclined shaft were simulated by using PLAXIS 3D software. A total of 9 measuring points are arranged at each inclined shaft, and the measuring points are distributed from bottom to top. Considering that the roadway crosses the epidote weak layer, one measuring point is arranged up and down the layer, as shown in Figure 8.

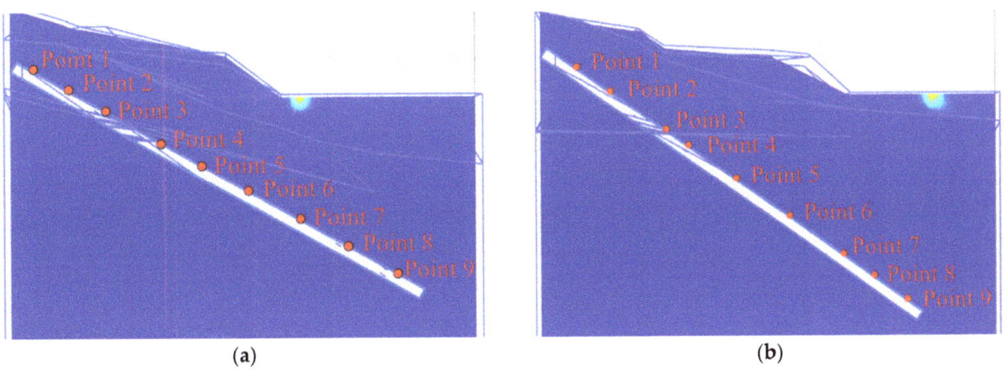

Figure 8. Schematic layout of measurement points. (**a**) East exhaust inclined shaft measurement point; (**b**) east sand injection inclined shaft measurement point.

3. Results and Discussion

3.1. Monitoring Result

The monitoring sampling frequency is 1000 Hz; the speed of the no-load train is 20 km/h; the weight is 30 t/section; and the speed of the full-load train is 15 km/h, and the weight is 60 t/section. Define the acceleration characteristic value: the maximum absolute acceleration, the maximum recorded absolute value of the acceleration at the same measuring point at each moment, and describe the maximum amplitude of the vibration acceleration at the measuring point when each train passes, and x, y, z direction is recorded as $|a_x|_{max}|a_y|_{max}|a_z|_{max}$. Based on the monitoring statistical results and data processing, the vibration acceleration reaction features were analyzed as follows: The surface vibration caused by three no-load trains and three fully loaded trains was monitored. Figure 9 shows the measured value of vertical and horizontal acceleration time curve at the surface of 1.5 m from the track under 60 t axle heavy-loaded train 15 km/h.

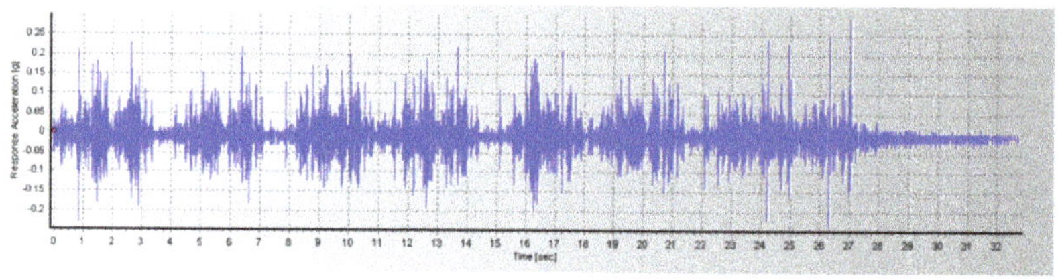

(a)

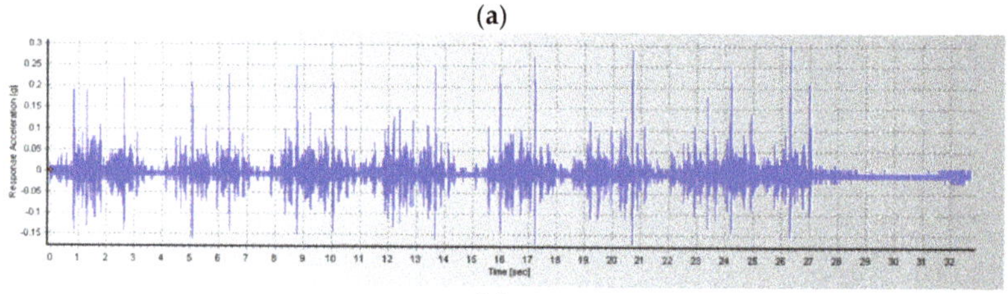

(b)

Figure 9. Ground vertical acceleration time. (**a**) Ground vertical acceleration time range curve at 1.5 m from the track; (**b**) ground horizontal acceleration time range curve at 1.5 m from the track.

The acceleration and the track distance curves under different operating conditions are shown in Figures 10 and 11.

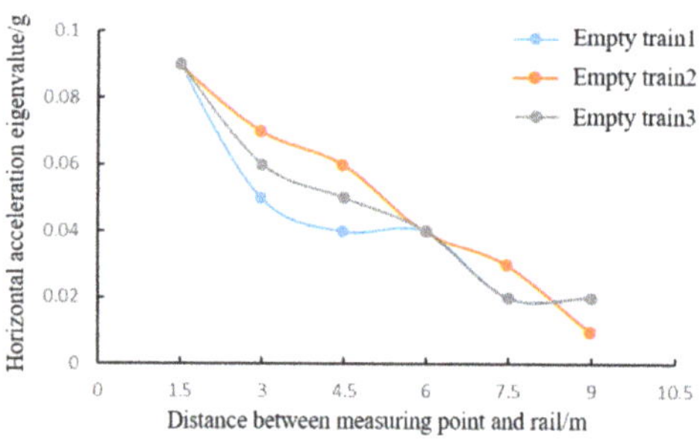

(a)

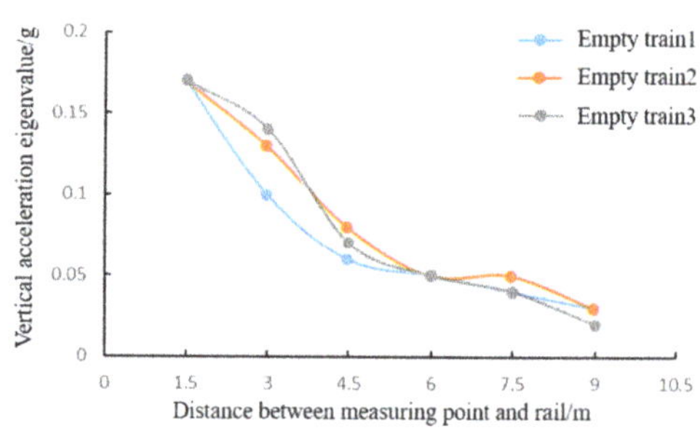

(b)

Figure 10. Acceleration change curve of each measuring point of the no-load train. (**a**) Horizontal acceleration change curve of each measuring point of the no-load train; (**b**) vertical acceleration change curve of each measuring point of the no-load train.

From the Figures 10 and 11, the characteristic values of horizontal and vertical acceleration decrease with the increase in distance from the measurement point, and the characteristic values of horizontal acceleration are less than 0.04 g, and vertical acceleration are less than 0.05 g at 9 m from the rail. The vibration response caused by a fully loaded train is obviously larger than that of an unloaded train; the decay speed of vibration caused by an unloaded train is obviously slower than that of a fully loaded train due to the faster driving speed. The vibration response caused by a fully loaded train is significantly larger than that of an unloaded train; the decay rate of vibration caused by an unloaded train is significantly slower than that of a fully loaded train due to its faster travel speed. This indicates that the train load and speed have an effect on the vibration response. Both horizontal and vertical acceleration eigenvalues decay significantly with increasing distance from the measurement point.

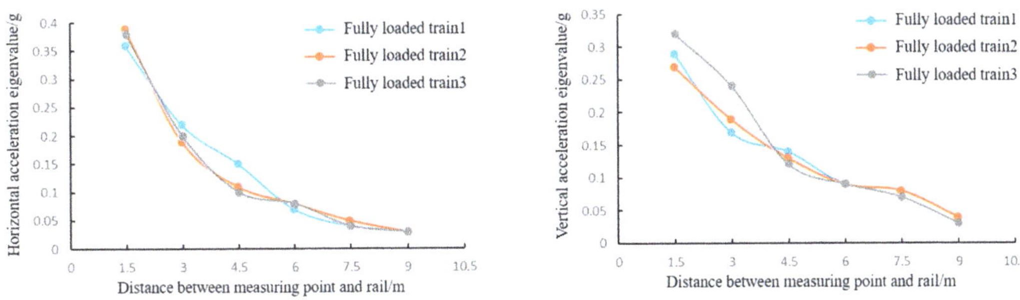

(a) (b)

Figure 11. Acceleration change curve of each measuring point of the fully load train. (**a**) Horizontal acceleration change curve of each measuring point of the fully load train; (**b**) vertical acceleration change curve of each measuring point of the fully load train.

3.2. Model Correctness Verification

To verify the finite element model, the numerical results of the train dynamic load model are compared with the field measurements of the fully loaded train, as shown in Figure 12. It can be seen from the figure that the horizontal and vertical acceleration eigenvalues of each measuring point in the numerical calculation model are basically the same as the measured value, indicating that the finite element model is feasible and correct.

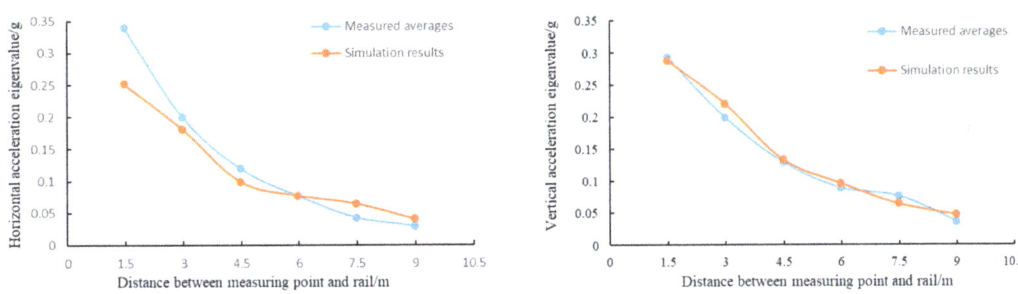

(a) (b)

Figure 12. Comparison of acceleration simulation of measuring points above the injection shaft and measured results. (**a**) Comparison of horizontal acceleration simulation of measuring points above the injection shaft and measured results; (**b**) comparison of vertical acceleration simulation of measuring points above the injection shaft and measured results.

To study the dynamic response characteristics of the inclined shaft under the train load, the surface horizontal and vertical acceleration features above the east exhaust inclined shaft are simulated by the finite element model. The simulation results are shown in Figure 13.

It can be seen from the figure that the decay law of the acceleration eigenvalues is roughly the same as the measured results, which both show a decreasing trend. The maximum horizontal acceleration of 0.23 g, vertical acceleration of 0.32 g is close to the measured results. It is considered that the establishment of this model is correct and feasible to analyze the dynamic response of the slope and inclined shaft.

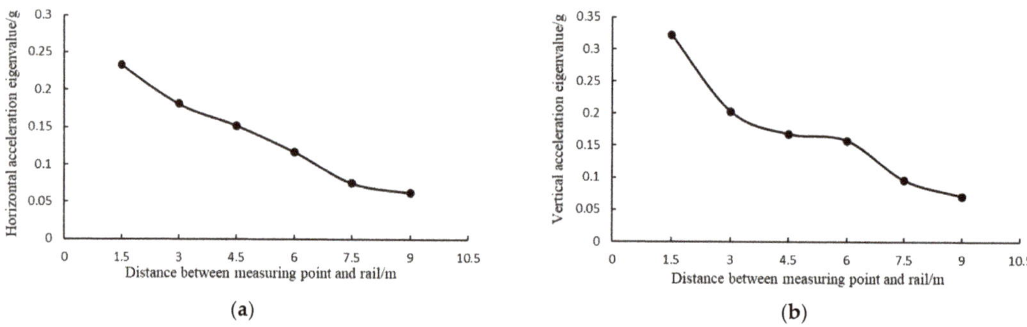

Figure 13. Simulation results of acceleration of east exhaust inclined shaft. (**a**) Simulation results of horizontal acceleration of east exhaust inclined shaft; (**b**) simulation results of vertical acceleration of east exhaust inclined shaft.

3.3. *Numerical Calculation Results and Analysis*

3.3.1. Slope Stability Analysis

After applying train load, the displacement of slope is shown in Figure 14 below.

The simulation results show that the maximum displacement of the slope is 5 mm and is directly below the train track; the miscellaneous fill layer on the side slope is only 0.5 mm. Considering that the distance between the slope and the dynamic load of the train is far, there is no landslide or damage in the soil. Both the internal dangerous epidote weak layer and the whole sandstone are less than 0.1 mm in displacement under the action of the train dynamic load, which can be ignored. Thus, the overall slope displacement under the dynamic load is very small, and the soil body does not suffer any damage.

The safety factor of the slope is derived from the strength-fold subtraction, which shaft unifies the strength reserve safety factor with the overall stability safety factor of the slope. The essence is that the initial shear strength τ of the slope decreases as the initial value, in order to find the shear strength value τ '(which just makes the slope in a critical unstable state and is referred to as critical shear strength). The critical shear strength value is:

$$\tau' = \frac{\tau}{F} = \frac{c + \sigma \cdot \tan\varphi}{F} = \frac{c}{F} + \sigma\frac{\tan\varphi}{F} \tag{4}$$

Correspondingly, the reduced shear strength parameters are: c' and φ'

$$c' = \frac{c}{F} \tag{5}$$

$$\varphi' = \arctan\left(\frac{\tan\varphi}{F}\right) \tag{6}$$

The slope safety factor calculated by this model is shown in Figure 15 below.

The safety coefficient of the slope finally converges to 1.201; that is, the safety coefficient of the slope is 1.201, and according to the Design Standard for Slope Engineering of Coal Industry Open-Pit Mine GB 51289-2018, the safety coefficient of the slope is not less than 1.05. Thus, it can be judged that the slope is stable under the action of traffic load.

3.3.2. Dynamic Response of East Exhaust Inclined Shaft

Under the action of the train dynamic load, the displacement of the inclined shaft and the acceleration response of the various measuring points on the inclined shaft are mainly considered. The displacement cloud diagram of the east exhaust inclined shaft is shown in Figure 16 below.

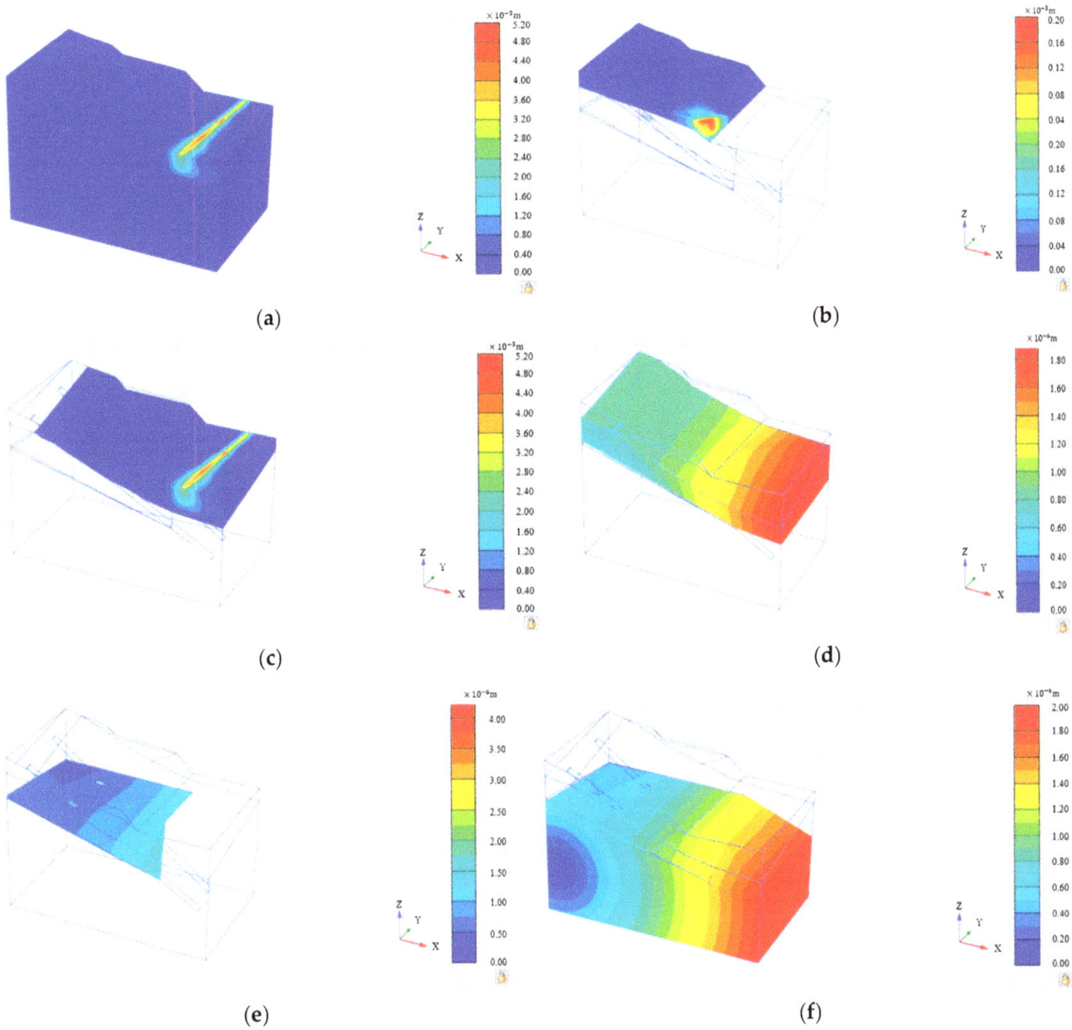

Figure 14. The displacement of slope. (**a**) Overall slope displacement; (**b**) mixed fill layer displacement; (**c**) gravelly soil displacement; (**d**) upper displacement of the epidote; (**e**) epidote weak layer displacement; and (**f**) sandstone displacement under the epidote.

Under the train load, the horizontal displacement peak of the east exhaust inclined shaft is 0.02 mm and located at the top of the east exhaust inclined shaft; the inclined shaft has the top-right (x positive) deformation trend under dynamic load; the vertical displacement peak of the east exhaust inclined shaft is 0.03 mm and is also at the top of the inclined shaft; and the bottom-left (x negative) deformation trend of the inclined shaft is under dynamic load. Overall, train load vibration has little impact on the east exhaust inclined shaft displacement. Its vertical and horizontal displacement are less than 0.1 mm, and the east exhaust inclined shaft is deflected negligibly under the influence of the dynamic train load.

The acceleration characteristic values of each measuring point of east exhaust inclined shaft are shown in Table 2.

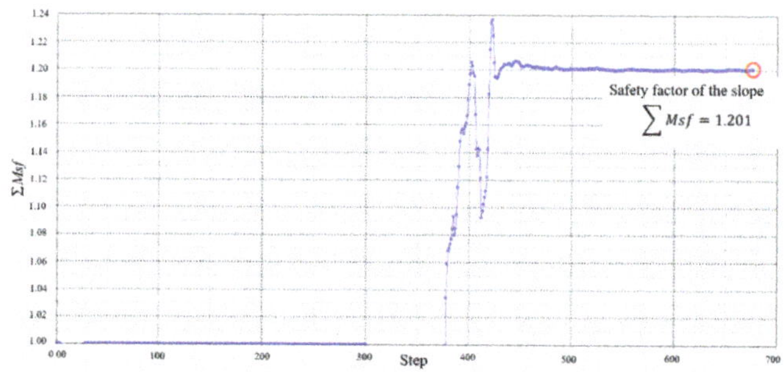

Figure 15. Slope safety factor.

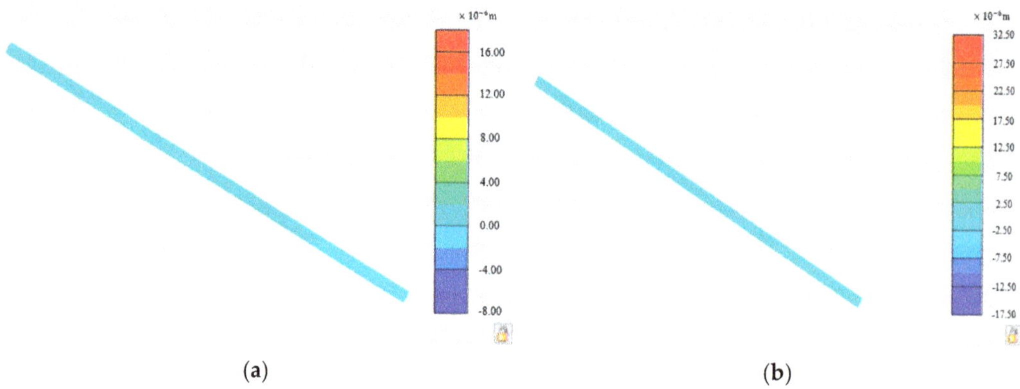

Figure 16. Displacement of east exhaust inclined shaft. (**a**) Horizontal displacement of east exhaust inclined shaft; (**b**) vertical displacement of east exhaust inclined shaft.

Table 2. Eigenvalue of acceleration of east exhaust inclined shaft.

Point	$\lvert a_x \rvert_{max}/g$	$\lvert a_z \rvert_{max}/g$
1	7.33×10^{-4}	1.16×10^{-3}
2	1.36×10^{-3}	1.75×10^{-3}
3	6.21×10^{-3}	8.19×10^{-3}
4	1.81×10^{-3}	8.02×10^{-3}
5	3.17×10^{-4}	4.75×10^{-4}
6	7.05×10^{-4}	1.05×10^{-3}
7	2.83×10^{-3}	3.54×10^{-3}
8	2.60×10^{-3}	4.08×10^{-3}
9	2.36×10^{-4}	2.70×10^{-3}

The acceleration characteristic value of the inclined shaft are shown in Figure 17.

As it can be seen from Table 2, in general, the acceleration characteristic values of each measurement point are small, and the maximum values of horizontal and vertical acceleration are 6.21×10^{-3} and 8.19×10^{-3} g, respectively, at measurement point 3, i.e., the upper part of the diagonal shaft going down through the weak layer of epidote. As can be seen from Figure 18, there are two significant increases in the characteristic value pattern of acceleration in each direction: the first one is a significant increase in the peak value when going down through the weak layer, and the second one is at the point directly below the load. Further, the horizontal acceleration of the rest of the inclined shaft measurement points is stable at 1×10^{-3} g, and the vertical acceleration is stable at

$1–3 \times 10^{-3}$ g. In general, the effect of train load on the safety of the east exhaust inclined shaft is relatively small.

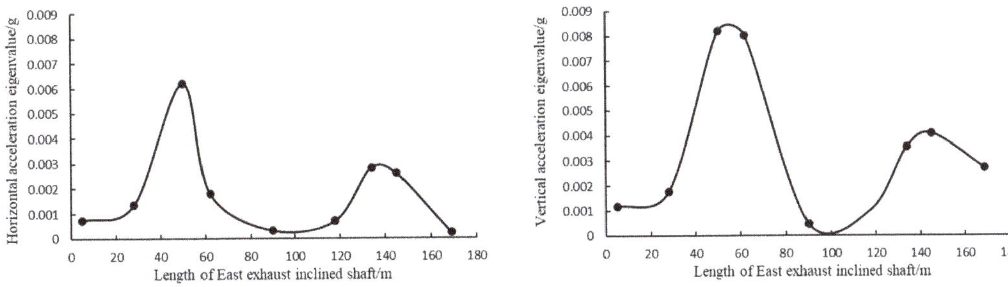

(**a**) (**b**)

Figure 17. Schematic diagram of relation between acceleration and inclined shaft length of east exhaust inclined shaft. (**a**) Schematic diagram of relation between horizontal acceleration and inclined shaft length of east exhaust inclined shaft; (**b**) schematic diagram of relation between vertical acceleration and inclined shaft length of east exhaust inclined shaft.

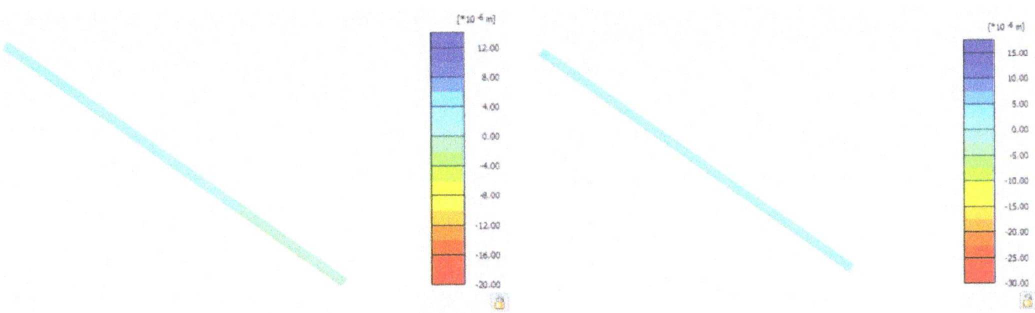

(**a**) (**b**)

Figure 18. Displacement of east injection inclined shaft. (**a**) Horizontal displacement of east injection inclined shaft; (**b**) vertical displacement of east injection inclined shaft.

3.3.3. Dynamic Response of East Injection Inclined Shaft

Under the action of the train dynamic load, the displacement of the inclined shaft and the acceleration response of the various measuring points on the inclined shaft are mainly considered. The displacement cloud map of the east injection shaft is shown in Figure 18 below.

Under the action of train load, the horizontal displacement of the east sand inclined shaft is 0.018 mm and located at the top of the inclined shaft, slightly offset (x direction) within 0.018 mm; given the vertical displacement of 0.028 mm, at the bottom of the inclined shaft, there is overall sinking trend, but the sinking amount is within 0.03 mm. In general, train load vibration has little impact on the displacement of the east injection inclined shaft, and its vertical and horizontal displacement are less than 0.1 mm and are negligible under the influence of train dynamic load.

The acceleration characteristic values of the east exhaust inclined shaft are shown in Table 3.

The acceleration characteristic value of the inclined shaft are shown in Figure 19.

In general, the acceleration characteristic values of each measurement point are small, and the maximum values of horizontal and vertical acceleration are 6.98×10^{-3} and 4.85×10^{-3} g, respectively, also at measurement point 3, which is the upper part of the inclined shaft down through the weak layer. The horizontal and vertical accelerations change in a similar pattern, and both have a significant increase at measurement point 3, i.e., measurement point 7, which is directly below the load. The horizontal and vertical acceleration time curves both increase to the maximum peak before 0.1 s. The peak value increases significantly through the weak layer, and the acceleration increases at the measurement point below the load, while the horizontal acceleration at the rest of the inclined shaft measurement points is stable at 1–2 × 10^{-3} g, and the vertical acceleration is stable within 1 × 10^{-3} g. In general, the effect of train load on the safety of the east sand injection inclined shaft is relatively small.

Table 3. Eigenvalue of acceleration of east injection inclined shaft.

Point	$\vert a_x \vert_{max}/g$	$\vert a_z \vert_{max}/g$
1	1.32 × 10^{-4}	1.80 × 10^{-3}
2	1.76 × 10^{-3}	3.08 × 10^{-4}
3	6.98 × 10^{-3}	4.85 × 10^{-3}
4	3.24 × 10^{-3}	3.02 × 10^{-3}
5	9.77 × 10^{-4}	6.90 × 10^{-4}
6	1.53 × 10^{-3}	1.99 × 10^{-4}
7	3.41 × 10^{-3}	1.95 × 10^{-3}
8	7.97 × 10^{-4}	8.75 × 10^{-4}
9	2.44 × 10^{-4}	1.95 × 10^{-4}

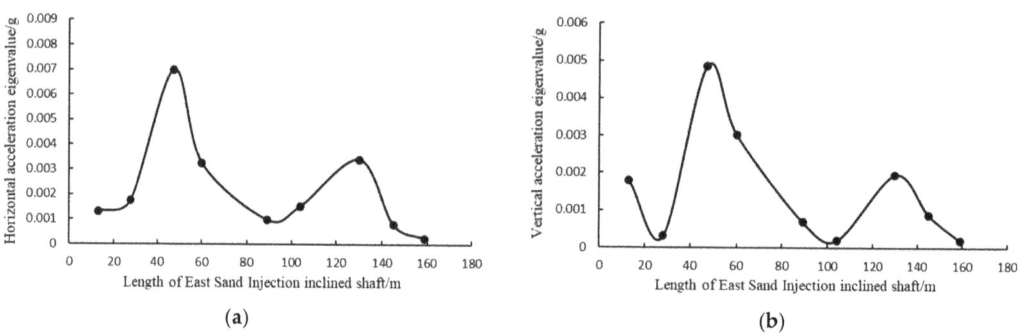

Figure 19. Schematic diagram of relation between acceleration and inclined shaft length of east injection inclined shaft. (**a**) Schematic diagram of relation between horizontal acceleration and inclined shaft length of east injection inclined shaft; (**b**) schematic diagram of relation between vertical acceleration and inclined shaft length of east injection inclined shaft.

4. Conclusions

Most of the previous studies by scholars on inclined shafts lie in underground mines, and there are fewer studies on open-pit mines. As one of the larger open-pit mines in Asia, the scale of the inclined shaft in Laohutai mine is also rare in the world. Unlike previous scholars, the dynamic load in this study is the dynamic load generated during the train travel, not the small-scale vehicle load. In this study, through site monitoring and power response analysis on the slope of the east sand injection inclined shaft and the east exhaust inclined shaft, the following conclusions are drawn:

1. The vibration response caused by the full-load trains is significantly greater than the no-load trains. The attenuation speed of the vibration caused by the no-load trains is significantly slower than the full-load trains, which shows that the train load and speed have an impact on the vibration response. Both the horizontal and vertical acceleration eigenvalues decay significantly with the distance of the measuring point;
2. Under the action of dynamic load, the maximum displacement of the slope appears in the gravel soil layer, 5 mm, and directly below the train track. The side slope and each

soil layer are less than 1 mm soil. The safety factor of the slope under the action of the dynamic load is 1.201.

3. The offset of east exhaust and east sand injection inclined shaft caused by train dynamic load is less than 0.1 mm and is far from the dynamic load, and the displacement caused by vibration is negligible;
4. The acceleration response law of east exhaust inclined shaft and east sand injection inclined shaft is consistent. Both the horizontal and vertical acceleration reached the maximum at the epidote weak layer, and the acceleration increased significantly directly below the load. The maximum horizontal acceleration of the east exhaust inclined shaft is 6.21×10^{-3} g, with a vertical acceleration of 8.19×10^{-3} g; the horizontal acceleration of the east injection inclined shaft is 6.98×10^{-3} g, with a vertical acceleration of 4.85×10^{-3} g. Overall, the inclined shaft acceleration response caused by the train vibration is small. Therefore, this test has a certain guiding significance for the selection of the through speed and load of trains on open-pit mines with inclined shafts and also provides some reference for the stability of the slopes of open-pit mines under the action of train loads.

Author Contributions: Funding acquisition, Y.W.; Project administration, Y.W.; Resources, Y.W., F.-W.Y., Z.-X.W., H.Z., K.M. and X.-J.L.; Software, S.-T.N.; Writing—original draft, Y.W. and S.-T.N.; Writing—review & editing, S.-T.N. All authors have read and agreed to the published version of the manuscript.

Funding: This study was funded by China Coal Technology Engineering Group (Grant No. wk004-2019 and No. 2018-2-ZD004).

Data Availability Statement: Data presented in this article are available at request from the corresponding author.

Acknowledgments: We wish to acknowledge the staff of Fushun Laohutai Mine, including Chief Engineer Fa-Wu Yang, and the team of Yu-Min Chen from Hohai University.

Conflicts of Interest: The authors declare no conflict of interest.

References

1. Valero, A.; Valero, A. Physical geonomics: Combining the exergy and Hubbert peak analysis for predicting mineral resources depletion. *Resour. Conserv. Recycl.* **2010**, *54*, 1074–1083. [CrossRef]
2. Haque, N.; Hughes, A.; Lim, S.; Vernon, C. Rare earth elements: Overview of mining, mineralogy, uses, sustainability and environmental impact. *Resources* **2014**, *3*, 614–635. [CrossRef]
3. Cheskidov, V.I.; Norri, V.K. Stripping with direct dumping in Kuzbass open pit mines: The current state and prospects. *J. Min. Sci.* **2016**, *52*, 725–731. [CrossRef]
4. Cheskidov, V.I.; Bobyl'sky, A.S. Technology and Ecology of Dumping at Open Pit Mines in Kuzbass. *J. Min. Sci.* **2017**, *53*, 882–889. [CrossRef]
5. Zaitseva, A.A.; Cheskidov, V.I.; Zaitsev, G.D. Effect of the mining sequence on the internal dump capacity in an open pit. *J. Min. Sci.* **2007**, *47*, 317–323. [CrossRef]
6. Meng, X.T.; Yu, S.Z.; Li, Z.F.; Wen, H.X. The slope monitoring method research. *Appl. Mech. Mater.* **2014**, *501–504*, 834–838. [CrossRef]
7. Selyukov, A.V. Technological significance of internal dumping in open pit coal mining in the Kemerovo Region. *J. Min. Sci.* **2015**, *51*, 879–887. [CrossRef]
8. Bakhaeva, S.P.; Gogolin, V.A.; Ermakova, I.A. Calculating stability of overburden dumps on weak bases. *J. Min. Sci.* **2016**, *52*, 454–460. [CrossRef]
9. Wajs, J. First experience with Remote Sensing methods and selected sensors in the monitoring of mining areas—A case study of the Belchatow open cast mine. *E3S Web Conf.* **2018**, *29*, 23. [CrossRef]
10. Selyukov, A.; Gerasimov, A.; Grishin, V. The Results of Mining and Geometric Analysis in Open Pit Mining of Promising Kuzbass Coal Deposits with Block System. *E3S Web Conf.* **2020**, *174*, 1020. [CrossRef]
11. Cao, C.; Feng, J.; Tao, Z. Start-up mechanism and dynamic process of landslides in the full highwaste dump. *Water* **2020**, *12*, 2543. [CrossRef]
12. Ye, H.; Wu, K.; Wang, J. Stability analysis about slope of Open pit Mine based on fuzzy synthetic evaluation. *Adv. Mater. Res.* **2012**, *396–398*, 2370–2374. [CrossRef]

13. Zhu, M.; Ma, C.; Tang, R.; Zhang, R.; Li, T.; Wang, Y.; Liu, J. Slope stability analysis of open pit mine based on finite difference method. In Proceedings of the 2nd International Conference on Information and Computing Science, Manchester, UK, 21–22 May 2009; pp. 208–211. [CrossRef]
14. Sun, S.; Yang, H.; Li, C.; Zhang, B.; Miao, A.; Wang, M. The study on key issues of slope instability during turning open-pit into underground mining. *Adv. Mater. Res.* **2012**, *594–597*, 70–75. [CrossRef]
15. Azhari, A.; Ozbay, U. Investigating the effect of earthquakes on open pit mine slopes. *Int. J. Rock Mech. Min. Sci.* **2017**, *100*, 218–228. [CrossRef]
16. Plotnikov, E.; Kolesnikov, V.; Šimková, Z.; Demirel, N. Features of new sites designing on quarry fields of the existing open pit mines of Kuzbass. *E3S Web Conf.* **2020**, *174*, 1003. [CrossRef]
17. Nunoo, S. Lessons Learnt from Open Pit Wall Instabilities: Case Studies of BC Open Pit Hard Rock Mines. *J. Min. Sci.* **2018**, *54*, 804–812. [CrossRef]
18. Petrović, B.; Vujić, S.; Čebašek, V.; Gajić, G.; Ignjatović, D. Predictive analysis of slope stability of internal dumps in Tamnava–West field mine after flooding. *J. Min. Sci.* **2016**, *52*, 110–114. [CrossRef]
19. Vyazmensky, A.; Stead, D.; Elmo, D.; Moss, A. Numerical analysis of block caving-induced instability in large open pit slopes: A finite element/discrete element approach. *Rock Mech. Rock Eng.* **2010**, *43*, 21–39. [CrossRef]
20. Segalini, A.; Giani, G.P. Numerical model for the analysis of the evolution mechanisms of the Grossgufer rock slide. *Rock Mech. Rock Eng.* **2004**, *37*, 151–168. [CrossRef]
21. Qian, Z.-W.; Jiang, Z.-Q.; Guan, Y.-Z.; Yue, N. Mechanism and Remediation of Water and Sand Inrush Induced in an Inclined Shaft by Large-Tonnage Vehicles. *Mine Water Environ.* **2018**, *37*, 849–855. [CrossRef]
22. Yan, Q.; Li, B.; Zhang, Y.; He, C. In situ automatic monitoring and working state assessment of inclined shafts in coal mines. *Proc. Inst. Civ. Eng. Geotech. Eng.* **2021**, *174*, 390–405. [CrossRef]
23. Voloshin, V.A.; Rib, S.V.; Petrova, O.A.; Belyaev, A.V. The choice of the site for inclined shaft and parameters of its stage-by-stage construction in the conditions of the Tersinsky coal region. *IOP Conf. Ser. Earth Environ. Sci.* **2018**, *206*, 012009. [CrossRef]
24. Zevgolis, I.E. Geotechnical characterization of mining rock waste dumps in central Evia, Greece. *Environ. Earth Sci.* **2018**, *77*, 566. [CrossRef]
25. Liu, C.; Gao, Y. Numerical simulation of slope stability analysis of open pit coal mine based on MIDAS/GTS. *Appl. Mech. Mater.* **2013**, *256–259*, 193–197. [CrossRef]
26. Huang, J.; Duan, T.; Lei, Y.; Hasanipanah, M. Finite Element Modeling for the Antivibration Pavement Used to Improve the Slope Stability of the Open-Pit Mine. *Shock Vib.* **2020**, *2020*, 6650780. [CrossRef]
27. Nie, L.; Li, Z.; Zhang, M.; Xu, L. Deformation characteristics and mechanism of the landslide in West Open-Pit Mine, Fushun, China. *Arab. J. Geosci.* **2015**, *8*, 4457–4468. [CrossRef]
28. Sakantsev, G.G.; Cheskidov, V.I.; Zyryanov, I.V.; Akishev, A.N. Validation of Slopes of Access Roads in Deep Open Pit Mining. *J. Min. Sci.* **2018**, *54*, 77–84. [CrossRef]
29. Abdellah, W.R.; Hussein, M.Y.; Imbabi, S.S. Rock slope stability analysis using shear strength reduction technique (SSRT)—Case histories. *Min. Miner. Depos.* **2020**, *14*, 16–24. [CrossRef]
30. Abdellah, W.; Beblawy, M.; Mohamed, M. Evaluation of open pit slope stability using various slope angles and element types. *Min. Miner. Depos.* **2018**, *12*, 47–57. [CrossRef]
31. Zairov, S.; Ravshanova, M.; Karimov, S. Intensification of technological processes in drilling and blasting operations during open-cut mining in Kyzylkum region. *Min. Miner. Depos.* **2018**, *12*, 54–60. [CrossRef]

metals

MDPI

Article

Experimental SHPB Study of Limestone Damage under Confining Pressures after Exposure to Elevated Temperatures

Lei Liu [1,2], Rui Li [1], Hao Qin [1] and Wei Sun [1,2,*]

[1] Faculty of Land Resources Engineering, Kunming University of Science and Technology, Kunming 650093, China; kgliulei@kust.edu.cn (L.L.); ruili@stu.kust.edu.cn (R.L.); 20172101051@stu.kust.edu.cn (H.Q.)
[2] Yunnan Key Laboratory of Sino-German Blue Mining and Utilization of Special Underground Space, Kunming 650093, China
* Correspondence: kmustsw@kust.edu.cn

Abstract: Studying the dynamic performance of rocks affected by high temperatures is a crucial theoretical foundation of mining engineering design and the construction of deep metallic mineral resources. More importantly, such studies can provide technical support for the green and low-carbon mining of these resources. However, systematic studies on the dynamic mechanical properties of rocks affected by both confining pressure and temperature during the mining of deep metallic mineral resources are lacking. Therefore, systematic research was conducted on the dynamic mechanical properties of limestone under confining pressure after high-temperature treatment, and a corresponding constitutive model was established. In this study, limestones were heated to 200 °C, 400 °C, 600 °C, and 800 °C, and the Split Hopkinson Pressure Bar impact test was conducted with confining pressures of 0.0 MPa, 0.5 MPa, 1.5 MPa, and 2.5 MPa. The test results show that the temperature has a significant effect on the dynamic compressive strength of limestone, and as the temperature rises, the strength tends to first increase and then decrease, reaching the turning point at a temperature of 400 °C. The dynamic compressive strength increases as the confining pressure increases. The constitutive equation of the dynamic damage to limestone under confining pressure after high-temperature treatment is consistent with the test results. Therefore, the established constitutive model can represent the dynamic behavior of limestone, providing a reference for evaluating the dynamic performance of this material, and serving as a theoretical basis for the green and low-carbon mining of deep metallic mineral resources.

Keywords: limestone; high temperature; confining pressure; SHPB; constitutive model

Citation: Liu, L.; Li, R.; Qin, H.; Sun, W. Experimental SHPB Study of Limestone Damage under Confining Pressures after Exposure to Elevated Temperatures. *Metals* **2021**, *11*, 1663. https://doi.org/10.3390/met11101663

Academic Editors: Srecko Stopic and Lijie Guo

Received: 25 August 2021
Accepted: 15 October 2021
Published: 19 October 2021

1. Introduction

With the depletion of shallow mineral resources, the mining depth for extracting underground metallic and non-metallic minerals has been growing [1]. The process of deep mining and construction is challenged by a complicated environment with a high ground stress, high ground temperature, high osmotic pressure, and strong mining disturbances of deep rock mass due to dynamic loads, such as blast waves and machine vibrations [2]. In order to provide a theoretical basis for the green and low-carbon mining of deep metallic mineral resources, it is essential to conduct research on the dynamic properties of rocks under a confining pressure and high temperature [3,4]. For example, shotcrete anchor supporting material and filling materials could be preserved. Recently, the continuous development of the Split Hopkinson Pressure Bar (SHPB) experimental technique has resulted in its wide application to the study of dynamic mechanical properties [5–7] and constitutive equations [8,9] of materials. Li X.B et al. [10] developed a dynamic and static loading test system for rocks at medium to high strain rates, and explored the dynamic mechanical properties of rocks under axial static pressure and confining pressure. Numerous scholars have conducted dynamic mechanical experiments for rocks under various conditions by utilizing SHPB, and great progress has been made. To research the effects of strain rates on the dynamic strength, deformation properties, and failure modes of rock, Ping Qi et al. [11] carried out numerous dynamic compression tests under passive confining pressure. A large number of experimental studies have shown that temperature has a significant impact on the mechanical properties of rock, which has attracted the attention of many scholars [12]. Li Ming [13] adopted XRD and SEM to research the meso-mechanics of the destruction of coal measure sandstone, and analyzed the impact of the temperature and impact load on the mechanical properties and behavior of the studied rock. Yin Tubing [14] developed a new experimental system and utilized it to conduct SHPB testing on siltstone under the coupling effect of temperature and pressure. The experimental results were used to develop a constitutive model of the rock under the coupling of temperature and pressure and dynamic disturbance. In addition to the influence of high temperatures, the deep rock mass is under triaxial stress in its natural state; the deeper the burial depth, the greater the stress of the surrounding rock, and the related indices of its mechanical properties will be affected accordingly. In the process of deep underground mining and underground engineering construction, the strength evaluation, stability control, and engineering design of an engineering rock mass subjected to high stress from the surrounding rock need to consider not only the physical and mechanical properties of rock blocks, but also the influence of confining pressure on its mechanical properties. Related studies have shown that the mechanical properties of the tested specimens change after confining pressure is applied in the SHPB experiment [15–17]. Therefore, to ensure that the study of the dynamic mechanical properties of rock materials is in line with engineering practice, the influence of confining pressure on these properties should not be ignored.

To date, many studies have been conducted on the dynamic mechanical properties of rock under normal temperature and dynamic coupling conditions, or on the uniaxial dynamic mechanical properties of rock subjected to high temperatures. However, few studies have used constitutive equations, which are more suitable for actual deep rock and soil engineering, to research dynamic mechanical properties under the combined effects of high temperature and confining pressure. As a result, it is necessary to conduct research on the dynamic mechanical properties of rocks affected by high temperatures under confining pressure so as to meet the needs of deep ground geotechnical construction.

In this research, the SHPB system was adopted to carry out dynamic compression experiments on limestone specimens after they were subjected to high temperatures (200 °C, 400 °C, 600 °C, and 800 °C) and different confining pressures (0.5 MPa, 1.5 MPa, and 2.5 MPa).

2. Specimen Preparation and Testing Procedure

2.1. Experimental Equipment

Figure 1 shows the Split Hopkinson Pressure Bar (SHPB) dynamic test system with a confining pressure device for testing the dynamic rock mechanics. The overall system consists of the SHPB test system and confining pressure device.

The Split Hopkinson Pressure Bar (SHPB) dynamic test system used in the experiment is composed of an electronic pulse loading device, incident rod, transmission rod, absorption rod, automatic speed monitoring device, signal acquisition and processing system, and other key components. In order to adjust the shape of the stress wave, the cross-section of the end of the incident rod is gradually changed. The material of the rod is 40Cr alloy steel, which has a Poisson's ratio of 0.2 and a longitudinal wave velocity of 5410 m/s. The diameters of the incident rod, transmission rod, and impact rod are 50 mm, and the lengths are 2400 mm, 1200 mm, and 800 mm, respectively.

Figure 2 shows the confining pressure loading system used in the dynamic mechanics experiment under the condition of confining pressure loading, which is mainly composed of the loading sleeve and the pressure oil pump. The pressure oil pump is used to change the oil pressure in the loading sleeve so as to adjust the confining pressure applied to the specimen.

Figure 1. SHPB test system.

Figure 2. Confining pressure device.

The heating device was manufactured by the Zhengzhou Xinghai Instrument Equipment Co., Ltd., used to heat the limestone specimen. The working voltage of this device is 220 V, and the rated power is 5000 W. The length, width, and height of the working chamber are 300 mm × 200 mm × 120 mm, respectively. The device is welded with high-quality high-temperature-resistant materials, and the maximum controllable temperature is 1200 °C.

After heating the rock for the high-temperature experiment, the storage device uses a WGLL-230BE electric heating blast drying oven made by Tianjin Tester. The working voltage is 220 V; the rated power is 3000 W; and its studio length, width, and height are 600 mm × 500 mm × 750 mm, respectively. The range of constant temperature fluctuation is ±1 °C.

2.2. SHPB Experimental Principle

In the process of impact loading with the SHPB impact system, due to the low brittleness of rock specimens, the experiment must not only decouple the effects of the one-dimensional stress wave and strain rate, but also meet the assumptions of one-dimensional stress and stress homogenization.

The shock wave is transmitted along the incident rod, and a part of the reflection occurs after contacting the specimen, which generates the reflection wave. The reflection wave continues to be transmitted after contacting the specimen, and the transmission wave is generated in the transmission rod, which is finally absorbed by the absorption rod. The electrical signals generated by the incident wave $\varepsilon_I(t)$, reflected wave $\varepsilon_R(t)$, and transmitted wave $\varepsilon_T(t)$ are collected and processed by a dynamic strain gauge and finally reflected in the oscilloscope. The dynamic stress $\sigma_s(t)$, dynamic strain $\varepsilon_s(t)$, and strain rate $\dot{\varepsilon_s}(t)$ of the specimen can be obtained by the three-wave method [18,19] (Equation (1)).

$$\begin{cases} \sigma_s(t) = \frac{EA}{2A_s}[\varepsilon_I(t) + \varepsilon_R(t) + \varepsilon_T(t)] \\ \dot{\varepsilon_s}(t) = \frac{C}{l}[\varepsilon_I(t) - \varepsilon_R(t) - \varepsilon_T(t)] \\ \varepsilon_s(t) = \int_0^t \dot{\varepsilon_s}(t)dt \end{cases} \quad (1)$$

2.3. Data Validity Verification

Similarly, when the experimental conditions satisfy the homogeneity assumption, the method shown in Section 2.2 (Equation (1)) can be used to test the validity of the experimental data. According to the relationship given in Equation (2), the reliability of the test data is verified by comparing the relationship between the transmitted wave, the incident wave, and the reflected wave, which is also called the stress balance test. In this experiment, all collected data were tested for stress balance. Figure 3 shows a typical test curve for stress equalization during the experiment.

$$\varepsilon_I(t) + \varepsilon_R(t) = \varepsilon_T(t) \quad (2)$$

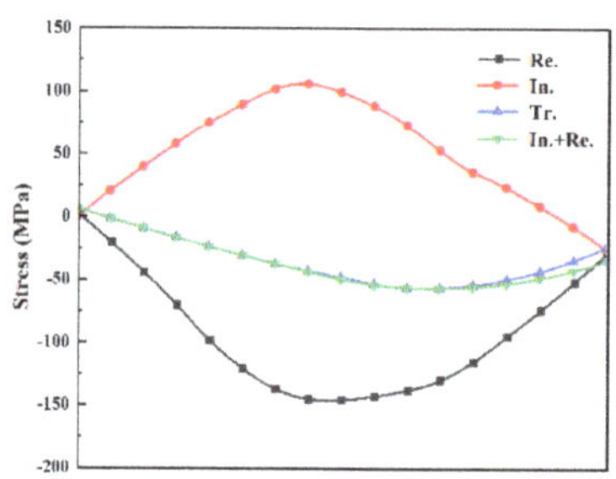

Figure 3. Typical test curve for stress equalization.

2.4. Experimental Materials and Scheme

The limestone specimens used in this work were taken from Katfang tin Mine, Honghe state, Yunnan Province. A cylindrical specimen with dimensions of Φ 50 mm × 40 mm (in Figure 4) was used, which is in accordance with the ISRM Suggested Methods [20].

Figure 4. Limestone specimen.

The heating rate is 10 °C/min. When the preset temperature of the device is reached, it remains constant for 2~3 h, and naturally cools down to room temperature. Figure 5 shows the heating temperature control curve. It is impossible to ensure that the heating rate of the rock specimen is completely consistent with the set heating rate in order to ensure that the specimen is damaged by the target temperature as a whole. Therefore, the heating furnace is maintained at the target temperature once it is reached. Before cooling the test piece, the infrared thermometer is used to measure the surface temperature of the test piece. As there will be some error between the surface and internal temperatures of the test piece, we generally set the temperature slightly higher than the target temperature.

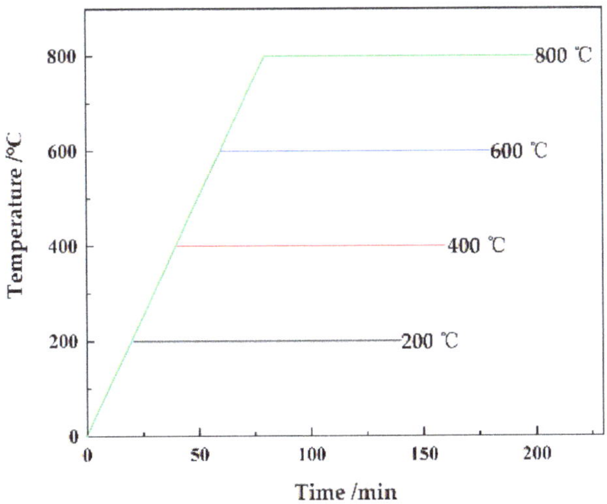

Figure 5. Heating temperature control curve.

In order to explore the influence of confining pressure and temperature damage on the dynamic mechanical properties of limestone specimens, the specimens were subjected to different temperatures and different confining pressures at loading impact velocities of 5.3 m/s and 8.6 m/s.

For comparison, the dynamic mechanical properties of the limestone specimens at room temperature were also tested at the corresponding impact rates without confining pressure. Three independent experiments were carried out under each working condition.

3. Experimental Results and Analysis

3.1. Experimental Results

The SHPB test system with a confining pressure device was utilized to study the effects of temperature (25 °C, 200 °C, 400 °C, 600 °C, and 800 °C), confining pressure (0.5 MPa, 1.5 MPa, and 2.5 MPa) and impact velocity (5.3 m/s and 8.6 m/s) on the dynamic mechanical properties of limestone.

The experimental results are shown in Table 1. In Table 1, a confining pressure of 0 MPa indicates that the dynamic impact experiments were carried out in a uniaxial state, and a temperature of 25 °C indicates that the dynamic impact experiments were performed at room temperature.

3.2. The Effects of High Temperature on Dynamic Mechanical Properties

Figure 6 presents the dynamic stress–strain curves of limestone specimens under different test conditions. At room temperature, the stress–strain curve can be divided into three stages: compaction stage, linear elastic stage, and destruction stage. The test results reported by Ping Qi et al. [21] also showed the same stages. The changing tendency of the uniaxial dynamic stress–strain curve for limestone after high-temperature treatment (200~800 °C) is similar to that at room temperature; the strength of the limestone first increases and then decreases with the increasing temperature.

Table 1. Mechanical properties of limestone under dynamic impact compression.

Impact Velocity (m/s)	Confining (MPa)	Temperature (°C)	Compressive Strength (MPa)	Peak Strain $\times 10^{-3}$	Elastic Modulus (GPa)
5.3	0	25	30.78	28.9	15.6
		200	33.68	28.1	16.83
		400	45.37	32.6	17.55
		600	38.56	34.5	22.03
		800	25.38	37.9	19.65
	0.5	25	33.35	2.72	7.65
		200	37.49	3.52	12.6
		400	46.42	4.66	14.38
		600	40.71	5.24	18.55
		800	32.72	5.51	14.7
	1.5	25	34.28	3.06	11.6
		200	39.13	5.48	20.11
		400	49.63	5.74	18.37
		600	41.06	4.62	28.5
		800	33.50	7.46	23.91
	2.5	25	46.06	4.11	11.39
		200	48.20	5.91	14.9
		400	53.56	7.88	24.16
		600	51.42	8.82	11.97
		800	37.39	8.89	13.66
8.6	0	25	39.63	48.8	19.72
		200	47.13	73.3	16.52
		400	48.20	48.8	16.33
		600	38.95	56.2	19.26
		800	39.73	54.2	15.24
	0.5	25	44.42	3.41	10.23
		200	53.20	6.03	14.44
		400	56.42	6.31	13.31
		600	49.99	8.88	15.33
		800	39.73	9.31	18.47
	1.5	25	55.70	4.5	16.32
		200	59.99	7.97	21.56
		400	66.42	8.2	14.77
		600	45.19	8.87	10.57
		800	43.63	10.91	23.11
	2.5	25	62.136	3.76	11.31
		200	68.778	5.31	18.78
		400	71.778	8.52	19.43
		600	49.280	6.16	17.18
		800	31.945	13.01	21.31

Figure 7 shows the relationship between the uniaxial dynamic compressive strength and temperature at an impact velocity of 5.3 m/s. Compared with the room temperature test condition, the dynamic compressive strength of limestone at 200 °C, 400 °C, and 600 °C is increased by 33.63%, 39.99%, and 19.87%, respectively. At 800 °C, the strength decreases by 19.54% and reaches the maximum value at 400 °C. This change is consistent with the pattern observed by Yin Tubing et al. when researching the impact of high temperatures on the dynamic compressive strength of granite specimens [22].

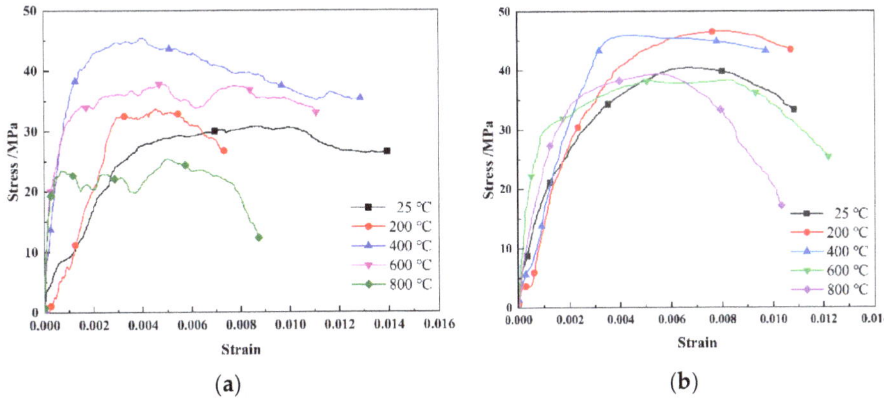

Figure 6. Dynamic stress–strain curves at different temperatures. (**a**) Impact velocity = 5.3 m/s; (**b**) Impact velocity = 8.6 m/s.

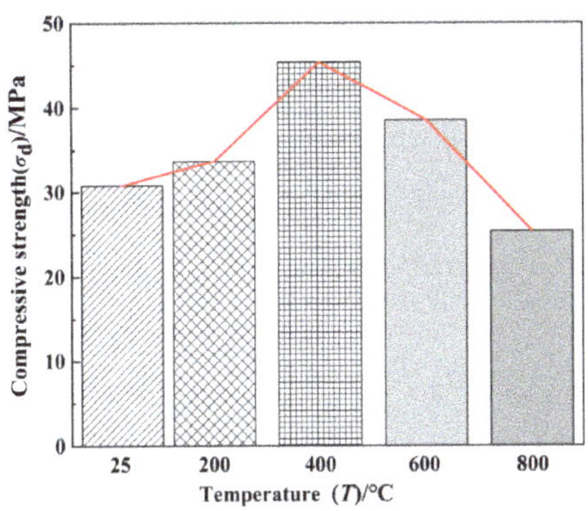

Figure 7. The relationship between uniaxial dynamic compressive strength and the temperature of limestone.

With the increase in temperature, the uniaxial dynamic compressive strength tends to first increase and then decrease, which shows that temperature has a significant effect on the uniaxial dynamic compressive strength of the limestone specimen. In the range of 25–400 °C, with the action of temperature, the water inside the specimen evaporates, the internal particles expand, and the internal pores close [23]. Hence, the temperature exerts a strengthening effect on the limestone. In addition, the increase in temperature contributes to the growth of limestone strength. When the temperature rises to 400 °C, the water inside the specimen completely evaporates, and the internal particles fully expand; at this point, the strengthening effect of the temperature on the uniaxial dynamic compressive strength of the limestone specimen is maximized. In the range of 400–600 °C, the internal particles of the limestone specimen continue to expand and squeeze each other, and the internal pores are basically closed. New pores and microcracks are generated inside the specimen [24]; at this time, the strengthening effect is still present (relative to room temperature of 25 °C), but it continuously decreases with the increasing temperature and reaches a minimum when the temperature is 400 °C. When the temperature exceeds 600 °C, the continuous development of internal pores and microcracks caused by the expansion of internal particles directly affects the dynamic compressive strength of the limestone specimen; at this time, the temperature generates a weakening effect on the uniaxial dynamic compressive strength of limestone (relative to room temperature of 25 °C). Furthermore, the weakening effect gradually increases as the temperature continues to rise.

3.3. The Effects of Confining Pressure on the Dynamic Mechanical Properties of Limestone after High-Temperature Treatment

Figures 8–10 show the dynamic stress–strain curves of limestone under different confining pressures at impact velocities of 5.3 m/s and 8.6 m/s after high-temperature treatment. Under the action of confining pressure, the peak stress in the limestone increases to different extents depending on the temperature. Moreover, the slope of the stress–strain curve changes accordingly. For temperatures lower than 400 °C, the peak of the curve shifts to the left as the temperature rises; in other words, the slope continuously increases. When the temperature exceeds 400 °C, the peak shifts to the right as the temperature rises; that is, the slope constantly decreases.

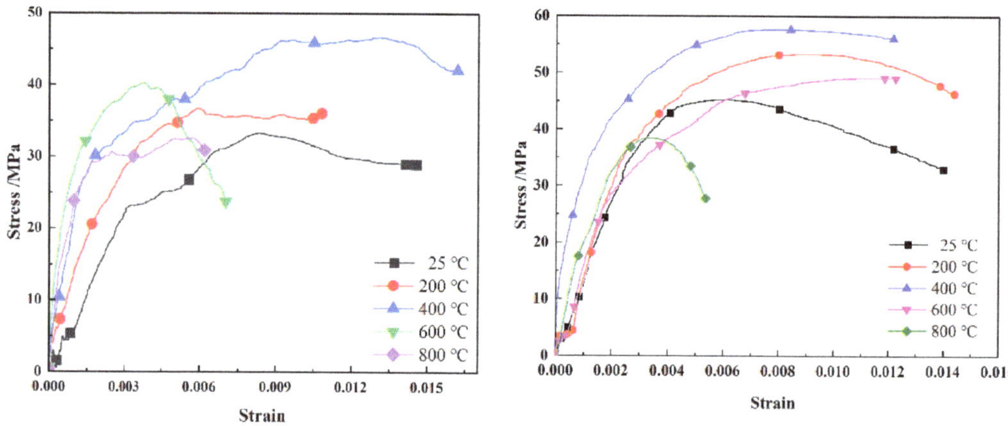

(a) (b)

Figure 8. Dynamic stress–strain curve for a confining pressure of 0.5 MPa. (**a**) Impact velocity = 5.3 m/s; (**b**) impact velocity = 8.6 m/s.

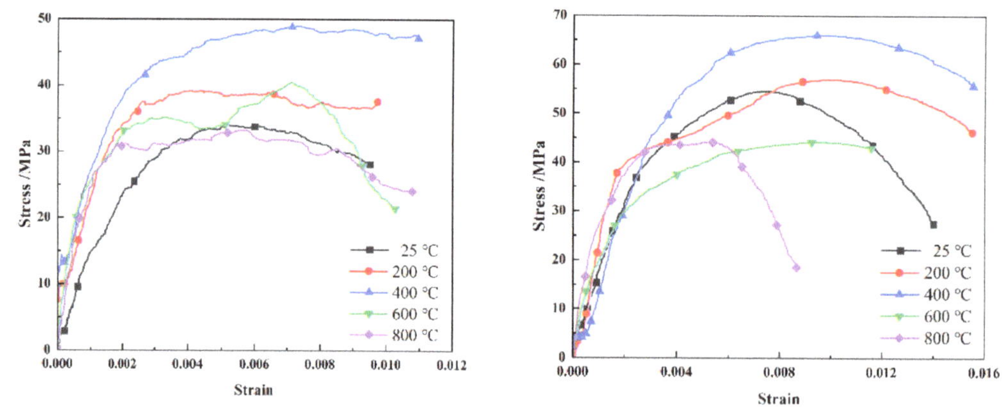

(a) (b)

Figure 9. Dynamic stress–strain curve for a confining pressure of 1.5 MPa. (**a**) Impact velocity = 5.3 m/s; (**b**) impact velocity = 8.6 m/s.

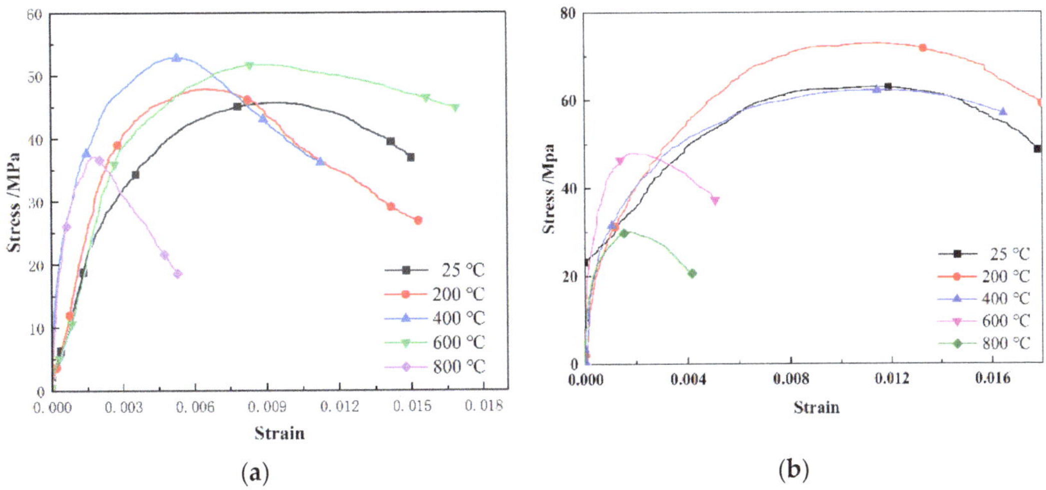

Figure 10. Dynamic stress–strain curve for a confining pressure of 2.5 MPa. (**a**) Impact velocity = 5.3 m/s; (**b**) impact velocity = 8.6 m/s.

The strength measurements of limestone subjected to different temperatures (25–800 °C) and different confining pressures (0 MPa, 0.5 MPa, 1.5 MPa, and 2.5 MPa) were used to draw the relationship curve between the dynamic compressive strength (σ_d) and confining pressure (σ_3) for different temperatures (T). Figure 11 shows that under different confining pressures, the effect of temperature on the strength of limestone is consistent with that without a confining pressure. As the temperature increases, the dynamic compressive strength of limestone first increases and then decreases. With the increase in confining pressure, the dynamic compressive strength also increases. A certain degree of positive correlation is observed, and it reaches the peak at a temperature of 400 °C and confining pressure of 2.5 MPa. According to the test results, high-temperature damage can affect the dynamic compressive strength by influencing the internal structure of the limestone specimen, while the confining pressure impacts the strength by affecting the tensile strength [25]. When both temperature and confining pressure act on the limestone specimen, the confining pressure enhances the temperature damage.

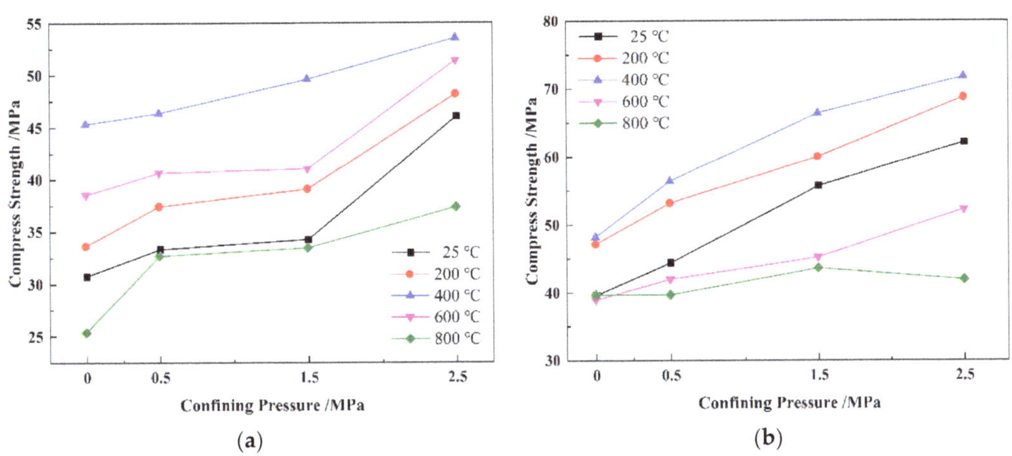

Figure 11. Dynamic compressive strength of limestone under different confining pressures after high-temperature treatment. (**a**) Impact velocity = 5.3 m/s; (**b**) impact velocity = 8.6 m/s.

3.4. The Dynamic Damage Constitutive Model for Limestone under the Action of Confining Pressure after High-Temperature Treatment

As the rock material, the limestone specimen can be regarded as a combination of a large number of differential elements; furthermore, it is also characterized by large numbers of pores and microcracks. Under naturally occurring conditions, these differential elements and internal pores and microcracks are distributed irregularly and unevenly. Therefore, the differential elements of rock can be assumed to be a parallel structure of viscous element λ and damage element D (Figure 12), and the following hypotheses can be made:

i. The impact of the acceleration of gravity on the constitutive relation of rock can be neglected;
ii. The damage and viscosity of the rock's differential element and the performance of the elastomer can be considered isotropous;
iii. Before the occurrence of damage, the differential element shows elastic properties and satisfies Hooke's Law;
iv. The constitutive relation of the viscous element satisfies the following:

$$\sigma = \eta \frac{d\varepsilon}{dt} \tag{3}$$

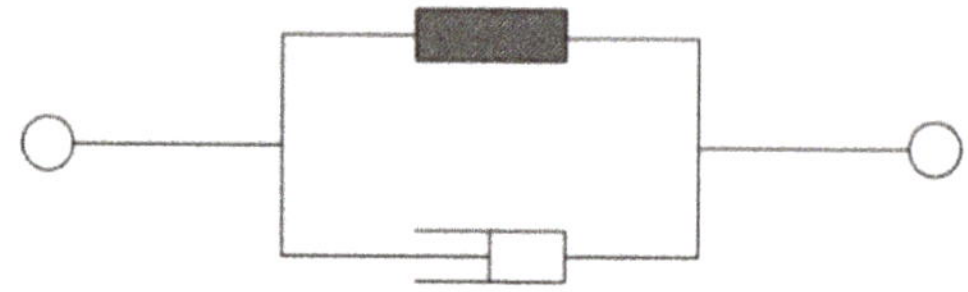

Figure 12. Differential element model.

v. The strength of the differential element has a Weibull distribution [26]; i.e., the density function satisfies the following:

$$\varphi\,(\mathrm{F}_0) = \frac{\mathrm{m}}{F_0}\left(\frac{F}{F_0}\right)^{m-1} \exp\left[-\left(\frac{F}{F_0}\right)^{m}\right] \tag{4}$$

The damage induced by the cumulative stress–strain and temperature is caused by effects on the internal structure of limestone. Therefore, when considering the internal damage of limestone due to the effects of the impact load and temperature, the impact load damage and the internal damage caused by temperature should be considered individually.

The analysis of the thermal damage of rock should factor in the influence of the internal material at the damaging temperature over a certain amount of time. From the perspective of wave velocity and modulus of elasticity, reference [27] presented a thermal damage model of different rocks. Given the need to include the stress–strain correspondence in this analysis, the elastic modulus representation was selected to characterize the thermal damage of limestone, as follows:

$$D_T = \frac{\Delta E}{E_0} = 1 - \frac{E_T}{E_0} \tag{5}$$

where D_T is the thermal damage to the limestone specimen under different temperatures, ΔE is the difference in elastic modulus among different temperatures and room temperature, and E_0 is the elastic modulus of limestone at room temperature.

The damage caused by the impact load effect is the result of the cumulative process of the internal stress–strain of the rock. The method in reference [28] expresses the damage of rock by statistically representing the degree of destruction of the internal differential element, as follows:

$$D_M = \frac{N_t}{N} \tag{6}$$

where D_M refers to the damage variable, and N_t and N indicate the failure number and sum, respectively, of differential elements in the rock under certain conditions.

According to the assumption of the density function of the differential element in the basic hypothesis, an interval of random variation of some differential element (assuming that $[\varepsilon, \varepsilon + d\varepsilon]$) is utilized to represent the tendency of the damage variable to change with the stress state. Then, the number of damaged differential elements in a certain state can be represented as follows (Equation (7)):

$$N_t(\varepsilon) = \int_0^{\varepsilon} N\varphi(\varepsilon)d\varepsilon \tag{7}$$

Equations (4) and (7) can be substituted in Equation (6) to obtain the damage equation of limestone due to the impact load, as follows:

$$D_M = \frac{N_t(\varepsilon)}{N} = 1 - \exp\left[-\left(\frac{F}{F_0}\right)^m\right]. \tag{8}$$

The relation among the total damage D, thermal damage D_T, and load damage D_M can be expressed according to Equation (9) [29]. D_T and D_M can be substituted accordingly to obtain the expression of the general damage variable D, which is expressed by Equation (10):

$$D = D_T + D_M - D_T D_M. \tag{9}$$

$$D = 1 - \frac{E_T}{E_0}\exp\left[-\left(\frac{F}{F_0}\right)^m\right]. \tag{10}$$

Based on the D-P model [30], the strength of the differential element of limestone can be expressed by Equation (11):

$$F = \alpha_0 I_1 + \sqrt{J_2} \tag{11}$$

where α_0 is the coefficient related to the internal friction angle of limestone. I_1 and J_2 relate to the stress state of the differential element, and the corresponding expressions are shown in Equations (12) and (13).

$$I_1 = \sigma_1' + \sigma_2' + \sigma_3'. \tag{12}$$

$$J_2 = \frac{1}{6}\left[(\sigma_1' - \sigma_2')^2 + (\sigma_1' - \sigma_2')^2 + (\sigma_2' - \sigma_3')^2\right]. \tag{13}$$

where $\sigma_1', \sigma_2',$ and σ_3'. are the effective stress received from different directions of various differential elements, as expressed by Equation (14):

$$\sigma_i' = \sigma_i(1 - D) \tag{14}$$

As shown in Figure 10, the stress obtained by the differential element of limestone is the sum of two parallel separations. The following relation should be satisfied for the two-fission strain of the differential element strain:

$$\sigma = \sigma_a + \sigma_b \tag{15}$$

$$\varepsilon = \varepsilon_a = \varepsilon_b \tag{16}$$

According to Hooke's Law, the strain of the elastic material and its corresponding stress state have a linear relationship [31,32]; furthermore, under confining pressure, the limestone receives equal radial force from all directions ($\sigma_2 = \sigma_3$). Then, ε_1 can be obtained as follows:

$$\varepsilon_1 = E^{-1}(\sigma_1' - 2\mu\sigma_2') \tag{17}$$

where μ is the Poisson's ratio of limestone.

Equation (14) can be substituted into Equation (17), resulting in the following:

$$\varepsilon_1 = E^{-1}(1-D)(\sigma_1 - 2\mu\sigma_2) \tag{18}$$

Then, the following can be deduced:

$$\sigma_1 = \varepsilon_1\, E(1-D)^{-1} + 2\mu\sigma_2 \tag{19}$$

The dynamic constitutive equation of the differential structure of limestone under confining pressure after high-temperature treatment can be obtained by combining Equations (10), (11), (15) and (19), which is as follows:

$$\sigma = \varepsilon \frac{{E_T}^2}{E_0} \exp\left[-\left(\frac{F}{F_0}\right)^m\right] + 2\mu\sigma_2 + \eta\frac{d\varepsilon}{dt} \tag{20}$$

In Equation (20), F_0 and m are unknown parameters in the Weibull distribution function. Under the uniaxial loading condition, that is, when the confining pressure is 0 MPa ($\sigma_2 = \sigma_3 = 0\ MPa$), the peak of the stress–strain curve, namely, σ_{max} and ε_{max}, can be obtained; then, the relation can be expressed in Equation (21):

$$\frac{d\sigma_{max}}{d\varepsilon_{max}} = 0 \tag{21}$$

By taking the derivative of Equation (22) with respect to ε and substituting Equations (11) and (21) accordingly to solve for F_0 and m, the following is obtained:

$$F_0 = \left(\alpha_0 + \frac{1}{\sqrt{3}}\right) E\varepsilon m^{\frac{1}{m}} \tag{22}$$

$$\mathrm{m} = \frac{1}{\ln\left(E_T^2\varepsilon_D\right) - \ln\left(E_0\sigma_D - E_0\eta\frac{d\varepsilon}{dt}\right)} \tag{23}$$

Figure 13 compares the experimental results obtained under the uniaxial loading condition (confining pressure = 0 MPa) and under a confining pressure of 2.5 MPa.

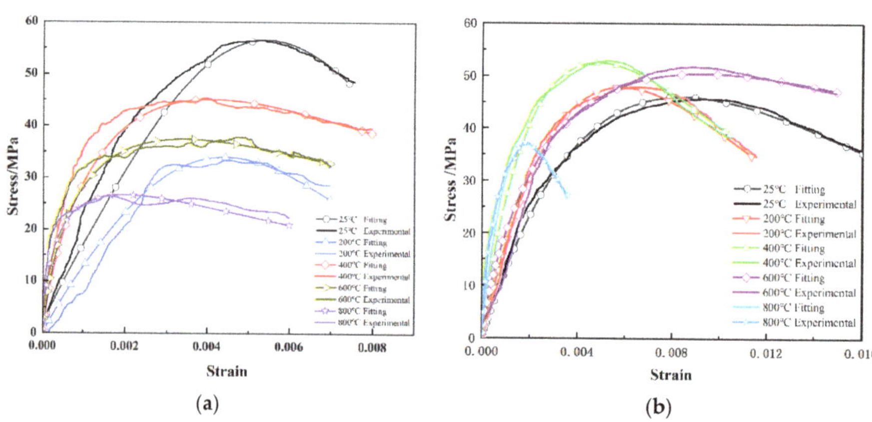

Figure 13. Fitting of the dynamic damage constitutive model. (**a**) Confining pressure = 0 MPa; (**b**) confining pressure = 2.5 MPa.

Figure 13 shows the dynamic stress–strain curve established using the dynamic damage constitutive model of limestone after high-temperature treatment compared with the SHPB experimental results for limestone under different experimental conditions. The changing tendency based on the model is almost identical to that of the actual laboratory experiment. The following characteristics apply to the fitted model:

The model can effectively express the change of "compaction" and "elastic" stages, in which failure occurs in the limestone specimen under the action of temperature, impact load, and confining pressure in various test conditions.

The model can effectively represent the dynamic compressive strength, peak stress, and elastic modulus of the limestone specimen under various test conditions.

In terms of the characterization of "failure" and "post-peak" stages in the dynamic failure process, the model results have certain differences from the laboratory test results. The model can be utilized to conduct research on the dynamic damage state of limestone under the action of confining pressure after high-temperature treatment. A laboratory experiment should be carried out with respect to the change features of the "post-peak" stage.

4. Conclusions

A high-temperature SHPB system was adopted to conduct dynamic compression tests on a limestone specimen under different confining pressures after being heated to different temperatures and naturally cooled to room temperature. Research on the SHPB test and dynamic damage constitutive model of limestone under the action of both confining pressure and temperature was carried out. The major conclusions are as follows:

The uniaxial dynamic stress–strain curve of limestone subjected to high temperatures has an almost identical changing pattern to the one at room temperature. The temperature can significantly impact the dynamic compressive strength of limestone, with a trend of first strengthening and then attenuating as the temperature rises.

After applying the confining pressure, the dynamic stress–strain curve of the limestone showed a consistent tendency with that in the uniaxial state. Through the limitation of lateral deformation, the confining pressure increased the dynamic compressive strength of the limestone. It also intensified the impact of temperature on the dynamic mechanical properties of limestone after high-temperature treatment. With the increase in confining pressure, the dynamic compressive strength also increased, reaching a maximum value at a temperature of 400 °C and confining pressure of 2.5 MPa.

A dynamic damage constitutive equation of limestone under the action of confining pressure after high-temperature treatment was established. According to comparisons with the experimental results, the model can reflect the relation among temperature, dynamic strength, strain, confining pressure, and strain rate under the action of confining pressure on the heated limestone.

Author Contributions: L.L. and W.S. designed the experiments; R.L. and H.Q. carried out the experiments; L.L. analyzed the experimental results. R.L. analyzed data and developed the analysis tools. W.S. and L.L. wrote the manuscript. All authors have read and agreed to the published version of the manuscript.

Funding: This research was funded by the National Natural Science Foundation of China (grant nos. 11862010 and 51964023).

Data Availability Statement: The data used to support the findings of this study are included within the article.

Conflicts of Interest: The authors declare no conflict of interest.

References

1. Xie, H.P.; Zhou, H.W.; Xue, D.J.; Wang, H.W.; Zhan, R.; Gao, F. Research and thinking on deep mining and limit mining depth of coal. *J. China Coal Soc.* **2012**, *37*, 535–542.
2. Xie, H.P. Research conception and prospect of "deep rock mechanics and mining theory". *Adv. Eng. Sci.* **2017**, *49*, 1–16.
3. Wang, S.M.; Sun, Q.; Qiao, J.W.; Wang, S.Q. Discussion on the geological guarantee of green mining of coal. *J. China Coal Soc.* **2020**, *45*, 8–15.
4. Zhao, G.Y.; Qiu, J.; Yuan, Z.; Pei, D.; Yang, L.; Pan, W. Discussion on evaluation method of green mining of metal. *Gold Sci. Technol.* **2020**, *28*, 169–175.

5. Mauko, A.; Fíla, T.; Falta, J.; Koudelka, P.; Rada, V.; Neuhäuserová, M.; Zlámal, P.; Vesenjak, M.; Jiroušek, O.; Ren, Z. Dynamic Deformation Behaviour of Chiral Auxetic Lattices at Low and High Strain-rates. *Metals* **2020**, *11*, 52. [CrossRef]
6. Fu, H.; Wang, X.; Xie, L.; Hu, X.; Umer, U.; Rehman, A.U.; Abidi, M.H.; Ragab, A.E. Dynamic Behaviors and Microstructure Evolution of Iron–nickel Based Ultra-high Strength Steel by Shpb Testing. *Metals* **2019**, *10*, 62. [CrossRef]
7. Klepaczko, J.R. Behavior of rock-like materials at high strain rates in compression. *Int. J. Plast.* **1990**, *6*, 415–432. [CrossRef]
8. Cherouat, A.; Borouchaki, H.; Jie, Z. Simulation of Sheet Metal Forming Processes Using a Fully Rheological-damage Constitutive Model Coupling and a Specific 3d Remeshing Method. *Metals* **2018**, *8*, 991. [CrossRef]
9. Lei, B.; Chen, G.; Liu, K.; Wang, X.; Jiang, X.; Pan, J.; Shi, Q. Constitutive Analysis on High-Temperature Flow Behavior of 3Cr-1Si-1Ni Ultra-High Strength Steel for Modeling of Flow Stress. *Metals* **2019**, *9*, 42. [CrossRef]
10. Li, X.B.; Gu, D.S.; Lai, H.H. Reasonable loading waveform in dynamic stress-strain map test of rock under impact load. *Explos. Shock. Waves* **1993**, *2*, 125–130.
11. Ping, Q.; Ma, Q.Y.; Lu, X.Y.; Yuan, W. Study on impact compression test of rock materials under passive confining pressure. *J. Vib. Shock.* **2014**, *33*, 55–59.
12. Xiao, M.; Ma, D.D.; Hu, D.W.; Zhou, H.; Chen, S.L.; Yu, Z.P.; Tan, X.F. Development and application of high temperature true triaxial test system in real time. *Chin. J. Rock Mech. Eng.* **2019**, *38*, 1605–1614.
13. Li, M.; Mao, X.; Cao, L.; Pu, H.; Lu, A. Influence of Heating Rate on the Dynamic Mechanical Performance of Coal Measure Rocks. *Int. J. Geomech.* **2017**, *17*, 04017020. [CrossRef]
14. Yin, T.; Li, X.; Xia, K.; Huang, S. Effect of Thermal Treatment on the Dynamic Fracture Toughness of Laurentian Granite. *Rock Mech. Rock Eng.* **2012**, *45*, 1087–1094. [CrossRef]
15. Fakhimi, A.; Azhdari, P.; Kimberley, J. Physical and Numerical Evaluation of Rock Strength in Split Hopkinson Pressure Bar Testing. *Comput. Geotech.* **2018**, *102*, 1–11. [CrossRef]
16. Mishra, S.; Chakraborty, T.; Matsagar, V. Dynamic Characterization of Himalayan Quartzite Using SHPB. *Procedia Eng.* **2017**, *191*, 2–9. [CrossRef]
17. Yin, Z.Q.; Li, X.B.; Jin, J.F.; He, X.Q.; Kun, D.U. Failure Characteristics of High Stress Rock Induced by Impact Disturbance Under Confining Pressure Unloading. *Trans. Nonferrous Met. Soc. China* **2012**, *22*, 175–184. [CrossRef]
18. Zhu, J.; Hu, S.; Wang, L. An Analysis of Stress Uniformity for Concrete-like Specimens During Shpb Tests. *Int. J. Impact Eng.* **2009**, *36*, 61–72. [CrossRef]
19. Ke, M.; Zhou, H. Dynamic Force Balance Analysis of SHPB System. *Met. Mine* **2010**, *39*, 27–30.
20. Zhang, Z.X.; Yu, J.; Kou, S.Q.; Lindqvist, P.A. Effects of High Temperatures on Dynamic Rock Fracture. *Int. J. Rock Mech. Min. Sci.* **2001**, *38*, 211–225. [CrossRef]
21. Ping, Q.; Wu, M.J.; Huan, Z.; Yuan, P. Experimental research on dynamic mechanical properties of sandstone under high-temperature condition. *Chin. J. Undergr. Space Eng.* **2019**, *15*, 691–698.
22. Yin, T.B.; Shu, R.H.; Li, X.B.; Pin, W.A.; Dong, L.J. Combined Effects of Temperature and Axial Pressure on Dynamic Mechanical Properties of Granite. *Trans. Nonferrous Met. Soc. China* **2016**, *26*, 2209–2219. [CrossRef]
23. Shu, R.H.; Yin, T.B.; Li, X.B.; Yin, Z.Q.; Tang, L.Z. Effect of heat-treatment on energy dissipation of granite under cyclic impact loading. *Trans. Nonferrous Met. Soc. China* **2019**, *29*, 158–169. [CrossRef]
24. Shi, L.; Xu, J. Effect of Strain Rate on the Dynamic Compressive Mechanical Behaviors of Rock Material Subjected to High Temperatures. *Mech. Mater.* **2015**, *82*, 28–38.
25. Yang, S.Q.; Yin, P.F.; Huang, Y.H.; Cheng, J.L. Strength, Deformability and X-ray Micro-ct Observations of Transversely Isotropic Composite Rock Under Different Confining Pressures. *Eng. Fract. Mech.* **2019**, *214*, 1–20. [CrossRef]
26. Kasyap, S.S.; Li, S.; Senetakis, K. Investigation of the Mechanical Properties and the Influence of Micro-structural Characteristics of Aggregates Using Micro-indentation and Weibull Analysis. *Constr. Build. Mater.* **2020**, *271*, 121509. [CrossRef]
27. He, A.; Wang, Z.; Bi, C. Experimental study on thermal damage characteristics and mechanism of Huashan granite. *Hydro-Sci. Eng.* **2018**, *1*, 95–101.
28. Krajcinovic, D.; Silva, M.A.G. Statistical Aspects of the Continuous Damage Theory—Science direct. *Int. J. Solids Struct.* **1982**, *18*, 551–562. [CrossRef]
29. Yin, T.; Wang, P.; Yang, J.; Li, X. Mechanical Behaviors and Damage Constitutive Model of Thermally Treated Sandstone Under Impact Loading. *IEEE Access* **2018**, *6*, 72047–72062. [CrossRef]
30. Zhou, J.; Yang, X.; Li, H. Numerical Solution for Mixed Mode Crack Propagation in Brittle Solids Combined with Finite Element Method and Failure Criteria. *Int. J. Mater. Prod. Technol.* **2012**, *45*, 96–107. [CrossRef]
31. Shi, X.; Zhou, W.; Chen, L.; Cai, Q.; Li, M.; Li, Z.; Luan, B. A New Unified Solution for Circular Opening Considering Different Strength Criteria and the Postpeak Elastic Strain Form. *Adv. Civ. Eng.* **2020**, *2020*, 1–21. [CrossRef]
32. Peng, J.; Tang, Z.C.; Hou, D. A Gsi-softening Model for Characterizing Strength Behavior of Thermally-damaged Rock. *Eng. Geol.* **2021**, *292*, 106251. [CrossRef]

metals

Article

Structural Parameter Optimization for Large Spacing Sublevel Caving in Chengchao Iron Mine

Yuye Tan [1,2,*], Mochuan Guo [1,2,*], Yimin Hao [1,2], Chi Zhang [1,2] and Weidong Song [1,2]

[1] School of Civil and Resource Engineering, University of Science and Technology Beijing, Beijing 100083, China; luobiha435@163.com (Y.H.); zhangchi9531@163.com (C.Z.); songwd@ustb.edu.cn (W.S.)

[2] Key Laboratory of Efficient Mining and Safety of Metal Mines, Ministry of Education, Beijing 100083, China

* Correspondence: tanyuye@ustb.edu.cn (Y.T.); guomochuan@163.com (M.G.)

Abstract: Non-pillar sublevel caving is beginning to use large structural parameters in China. Appropriate structural parameters can effectively control the loss and dilution of stope and improve ore drawing efficiency. In this study, taking Chengchao Iron Mine as the engineering background, a theoretical calculation, a numerical simulation, and physical similarity experiments were combined to optimize sublevel height, production drift spacing, and drawing space. The optimal structural parameter range, based on the ellipsoid ore drawing theory, was obtained as a theoretical reference for subsequent studies. A "two-step" strategy was used, in which PFC2D software (Itasca Consulting Group, Minneapolis, MN, USA) was used to numerically simulate 20 groups of different sublevel heights and production drift spacing parameters were used to determine the appropriate sublevel height and production drift spacing for the project. Subsequently, the optimization of the ore drawing space was studied using PFC3D (Itasca Consulting Group, Minneapolis, MN, USA) particle unit software, numerical simulation analysis, and similar physical experiments. The results showed that safe and efficient mining can be achieved when the structural parameters of the stope are 17.5 m sublevel height, 20 m production drift spacing, and 6 m drawing space. The findings of this study can further the goal of green and efficient mining, and provide a theoretical reference for the popularization and application of pillarless sublevel caving with large structural parameters at home and abroad. It is an effective measure for the green mining of caving mines.

Keywords: sublevel caving; numerical simulation; physical model; structural parameter; green mining

Citation: Tan, Y.; Guo, M.; Hao, Y.; Zhang, C.; Song, W. Structural Parameter Optimization for Large Spacing Sublevel Caving in Chengchao Iron Mine. *Metals* **2021**, *11*, 1619. https://doi.org/10.3390/met11101619

Academic Editor: Lijie Guo

Received: 14 September 2021
Accepted: 5 October 2021
Published: 12 October 2021

1. Introduction

Non-pillar sublevel caving offers the advantages of simple operation, high-intensity mining, high mechanization degree, safety and reliability, and relatively low mining costs [1–4]. Therefore, it has been widely used in ore mining at home and abroad [5]. The purpose of mining is to discharge ore under the overburden, leading to a large loss coefficient and dilution ratio [6,7]. Many scholars at home and abroad have conducted in-depth research on reducing loss and dilution. They have successively proposed various ore drawing theories, including ore drawing ellipsoidal and stochastic medium drawing theories [8]. These theories have been widely applied to guide the production of mines in sublevel caving [9].

The three important parameters used in sublevel caving are sublevel height, production drift spacing, and drawing space. Several researchers have studied the adjustment and optimization of mine parameters and mining management to control the recovery effect. Kvapil, R., et al. [10] optimized structural parameters through laboratory experiments on similar materials. Janelle, I. and Kvapil, R. [11] performed an experimental field study on a prototype size based on tracer recovery. David, J., et al. [12] conducted a numerical simulation experiment based on ore drawing theory. These studies focused on the flow principle of ore and rock particles [13]. For the optimization principle of structural parameters, the difference between recovery and dilution rates in the actual production of mines is used as the criterion for determining the optimal structural parameters. In particular, high recovery and low dilution rates indicate large differences and a good recovery effect [14].

Because the structural parameters of the sublevel caving mining method can directly affect the ore recovery ratio and recovery efficiency, and then affect the overall technical and economic indicators and benefits of the mine, many scholars use different methods to optimize the structural parameters of sublevel caving stope [15,16]. To determine the appropriate stope structural parameters for the sublevel caving method, Wu et al. [17] compared and analyzed nine different stope structural parameters by laboratory test and computer numerical simulation. Zhao et al. [18] obtained the optimal structural parameters of an iron ore stope through a single-ore drawing experiment and MATLAB software (MathWorks, Natick, MA, USA) fitting analysis, combined with industrial eccentricity. Using the three ore drawing methods, the no-dilution caving method, the low-dilution caving method, the and the current cutoff caving method under large structural parameters, similar simulation experiments and PFC2D (Particle Flow Code 2D) (Itasca Consulting Group, Minneapolis, MN, USA) are used to simulate ore recovery and rock mixing. The optimal caving step of the three ore drawing methods was obtained by Jin et al. [19]. An improved flow diversion drawing technology can reduce the ore loss rate and dilution ratio [20].

In this study, the exploitation of the Chengchao Iron Mine through sublevel caving without a sill pillar was used as the engineering background. The range of optimal structure parameters was calculated based on ellipsoid drawing theory and the intersection degree of the drawn-out orebody, and a "two-step" strategy was adopted. PFC2D software (Itasca Consulting Group, Minneapolis, MN, USA) simulated the sublevel caving with different sublevel heights and production drift spacing to determine the optimal solution while maintaining the existing drawing space. Next, PFC3D three-dimensional discrete element software (Itasca Consulting Group, Minneapolis, MN, USA) was used to adopt the finite difference method. Combined with the physical ore drawing similar simulation experiment [21,22] and theoretical calculation results with the experimental model ratio of 1:100, and considering the relationship between drawing step and ore loss and dilution, the optimal drawing step was determined by mutual verification from the numerical simulation, physical experiment, and theoretical analysis [23].

2. Theoretical Calculation of Optimal Structural Parameters

At present, the generally accepted criterion for determining the optimal structural parameters of sublevel caving without a sill pillar is that "the morphology of the caving body should be consistent with that of the drawn-out ore body". Jin et al. [24] modified this criterion, where the optimization of structural parameters refers to optimizing the spatial arrangement of the drawn-out ore body, and the degree of compaction is the best. Theoretically, the drawn-out ore body can be divided into two equivalent and optimal arrangements: high sublevel and large space. The arrangement of large space is advantageous [25] considering mines' current technical equipment conditions in China, as shown in Figure 1.

The relationship between sublevel height and production drift spacing is expressed as:

$$\frac{H}{B} = \frac{\sqrt{3}}{6}\frac{a}{b} \tag{1}$$

$$\frac{L}{H} = \frac{(1+\sin\theta)(c+\cos\theta)}{2a} \tag{2}$$

In this equation, H is the sublevel height, m; B is the production drift spacing, m; a is the long axis of the discharged ellipsoid, m; b is the short axis of the discharged ellipsoid, m; L is the step distance of the ore drawing, m; θ is the included angle (2°–4°, θ = 2°) between the long axis of the quasi-ellipsoid and the end wall; c is the short semiaxis of the quasi-ellipsoid in the direction of vertical access, m.

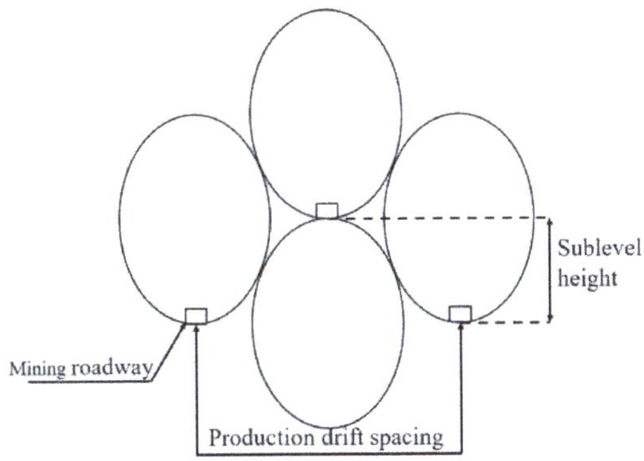

Figure 1. Arrangement of large space of discharged ellipsoid.

In actual ore drawing, the morphology change of the drawn-out ore body is dynamic, and its actual morphology is close to that of Figure 2. Chengchao Iron Mine adopts low dilution ore drawing of sublevel caving without a sill pillar. The optimal structural parameters [26] can be theoretically determined, including the sublevel height, production drift spacing, and the interval of caved ore, based on the geometric relation of the optimal spatial arrangement of the discharged ellipsoid.

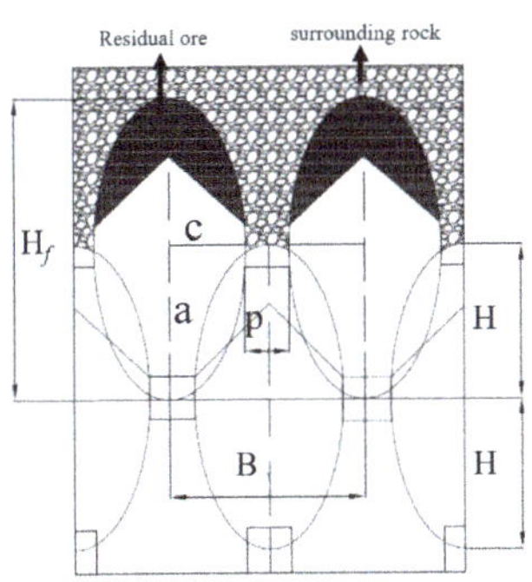

Figure 2. The ellipsoid arrangement is released in the direction of the vertical access.

According to the drawing theory, the drawn-out ore body is a quasi-ellipsoid, and its size can be determined by the a, b, and c axes of the ellipsoid. The ellipsoidal arrangement in Figure 2 allows as much ore to be recovered as possible and can be expressed in the vertical direction as follows:

$$B = 2c + p \tag{3}$$

$$H_f = 2H = 2a \tag{4}$$

In this equation, H_f is the ellipsoid discharge height, m; p is the access width, m.

The caving step L can be determined through geometric analysis, using the following equation:

$$L = b\cos\theta + a\sin\theta \tag{5}$$

The eccentricity of the ellipsoid is expressed as:

$$b = a\sqrt{1 - {\varepsilon_b}^2} \tag{6}$$

$$c = a\sqrt{1 - {\varepsilon_c}^2} \tag{7}$$

According to the actual situation of the mine, at the level of −500~−570 m, the sublevel height H is 17.5 m, that is, a = 17.5 m, and the access width p = 3.8 m, through the industrial experiment eccentricity, $\varepsilon_b = 0.95$, $\varepsilon_c = 0.90$. These values are substituted in Equations (6) and (7) to yield b = 5.46 m and c = 7.62 m. Next, the production drift spacing B_1 is 18.91 m by substituting the values of a and b into Equation (1). Production drift spacing B_2 is 19.04 m by substituting the values of b and c into Equation (3). Thus, the theoretical production drift spacing ranges from 18.91 m to 19.04 m. The interval of caved ore L_1 is 4.46 m by substituting a, c, and θ into Equation (2), and the interval of caved ore L_2 is 6.07 m by substituting a, b, and θ into Equation (5). Therefore, the interval of caved ore ranges is 4.46~6.07 m.

3. PFC Numerical Simulation Study

3.1. Optimization of Sublevel Height and Production Drift Spacing

3.1.1. Simulation Scheme

Sublevel caving without a sill pillar was adopted for the mining stage −500~−570 m in Chengchao Iron Mine. The ore body was about 1500 m long and 296° in the overall strike; the direction of dip was 214°, the amount of inclination was 0°–45°, and the average thickness was 72.95 m. The ore body mainly comprised marble, with good stability. The mine production scale was 3.4 million t/a, and the ore geological grade was 48%. The sublevel height was 17.5 m, and the production drift spacing was 15 m. Compared with the advanced level at home and abroad, the stope structure parameters can be optimized. Through PFC2D (Itasca Consulting Group, Minneapolis, MN, USA), the numerical simulation was carried out with sublevel heights of 17.5 m, 19 m, 21 m, 23.5 m, and 26 m, and production drift spacing of 15 m, 18 m, 20 m, and 22 m to study the influence of sublevel height and production drift spacing on ore recovery and dilution index.

In this study, the vertical production drift direction profile was selected to establish a calculation model. An ore drawing scheme, with a sublevel height of 21 m and production drift spacing of 20 m, was taken as an example, as shown in Figure 3a. The model's width was 60 m, its height was 126 m, the upper overburden was 42 m, the lower ore body was 84 m (four sublevels), and the edge angle of the middle and deep holes was 60°. The equivalent radius R of the ore unit in the model was 200 mm, and the equivalent radius R of the overburden unit was 300 mm, using a volume equivalent method. Gravity ore dropping was adopted to make the corresponding collapsed ore move under the action of gravity.

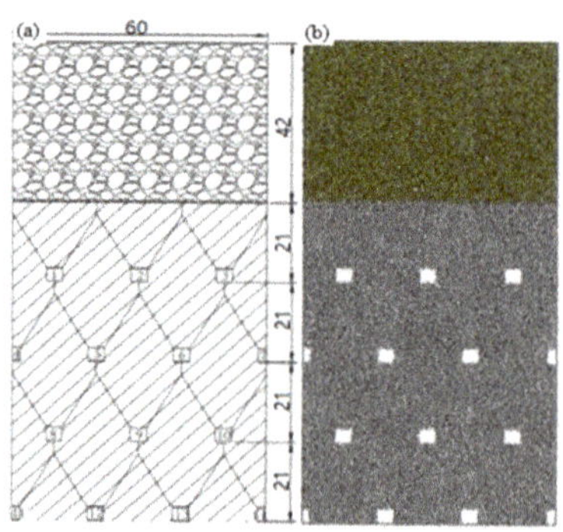

Figure 3. Schematic of the model: (**a**) schematic of model size, m; (**b**) 2D numerical model diagram.

A total of 41,331 balls are found in the model, where 35,093 red balls represent the ore, and 6238 yellow balls represent the waste rock. The particle unit diagram of the 2D numerical simulation of the ore drawing scheme is shown in Figure 3b.

PFC2D modeling requires debugging micromechanical parameters. The micromechanical parameters assumed by the model are first assigned to conduct numerical tests and then are matched with the macro test parameters obtained from laboratory tests to enable continuous debugging. These micromechanical parameters can be applied to the numerical calculation when the calculated results are consistent with the laboratory test results [26,27]. The size of the 3D ore drawing numerical model adopted in this study was consistent with the laboratory test, and the contact-stiffness model was adopted for the particle interaction model. The modeling parameters of this model mainly included ore and rock particle radius, normal stiffness, shear stiffness, friction coefficient, density, and color [28,29]. The micromechanical parameters in Table 1 are consistent with the macroscopic mechanical parameters after repeated debugging.

Table 1. Micromechanical parameters of the model.

Particle Type	Average Particle Size/m	Normal Stiffness/N·m^{-1}	Tangential Stiffness/N·m^{-1}	Friction Coefficient	Density/ kg·m^{-3}	Color
Ore	0.2	1.2×10^8	1.2×10^8	0.1	4000	Red
rock	0.3	1.0×10^8	1.0×10^8	0.2	2700	yellow

Before ore drawing, the cutoff condition of simulated ore drawing was determined based on the principle of equal ore volume dilution ratio. The specific calculation process of the ore drawing cutoff condition is expressed as follows:

$$C_i = \frac{C_k}{W} \times W_k + \frac{C_y}{W} \times W_y \tag{8}$$

$$W = W_k + W_y \tag{9}$$

$$W_y = 1.67W_k \tag{10}$$

In this equation, according to the actual situation of the mine, C_i is the cutoff ore grade, 18%; C_k is the ore geological grade, 48%; C_y is the surrounding rock grade, 0; W_k is the current ore weight; and W_y is the current rock weight.

The ore is mined by free falling. Equation (10) can be obtained from Equations (8) and (9), when the ratio of the waste rock mass to the ore mass reached 1.67, the ore was stopped. The drawing process with sectional height and route spacing of 19 m × 20 m was selected as an example, as shown in Figure 4.

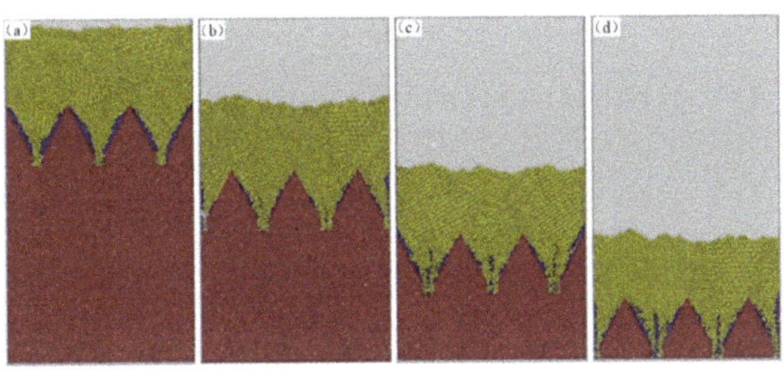

Figure 4. Numerical simulation of ore drawing process: (**a**) subparagraph 1; (**b**) subparagraph 2; (**c**) subparagraph 3; (**d**) subparagraph 4.

3.1.2. Analysis of Simulation Results

The ore drawing statistical results of 20 simulation schemes are shown in Table 2. From the statistical results of ore drawing in each sublevel, only the sublevel height × production drift spacing of 19 m × 20 m are presented as an example because of space limitations, as shown in Table 3.

Table 2. Statistical table of ore drawing results of each simulation scheme.

Analog Scheme	Segment Height/m	Approach Space/m	Difference between Recovery and Dilution Ratio/%
A 1	17.5	15	82.5
A 2		18	83.7
A 3		20	85.16
A 4		22	84.3
A 5	19	15	83
A 6		18	84.25
A 7		20	86.54
A 8		22	85.39
A 9	21	15	81.8
A 10		18	83.31
A 11		20	84.51
A 12		22	83.8
A 13	23.5	15	80
A 14		18	81.64
A 15		20	83.34
A 16		22	82.2
A 17	26	15	78.2
A 18		18	80.6
A 19		20	82.29
A 20		22	81.5

Table 3. Statistical data of 19 m × 20 m ore discharge results.

Discharge Section	Mine Release/ Mg	Total Release Ore /Mg	Total Release Ore /Mg	Recovery Ratio/%	Dilution Ratio/%
First segment	120.62	23.91	144.53	78.47	16.54
Second segment	225.56	25.91	251.47	105.61	10.30
Third segment	276.25	24.07	300.32	96.19	8.01
Fourth segment	206.12	23.81	229.93	97.18	10.35
Average	207.14	22.90	231.56	96.43	9.89

(1) Single-factor analysis

According to the numerical simulation drawing results, the single factor analysis of 20 schemes was carried out [30]. The difference between recovery and dilution ratio Y and sublevel height H and production drift spacing B is shown in Figure 5.

By analyzing the changing trend of the curve in Figure 5a, it can be seen that the difference between recovery and dilution ratio at each production drift spacing first increased and then decreased with the increase in sublevel height. The difference between recovery and dilution ratio when the production drift spacing was 20 m constantly remained the largest compared with other production drift spacings with the increase in sublevel height, indicating that the mining effect was optimal when the production drift spacing was 20 m, and the maximum value was 86.54% at the sublevel height of 19 m. The difference between recovery and dilution ratio increased from 85.16% to 86.54%, with a growth ratio of 1.62% and a small change when the production drift spacing was 20 m and the sublevel height increased from 17.5 m to 19 m, indicating that the values ranging from 17.5 m to 19 m were the appropriate sublevel heights.

By analyzing the changing trend of the curve in Figure 5b, it can be observed that the difference between recovery and dilution ratio at each sublevel height first increased and then decreased with the increase in production drift spacing. The difference between recovery and dilution ratio when the sublevel height was 19 m constantly remained the largest compared with other sublevel heights with the increase in production drift spacing, indicating that the stope had the best mining effect when the sublevel height was 19 m.

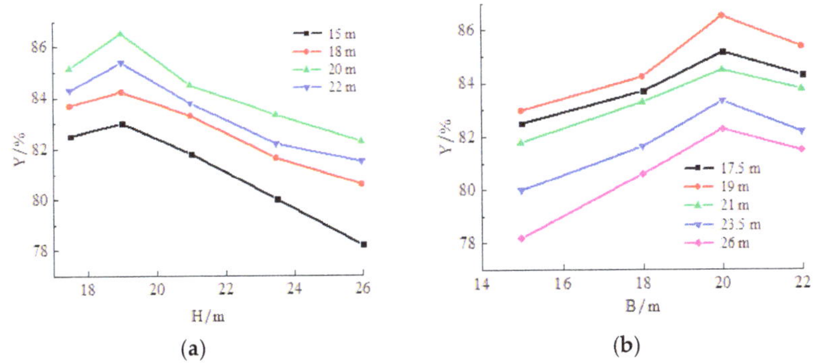

Figure 5. Relationship of the difference between recovery and dilution ratio Y and sublevel height H and production drift spacing B: (**a**) relationship curve of sublevel height H and the difference between recovery and dilution ratio Y; (**b**) relationship curve of production drift spacing B and the difference between recovery and dilution ratio Y.

(2) Comprehensive analysis

Table 2 demonstrates a quadratic polynomial with MATLAB (MathWorks, Natick, MA, USA), and the results are shown in Figure 6. The vertex of the surface in the figure was at a sublevel height of approximately 19 m, and the production drift spacing was approximately 20 m. The quadratic function relation of the difference between recovery and dilution ratio Y concerning sublevel height H and production drift spacing B can be obtained through fitting analysis, which can be expressed as:

$$Y = -0.13H^2 - 0.157B^2 + 5.594H + 6.517B - 0.018HB - 36.172(R^2 = 0.8756) \quad (11)$$

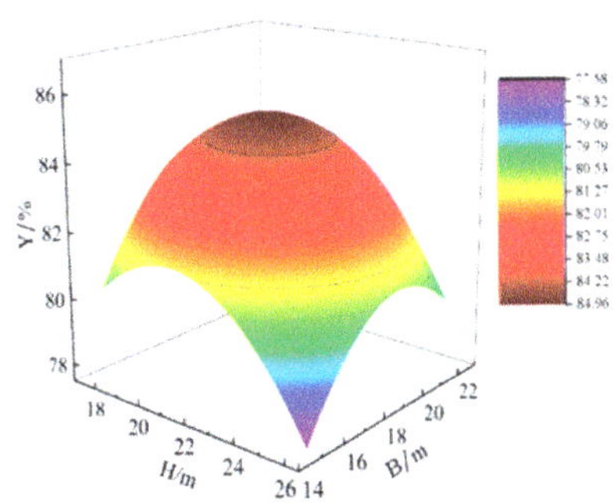

Figure 6. Relation curve of the difference between recovery and dilution ratio Y and sublevel height H and production drift spacing B.

In this fitting function, partial derivatives of H and B were calculated, and the maximum difference between recovery and dilution ratio was 84.15% when H = 20.18 m and B = 19.61 m. The results of theoretical calculation, numerical simulation, and function prediction are shown in Table 4.

Table 4. Statistical table of the range of structural parameters obtained by each method.

Bottom Structure	Theoretical Calculation	Numerical Simulation	Function Prediction
Segment height/m	17.5	19	20.18
Approach space/m	18.91~19.04	20	19.61

Combined with the occurrence of ore bodies (Figure 7) ranging from −500 m to −570 m, the changes in size and morphology of the ore body were large with the increase in sublevel height and production drift spacing. The edge of the ore body was difficult to control, resulting in a decrease in the ore recovery ratio and an increase in the dilution ratio. The sublevel height of 17.5 m in the development stage was defined as the mine design and was difficult to adjust. Thus, the sublevel height was kept constant at 17.5 m. As shown in Table 4, the optimal production drift spacing ranged from 18.91 m to 20 m. The crossing space of the −500 m level in the east of Chengchao Iron Mine is 60 m. The crossing space should be an integral multiple of the production drift spacing. The production drift spacing was set to 20 m based on the ellipsoid drawing theory [31] to ensure that the production drift spacing of the upper and lower sublevels formed a diamond layout. In particular, the optimal structure parameters of the stope were 17.5 m of sublevel height and 20 m of production drift spacing.

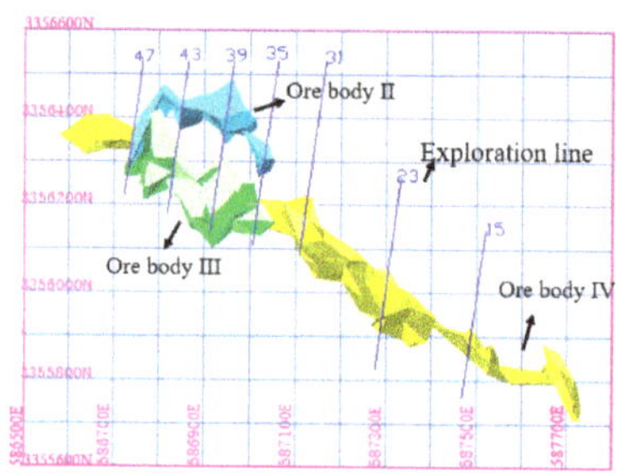

Figure 7. Top view of 3D visual model of No. II, III, and IV ore bodies from −500 m to −570 m.

3.2. Optimization of the Caving Step

3.2.1. Simulation Scheme

The interval of the caved ore of the stope was 3.0 m, and the caving step was approximately 4.0 m. For the simulation study, nine structural parameters, namely, 3.0, 3.5, 4.0, 4.5, 5.0, 5.5, 6.0, 6.5, and 7.0 m, were selected to determine the optimal caving step. The drawing model parameters are shown in Table 5, and the micromechanical parameters were the same as in Table 1.

Table 5. Macro model simulation parameter table.

Analog Number	Ore Height/m	Width/m	Caving Step/m	Current Cutoff Grade/%
B 1	33	20	3.0	18
B 2	33	20	3.5	18
B 3	33	20	4.0	18
B 4	33	20	4.5	18
B 5	33	20	5.0	18
B 6	33	20	5.5	18
B 7	33	20	6.0	18
B 8	33	20	6.5	18
B 9	33	20	7.0	18

Figure 8 shows the diagram of a single model of 17.5 m × 20 m. The actual drawing height was the sum of the sublevel and ridge heights formed through sublevel blasting. After the calculations, the actual drawing height was set to 33 m.

Figure 9 shows the initial diagram of the single model. The production drift direction is the x-axis, the vertical production drift is the y-axis, and the direction of the model height is the z-axis. The yellow particles are the overburden, the blue particles are the frontal waste rock, and the red particles are the ore. Under the action of the initial gravity field, the ore and rock flow from the lower ore discharge port to form recovery mining and stop when the cutoff grade is reached.

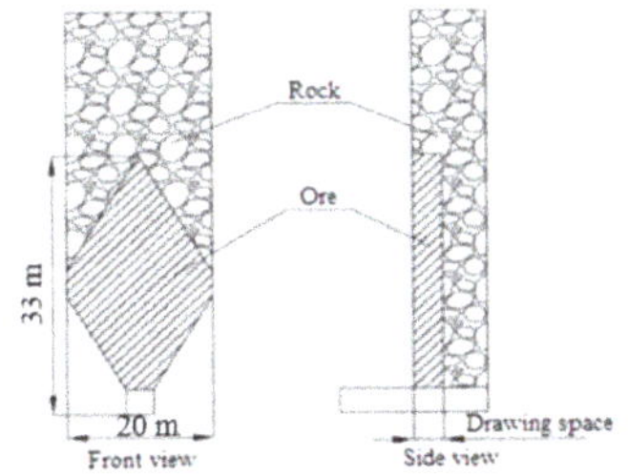

Figure 8. Single model design.

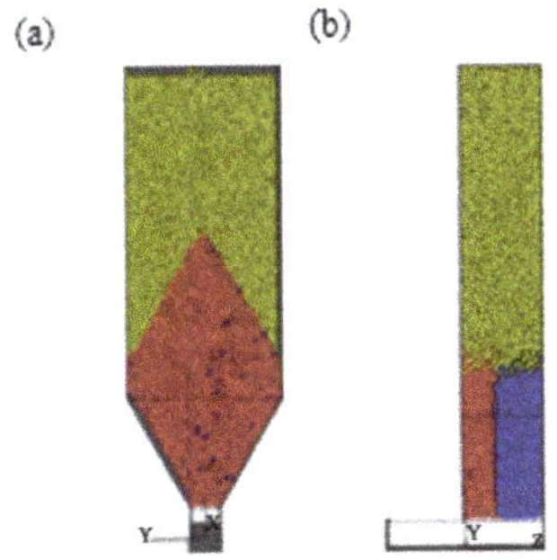

Figure 9. Initial diagram of the single model: (**a**) front view; (**b**) side view.

The cutoff grade of the 3D simulated ore drawing was the same as that of the 2D simulated ore drawing. Considering the model's limitations and shovel simulation, ore drawing in the three-dimensional single simulation experiment should be stopped when the mass ratio of waste rock and ore reaches 1.67 two consecutive times to ensure complete ore discharge.

3.2.2. Analysis of Simulation Results

Figure 10 shows the displacements of nine simulation schemes. The experimental results of each simulation scheme are shown in Table 6.

As shown in Figure 10a, the height of the drawn-out ore body increased with the increase in the caving step. The drawn-out ore body remained constant when the caving step reached 5.0 m, indicating that the height of the drawn-out ore body remained unchanged. The overall shape resembled an ellipsoid when the caving step increased to a certain degree.

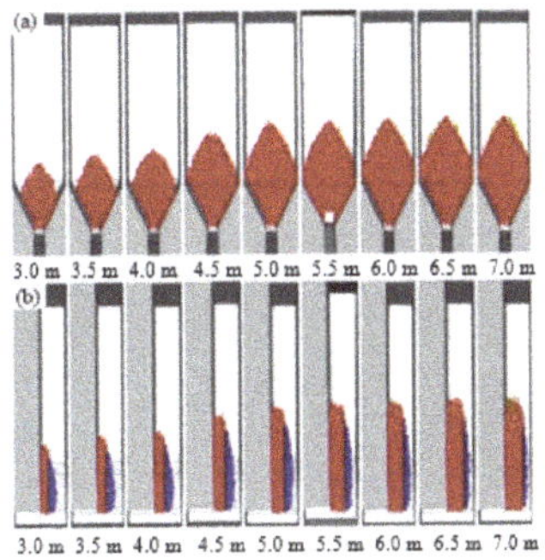

Figure 10. Schematic of the drawn-out ore body: (**a**) front view contrast of the drawn-out ore body; (**b**) side view contrast of the drawn-out ore body.

Table 6. Results of recovery indexes in the optimization simulation of caving step.

Analog Number	Caving Step/m	Recovery Ratio/%	Dilution Ratio/%	Difference between Recovery and Dilution Ratios/%
B 1	3	40.35	24.41	15.94
B 2	3.5	50.78	20.25	30.52
B 3	4	57.43	18.02	39.41
B 4	4.5	77.36	17.35	60.02
B 5	5	89.64	15.38	74.26
B 6	5.5	93.94	12.91	81.04
B 7	6	93.99	9.79	84.20
B 8	6.5	92.02	8.12	83.90
B 9	7	91.10	7.38	83.72

As shown in the analysis of Figure 10b, the thickness of the drawn-out ore body increased with the increase in the caving step. The thickness of red ore gradually increased, and the waste rock with the blue face slightly decreased. The body's morphology developed from "short and thick" to "long and thin" with the increase in the ore caving step.

As shown in Figure 10a,b, the morphology of the drawn-out ore body of each scheme resembled an ellipsoid, and the ore drawing process reflected the actual situation, indicating the rationality of the simulation study.

The ore recovery indexes, such as ore caving step, recovery ratio, dilution ratio, and the difference between recovery and dilution ratio, were obtained through sorting and summarizing. The data were fitted on MATLAB software (MathWorks, Natick, MA, USA), and the results are shown in Figures 11–13.

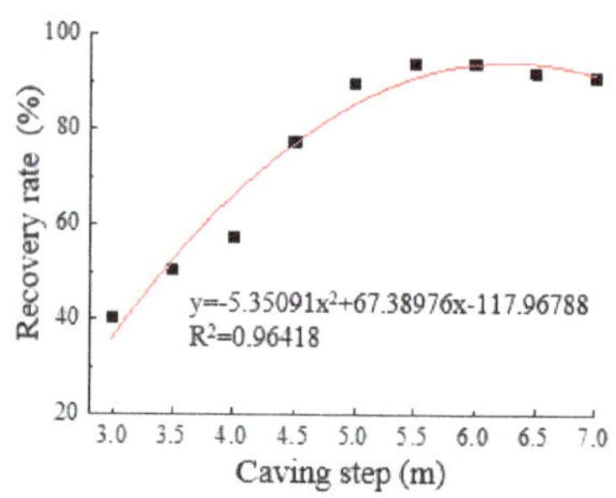

Figure 11. Relationship between ore recovery ratio and caving step.

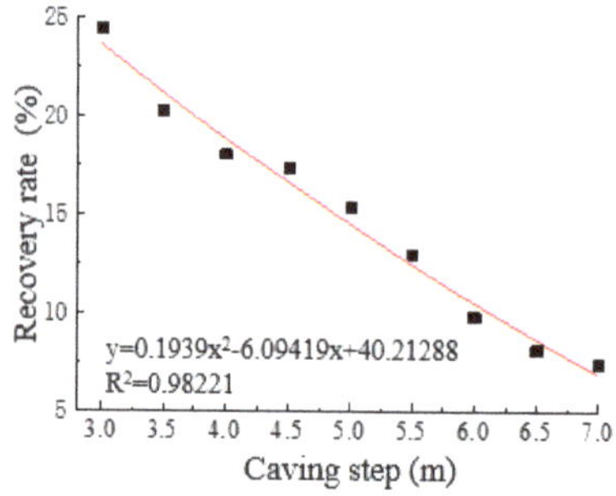

Figure 12. Relationship between ore dilution ratio and caving step.

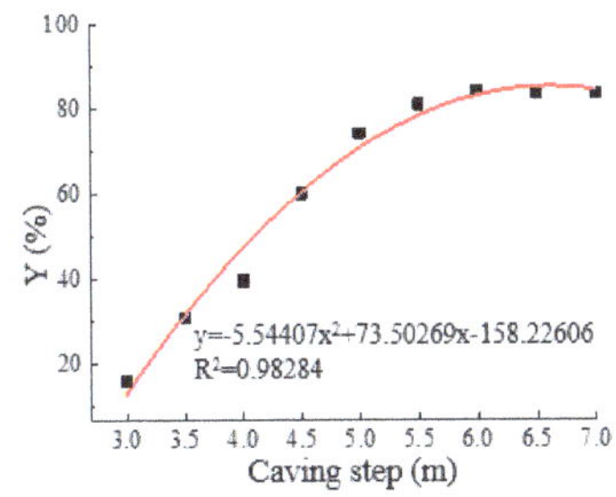

Figure 13. Relationship of the difference between recovery and dilution ratio Y and caving step.

(1) Recovery ratio

Figure 11 shows that the ore recovery ratio first increased and decreased with the increase in the caving step. The fitting curve results show that their relationship was quadratic, and the determination coefficient reached 0.96418. The recovery ratio was 40–60% when the caving step was 3–4 m, indicating that a large amount of waste rock in the front was mixed into the ore to intercept the normal outflow because of the small caving step, leading to premature dilution. The ore recovery ratio reached approximately 90% and remained unchanged when the ore drawing interval was 5–7 m. The recovery slightly declined when the caving step was 6 m, with an average reduction ratio of 1.55%. From the perspective of the fitting function, the recovery reached its maximum when the ore caving step was 6.3 m. Therefore, the recovery ratio showed that an ore caving step of 6–6.3 m was appropriate.

(2) Dilution ratio

As shown in Figure 12, the dilution ratio gradually decreased with the increase in caving step. The fitting curve results show that their relationship was quadratic, and the determination coefficient reached 0.98221. The dilution ratio was higher than 20% when the caving step was 3–3.5 m, indicating that a large amount of waste rock was mixed at this time, which would lead to an increase in mining production costs. The dilution ratio decreased to less than 10% when the caving step was 6–7 m, and the change was small. The average reduction ratio was 13.09%, indicating that the increase in waste rock mixing amount was less than that of the increase in total ore, leading to the decrease in dilution ratio. From the perspective of the fitting function, the dilution ratio was at its minimum value when the caving step was 15.71 m. Therefore, a caving step of 6–7 m was reasonable based on the dilution ratio.

(3) Difference between recovery and dilution ratio

The difference between recovery and dilution ratio is an important indicator for evaluating the effects of mining methods [32]. The greater the difference between the recovery and the dilution ratio, the better the economic effect. As shown in Figure 13, the difference between recovery and dilution ratio increased and decreased with the increase in the caving step. The fitting curve results show that their relationship was quadratic, and the determination coefficient reached 0.98284. The difference between recovery and dilution ratio gradually increased when the caving step was 3–6 m. The caving step was greater than 6 m, and the difference between recovery and dilution ratio showed a downward trend. From the perspective of fitting function, the difference between recovery and dilution ratio was at its maximum when the caving step was 6.63 m. Therefore, the appropriate range of the caving step was 6–6.63 m.

The appropriate range of the caving step of 6–6.3 m was obtained by considering the relationship between the recovery and the dilution ratio and the difference between the recovery ratio, the dilution ratio, and the caving step.

4. Experimental Study on Physical Ore Drawing

4.1. Experimental Scheme

In this experiment, the caving steps of 3.0, 4.0, 5.0, 6.0, and 7.0 m were taken under sublevel height and production drift spacing of 17.5 m × 20 m to conduct a simulation study of porous ore drawing. Appropriate suggestions were provided through thorough analysis and comparison of the experimental results.

The multi-segment three-dimensional ore drawing model was used in the experiment. According to the similarity theory, the stope structure parameters, ore and rock particles, and the total amount of ore and rock at the time of mining were simulated. The model size was length × width × height = 800 mm × 140 mm × 1100 mm. The experimental ore rock was Chengchao iron ore field ore rock. Compared with the field block, the particle size was ground at a 1:100 ratio, as shown in Figure 14a. The cross section of the drawing roadway was 38 mm × 36 mm (equivalent to 3.8 m × 3.6 m on site), and the roadway was arranged in a rhombic staggered arrangement. As shown in Figure 14b, the ore flowed downward by weight.

(**a**) Ore and waste rock particles (**b**) Physical drawing model framework

Figure 14. Physical ore drawing model. (**a**) Ore and waste rock particles; (**b**) Physical drawing model framework.

Each group of ore drawing test waste rock and ore was loaded at a 1:3 ratio, and three to four production drift roads were set for each sublevel. The ore drawing process is shown in Figure 15.

4.2. Results Analysis

As shown in Figure 15, the discharged bodies displayed a quasi-ellipsoid morphology and conformed to the ellipsoid theory, thereby confirming that this experiment was theoretically reasonable and feasible. The simulation experiment results of five ore caving steps of 3.0, 4.0, 5.0, 6.0, and 7.0 m were calculated under a sublevel height of 17.5 m and a production drift spacing of 20 m, as shown in Figure 16.

From Figure 16a,c, it can be seen that the variation trend of the recovery ratio and the difference between the recovery and the dilution ratio of the ore in each sublevel with different structural parameters were similar under the same ore drawing method. The residual bodies and recovery indexes in the discharged bodies gradually stabilized with the ore drawing sublevel. These findings indicate that each ore sublevel can be fully recovered under the existing structural parameters [33]. For the structural parameters of 17.5 m × 20 m × 5 m at sublevel II, the recovery ratio and the difference between recovery and dilution ratio were higher than the other structural parameters.

According to Figure 16b, the rock mixing ratio of each sublevel was significantly affected by the structural parameters under the same ore drawing method. Rock with structural parameters of 17.5 m × 20 m × 3 m had the highest mixing ratio. The actual caving step of the mine was approximately 3.5 m, indicating that the caving step of the stope should be increased at the same rate to optimize the recovery indexes.

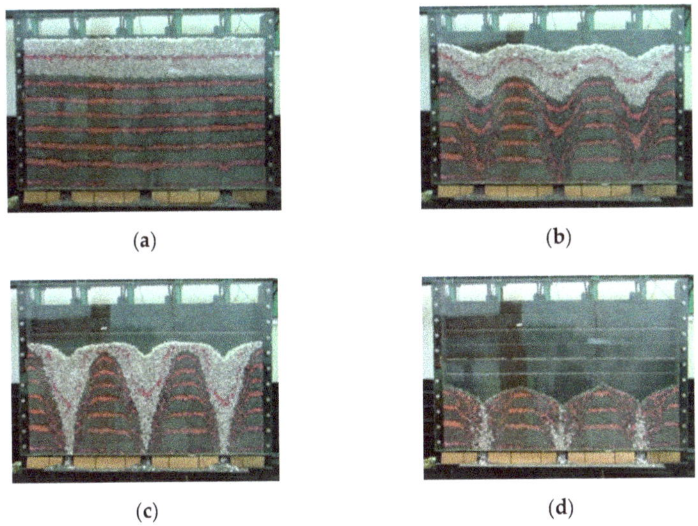

Figure 15. Drawing process diagram (**a**) before drawing, (**b**) at the initial drawing stage, (**c**) at the middle drawing stage, and (**d**) at the end of ore drawing.

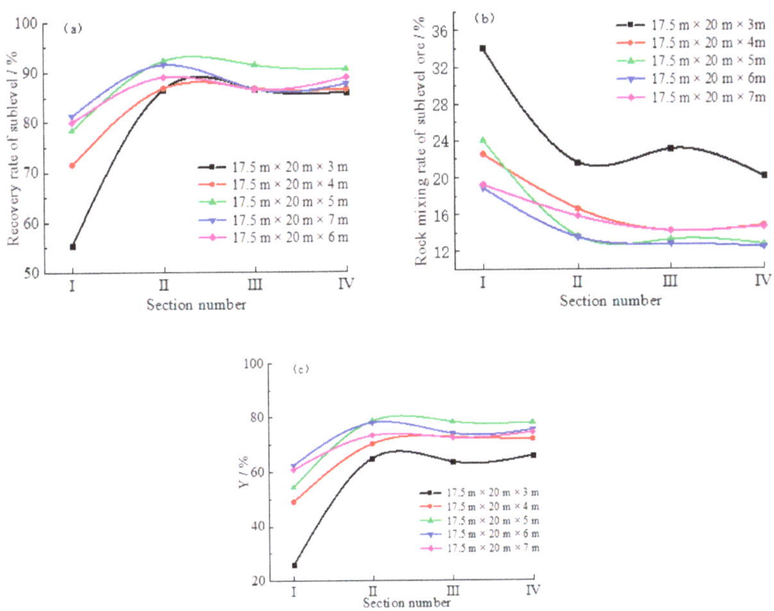

Figure 16. Curves of the recovery indexes of sublevel ore in each structural parameter scheme: (**a**) recovery ratio curve of sublevel ore; (**b**) rock mixing ratio curve of sublevel ore; (**c**) curve of the difference between recovery and dilution ratio of sublevel ore y.

Figure 17 presents the relationship between recovery indexes and ore interval drawing from an overall perspective without considering sublevels. The analysis of the curve showed that the dilution and ore recovery ratio first decreased and then increased. The change law of the difference between the recovery and the dilution ratio was similar to the recovery ratio when the caving step was 5.0 m. The maximum value of the difference between the recovery and the dilution ratio was obtained. Therefore, the caving step of 5.0 m (loose coefficient of 1.3, equivalent to 3.8 m or so of the interval of caved ore) and the recovery effect were optimal when the sublevel height and production drift spacing were 17.5 m × 20 m, based on the difference between the recovery and the dilution ratio.

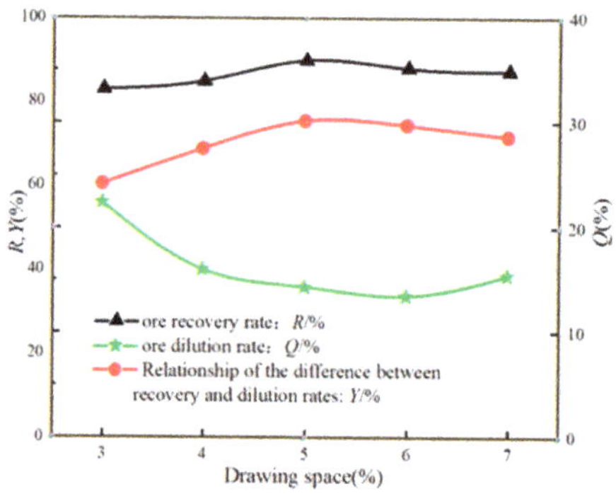

Figure 17. Relationship between recovery indexes and drawing space from an overall perspective.

4.3. *Results Comparison between Numerical Simulation and Physical Experiment*

PFC3D software (Itasca Consulting Group, Minneapolis, MN, USA) was used to conduct the numerical simulation research on nine drawing plans and the fitting analysis of each recovery index and ore drawing interval to determine the optimal caving step. A similar experiment of physical ore drawing in the laboratory was designed and combined with the theoretical calculation range of the caving step. The optimal caving step determined by each method is shown in Table 7.

Table 7. Various methods are used to obtain the optimal step range.

Bottom Structure	Theoretical Calculation	Numerical Simulation Research	Physical Similarity Simulation Experiment
Optimal caving step/m	4.46~6.07	6~6.3	5

Table 7 shows that the optimal caving step ranged from 6 m to 6.07 m through theoretical calculation and numerical simulation. The results only serve as a reference because similar physical experiments in practice inevitably involve human error. The optimal caving step, 6 m, was determined (extrusion blasting, loose coefficient of 1.3, equivalent to approximately 4.5 m of the interval of caved ore) by combining the actual effect of mine blasting.

5. Conclusions

In this paper, the sublevel caving method was studied through theoretical calculation, numerical simulation, and a laboratory test, and the following conclusions were drawn.

(1) The theoretical ranges of the optimal drift interval and caving step were calculated as 18.91~19.04 m and 4.46~6.07 m, respectively, based on the optimal arrangement and intersection degree of the discharged ellipsoid.
(2) Twenty groups of structural parameters were designed for simulation research. The binary quadratic function relation with sublevel height and production drift pace was fitted with the difference between the recovery and the dilution ratio as the objective function, on PFC2D software (Itasca Consulting Group, Minneapolis, MN, USA). The sublevel height and drift interval were 17.5 m × 20 m, based on the theoretical calculation results and the actual situation on site.

(3) The optimal caving step was investigated through a similar physical experiment and the theoretical calculation was performed on PFC3D software (Itasca Consulting Group, Minneapolis, MN, USA). The optimal caving step, 6 m, was determined through numerical simulation, physical experiment, and theoretical analysis.
(4) Through the optimization of bottom structure parameters, the loss and dilution of stope are effectively controlled, and energy consumption is significantly reduced. It is an effective measure for green mining and improves the production efficiency of the pillarless sublevel caving method.

Author Contributions: Conceptualization, Y.T. and M.G.; methodology, Y.T. and W.S.; software, M.G. and C.Z.; validation, Y.T., M.G. and Y.H.; formal analysis M.G.; investigation, Y.T.; resources, Y.T., data curation, M.G.; writing—original draft preparation, Y.T.; writing—review and editing, W.S.; visualization, Y.H. All authors have read and agreed to the published version of the manuscript.

Funding: This research was funded by the National Natural Science Foundation of China, grant number 52004019.

Institutional Review Board Statement: Not applicable.

Informed Consent Statement: Not applicable.

Data Availability Statement: Exclude this statement.

Acknowledgments: Financial support from the National Key R&D Program of China 2017YFC0602900 to the first author is gratefully acknowledged. The research is also partially supported by the Fundamental Research Fund for the Central Universities (FRF-TP-19-014A3). The authors would also like to thank Tan Y. Y. and Song W. D. for the discussions of the mining method.

Conflicts of Interest: The authors declare no conflict of interest.

References

1. An, L.; Xu, S.; Li, Y.H.; Peng, J. Optimization of rate of advance during ore breaking of caving method based on multi-method joint application. *Chin. J. Rock Mech. Eng.* **2013**, *32*, 754–759.
2. Liu, X.G.; Zhang, Z.T. Basis of Ore-drawing without Dilution for Sublevel Caving Method (First Part). *Metal Mine* **1995**, *11*, 5–9.
3. Wang, Y.P.; Yu, J. Optimization of breaking interval in non-pillar sublevel caving mining. *J. Cent. South Univ.* **2014**, *45*, 603–608.
4. Kaicheng, D.; Guihai, P.; Dan, H. Transition sublevel drawing optimization for large structural parameters of pillarless sublevel caving method. *China Min. Mag.* **2020**, *29*, 356–371.
5. Sun, G.; Lv, G. The development direction of no-pillar sublevel caving mining in China. *J. Inst. Technol.* **2007**, *29*, 4–6.
6. Brunton, I.D.; Fraser, S.J.; Hodgkinson, J.H.; Stewart, P.C. Parameters influencing full scale sublevel caving material recovery at the Ridgeway gold mine. *Int. J. Rock Mech. Min. Sci.* **2010**, *47*, 647–656. [CrossRef]
7. Jin, A.; Sun, H.; Wu, S.; Gao, Y. Confirmation of the upside-down drop shape theory in gravity flow and development of a new empirical equation to calculate the shape. *Int. J. Rock Mech. Min. Sci.* **2017**, *92*, 91–98. [CrossRef]
8. Yu, J. *Research on the Prediction of Ore Dilution and Loss Rate and the Optimization of Structural Parameters of High Sublevel Large Space Drive Interval Pillarless Sublevel Caving*; Center South University: Changsha, China, 2007.
9. Zhang, Z.G.; Liu, X.G.; Yu, G.L. *Bottom-Free Sublevel Caving Method without Depleted Ore Deposit: Non-Depleted Ore-Concentration Theory and Its Practice in Mines*; Northeastern University Press: Shenyang, China, 2007; pp. 6–7.
10. Kvapil, R. Gravity flow of granular materials in hoppers and bins. *Int. J. Rock Mech. Min. Sci.* **1965**, *2*, 35–41. [CrossRef]
11. Janelid, I.; Kvapil, R. Sublevel caving. *Int. J. Rock Mech. Min. Sci.* **1966**, *3*, 129–153. [CrossRef]
12. David, J. Computer simulation of the movement of ore and waste inan underground mining pillar. *Can. Min. Metall. Bull.* **1968**, *61*, 854–859.
13. Xu, S.; An, L.; Feng, X.T.; Dong, J.K.; Li, Y.H. Research on granular flow laws of caved ore and rock for steeply dipping thin vein. *J. Min. Saf. Eng.* **2013**, *30*, 512–517.
14. Xu, S.; An, L.; Li, Y.H.; Dong, J.K.; Li, Y.H. Optimization of caving space for different angles of end-wall during pillarless sublevel caving. *J. Northeast. Univ. Nat. Sci.* **2012**, *33*, 120–123.
15. Li, X.; Sheng, J.; Cheng, A.; Huang, Z. Study on structural parameter for non-pillar sublevel caving. *Ind. Miner. Process.* **2011**, *40*, 21–23.
16. Guo, H.; He, Z.; Zhang, Z.; Su, H.; Zhu, Q. Influence analysis of stope structure parameters on the drift stability in non-pillar sublevel caving under backfill. *Min. Res. Dev.* **2020**, *40*, 12–18.
17. Wu, A.; Wu, L.; Liu, X.; Sun, X.; Zhou, Y.; Yin, S. Study on structural parameters of sublevel caving. *J. Cent. South Univ.* **2012**, *43*, 1845–1850.

18. Zhao, Y.; Chen, Y.; Li, C.; Chen, J. Research on structural parameter optimization of sub level caving method. *Ind. Miner. Process.* **2019**, *48*, 16–21.
19. Jin, A.; Sun, H.; Meng, X.; Gao, Y.; Wu, Q.; Zhang, G. Study of best caving steps under different ore methods of sublevel caving. *J. Cent. South Univ.* **2017**, *48*, 3037–3043.
20. Tao, G.; Lu, M.; Zhang, X.; Zhang, R.; Zhu, Z. A new diversion drawing technique for controlling ore loss and dilution during longitudinal sublevel caving. *Int. J. Rock Mech. Min. Sci.* **2019**, *113*, 163–171. [CrossRef]
21. Li, T.; Fu, J.X.; Song, W.D. Study on law of surrounding rock movement induced by mining large and thick ore body. *J. Min. Saf. Eng.* **2018**, *35*, 978–983.
22. Fu, J.X.; Du, J.H.; Tan, Y.Y. The falling process and mechanism of concealed gob roof during the caving mining of the gently inclined heavy ore. *J. Min. Saf. Eng.* **2017**, *34*, 891–898.
23. Meng, Q.B.; Wang, C.K.; Han, L.J. Analysis on stope surrounding rock stability and mining entrance repair technology in sublevel caving method. *J. Min. Saf. Eng.* **2019**, *36*, 364–372.
24. Jin, C.; Dong, Z.M.; Gong, S.G. Study on the enlarged-structure parameters of sublevel caving in Meishan Iron Mine. *Metal Mine* **2000**, *4*, 17–19.
25. Huang, Z.; Sheng, J.L.; Li, X. Parametric analysis of high-segment and large-pitch structural parameters of sublevel caving without pillars. *Min. Technol.* **2011**, *11*, 11–12.
26. Zhu, H.C. PFC and application case of caving study. *Chin. J. Rock Mech. Eng.* **2006**, *25*, 1927–1931.
27. Zhou, C.B.; Yao, Y.K.; Guo, L.W.; Yin, X.P.; Fan, X.F.; Shang, Y. Numerical simulation of independent advance of ore breaking in the non-pillar sublevel caving method. *J. China Univ. Min. Technol.* **2007**, *17*, 295–300. [CrossRef]
28. Itasca Consulting Group Inc. *Particle Flow Code (Version 3.0)*; ICG: Minneapolis, MN, USA, 2004; pp. 22–26.
29. Brady, B.H.G.; Brown, E.T. *Rock Mechanics for Underground Mining*; Allen & Unwin: London, France, 1985; pp. 334–339.
30. Zhao, Y.L.; Cheng, Y.M.; Li, C.; Chen, J. Structural parameters optimization of sublevel caving method without pillars based on PFC3D. *Nonferrous Met. (Min. Sect.)* **2019**, *71*, 10–16.
31. Wang, C.H. *Theory of Ore Flow*; Metallurgical Industry Press: Beijing, China, 1982; pp. 5–10.
32. Zhang, H.F.; Chi, X.W. Application of particle flow method in structural parameter optimization of sublevel caving method without bottom column. *Min. Technol.* **2018**, *18*, 19–22.
33. Tan, B.H.; Zhang, Z.G.; He, R.X.; Zhu, Q. Discussion on the rationality and experimental research of the ore-drawing ellipsoid arrangement theory. *J. Northeast. Univ. Nat. Sci.* **2019**, *40*, 1014–1019.

metals

MDPI

Article

Enhancing Arsenic Solidification/Stabilisation Efficiency of Metallurgical Slag-Based Green Mining Fill and Its Structure Analysis

Wei Gao [1,2,3], Zifu Li [1,2,3], Siqi Zhang [2,3,4,*], Yuying Zhang [2,3,4], Pingfeng Fu [2,3,4], Huifen Yang [2,3,4] and Wen Ni [2,3,4,*]

1 School of Energy and Environmental Engineering, University of Science and Technology Beijing, Beijing 100083, China; 15201455635@163.com (W.G.); zifulee@aliyun.com (Z.L.)
2 Key Laboratory of High-Efficient Mining and Safety of Metal Mines, Ministry of Education, Beijing 100083, China; yuying.zhang@polyu.edu.hk (Y.Z.); pffu@ces.ustb.edu.cn (P.F.); yanghf@ustb.edu.cn (H.Y.)
3 Key Laboratory of Resource-Oriented Treatment of Industrial Pollutants, Beijing 100083, China
4 School of Civil and Resource Engineering, University of Science and Technology Beijing, Beijing 100083, China
* Correspondence: zsq2017@ustb.edu.cn (S.Z.); niwen@ces.ustb.edu.cn (W.N.)

Abstract: To dispose of arsenic-containing tailings with low carbon and high efficiency, sodium sulphate (Na_2SO_4), sodium hydroxide (NaOH), calcium nitrate $Ca(NO_3)_2$ and calcium hydroxide $Ca(OH)_2$ were independently added to metallurgical slag-based binder (MSB) solidification/stabilisation (S/S)-treated tailings (MSTs) to enhance the MST arsenic S/S performance. Results showed that only $Ca(OH)_2$ could increase the unconfined compressive strength of MST from 16.3 to 20.49 MPa and decrease the leachate As concentration from 31 μg/L to below 10 μg/L. $Na_3AsO_4{\cdot}12H_2O$ and $NaAsO_2$ were used to prepare pure MSB paste for mechanism analysis. The results of microstructure analyses showed the high specific surface area and amorphous properties of calcium–sodium aluminosilicate hydrate facilitated the adsorption or solid-solution formation of As(V) and As(III). As(V) formed an inner-sphere complex in ettringite, whereas As(III) formed an outer-sphere complex, and the relatively larger size and charge of As(V) compared with SO_4^{2-} restrict substitution inside channels without affecting the ettringite structure under high loading of As(V). The added $Ca(OH)_2$ promoted the hydration reaction of MSBs and facilitated the formation of a Ca–As(V) precipitate with low solubility, from $Ca_4(OH)_2(AsO_4)_2{\cdot}4H_2O$ (K_{sp} = $10^{-27.49}$) to $Ca_5(AsO_4)_3(OH)$ (K_{sp} = $10^{-40.12}$). This work is beneficial for the application of cement-free MSB in the S/S process.

Keywords: metallurgical slag-based binders; solidification/stabilisation; As(III); As(V); calcium hydroxide

Citation: Gao, W.; Li, Z.; Zhang, S.; Zhang, Y.; Fu, P.; Yang, H.; Ni, W. Enhancing Arsenic Solidification/Stabilisation Efficiency of Metallurgical Slag-Based Green Mining Fill and Its Structure Analysis. *Metals* **2021**, *11*, 1389. https://doi.org/10.3390/met11091389

Academic Editor: Lijie Guo

Received: 8 August 2021
Accepted: 27 August 2021
Published: 1 September 2021

1. Introduction

Arsenic (As) is a toxic and carcinogenic element, which can cause severe effects on human health, such as skin, lungs, kidney and liver cancers, even at a low concentration [1]. The mining, processing and smelting industry of nonferrous metals is the most widespread source of As emissions in China, accounting for 61.8% [2]. Mining wastewater, smelting slag, dust and tailings are piled up in tailings ponds, which have potential environmental risks and occupy a large percentage of soil [3]. Heavy metal pollution from a large number of tailings poses a major threat to the environment [4,5]. The treatment of heavy metal pollution has received increasing attention since the implementation of the Soil Pollution Prevention and Control Law of the People's Republic of China in 2019 [6]. It is imperative to develop effective and economically viable technologies to reduce pollution from the nonferrous metal industry.

Solidification/stabilisation (S/S) technology has been used to dispose of metal(loid)-containing waste by forming a less soluble, mobile or toxic product through adsorption, co-precipitation and physical encapsulation [7]. Ordinary Portland cement (OPC) is a traditionally used binder in S/S processes, while the high carbon footprint of cement clinker manufacture is of concern [8,9]. To achieve an environment-friendly binder for the S/S of As-containing waste, many studies have been conducted to replace cement in part or whole using materials, such as red mud, blast furnace or metakaolin-assisted cement [1,10], cement-free clay-based binders [11], red mud-enhanced magnesium phosphate cement [12], pine sawdust biochar [13], coal fly ash [14], zero-valent iron or magnetic biochar [15]. However, there have been few reports on the use of cement-free metallurgical slag-based binders (MSB) [16,17], which consist of steel slag powder (SSP), ground-granulated blast furnace slag (GGBFS) and flue gas desulphurisation gypsum (FGDG). Steel slag is a by-product of a steel-making process, known as overburnt clinker because its mineral phase composition is similar to that of cement clinker [18]. In 2019, 149.45 million tonnes of it were produced in China; however, the usage rate is less than 40% [19]. A large accumulation of steel slag pressurises the environment [20]. GGBFS is a by-product of the steel-making industry and is used as a supplementary cementitious material. In the hydration process of MSB, the relatively high alkalinity of SSP facilitates the breakage of Si–O and Al–O bonds in the GGBFS vitreous structure to form SiO_4^{4-} and AlO_4^{5-} [21,22], which react with dissolved Ca^{2+} to form calcium silicoaluminate hydrate (C–A–S–H) gels [11,23]. The AlO_4^{5-} dissolved in solution transforms from four coordinated aluminium ions to six coordinated aluminium ions $[Al(OH)_6]^{3-}$ in alkaline environments, which reacted with the Ca^{2+} and SO_4^{2-} dissolved from gypsum to form ettringite [24].

In this study, As-containing tailings were chosen as the S/S objects, as other hazardous elements' leaching risk is very low after S/S treatment (Table S1). In order to dispose of As-containing tailings with low carbon and high efficiency, the MSB (60 wt% GGBFS, 30 wt% SSP, and 10 wt% FGDG) was chosen as the binder for the As-containing tailings, based on a previous study [25]. SO_4^{2-} (Na_2SO_4), OH^- (NaOH), Ca^{2+} ($Ca(NO_3)_2$ or Ca^{2+} and OH^- ($Ca(OH)_2$) was added to the MSB-S/S treated tailings (MST) to enhance the MST-As-S/S performance. Arsenic in tailing occurs in various states in the natural environment, and its main forms in regard to leaching and migration characteristics are soluble arsenite As(III) and arsenate As(V) compounds [26]. The As-S/S mechanism of MSB and the mechanism of efficiency improvement are both unclear; therefore, $Na_3AsO_4 \cdot 12H_2O$ and $NaAsO_2$ were used to study the S/S mechanisms through X-ray diffraction (XRD), Fourier-transform infrared spectroscopy (FTIR) and scanning electron microscope-energy dispersive spectrometer (SEM-EDS) analysis. This study will improve the engineering application prospects of MSB and reusability of MST.

2. Materials and Methods

2.1. As-Containing Tailings and Metallurgical Slag-Based Binders

As-containing tailings were collected from a closed lead-zinc tailings pond in Hechi Nandan, Guangxi (China). The main components of the tailings were quartz, calcite and fluorite, of which the D_{50} and D_{90} particle sizes were 42 and 188 μm, respectively (Figure S1). The total concentration of As in the tailings was 2098 mg/kg, as determined using an Agilent 7500a ICP-MS (Agilent Technologies, Santa Clara, CA, USA) after aqua regia digestion (1/3 concentrated HNO_3 and HCl (v/v)) based on Chinese standard method HJ 803-2016. The leaching concentration of As in the tailings after horizontal oscillation-leaching tests was 0.57 mg/L, based on the Chinese standard method HJ 557-2010 (Table 1). The MSB consisted of GGBFS, SSP and FGDG. The chemical components of raw materials are shown in Table 1, and the XRD spectra of the binders are shown in Figure S2. The analytical reagents, Na_2SO_4, NaOH, $Ca(NO_3)_2$ and $Ca(OH)_2$, were used as added additive for the binders, respectively. A polycarboxylic acid water-reducer (WR) was used to achieve an acceptable flowability of mortar during S/S treatment. The analytical reagents, $Na_3AsO_4 \cdot 12H_2O$ and $NaAsO_2$, were used to prepare the pure MSB paste to study the S/S mechanisms of As(III) and As(V).

Table 1. Raw material chemical components, blaine fineness, component concentrations and leaching concentration of As.

Inspected Item		Tailings	GGBFS	SSP	FGDG
Chemical composition Oxide (wt%)	SiO_2	53.23	33.32	17.05	2.03
	TiO_2	0.12	0.85	0.91	0.04
	Al_2O_3	5.07	15.43	5.73	0.78
	Fe_2O_3	1.65	1.01	22.33	0.48
	MnO	0.70	0.52	3.63	0.03
	MgO	1.17	10.78	9.01	1.04
	CaO	25.49	36.89	38.42	30.01
	Na_2O	-	0.49	0.16	0.06
	K_2O	1.38	0.38	0.09	0.15
	SO_3	1.12	-	-	44.97
	As_2O_3	0.18	-	-	-
	LOI	10.61	0.12	1.42	22.07
Blaine fineness (m^2/kg)	-	-	435	440	325
pH	-	7.64	11.78	11.96	7.68
Total As concentration (mg/kg)		2098	-	-	-
As leaching concentration (mg/L)		0.57	-	-	-

2.2. Cement-Free S/S Treatment

For the preparation of MST mortar samples, the different constituents of binders were mixed with tailings at specific ratios (Table 2). MSB with 60 wt% GGBFS, 30 wt% SSP and 10 wt% FGDG was chosen due to its high compressive strength, according to the findings of previous studies [25]. Four additives (Na_2SO_4, NaOH, $Ca(NO_3)_2$ and $Ca(OH)_2$) were incorporated into the MSB mass individually at 5 wt%, 10 wt% and 15 wt%. The binders-to-tailings (B/T), water-to-solids (W/S) and WR-to-solids (WR/S) mass ratios were maintained at 0.25, 0.2 and 0.008, respectively, in all MST samples. After the tailings and binders were mixed for 1 min using a standard cement mortar mixer, distilled water used to dissolve the WR was added, and the blended mortar was mixed for another 5 min, before being cast into steel moulds (40 mm × 40 mm × 160 mm) to prepare the MST samples. To prepare MSB paste samples, 5 wt% concentrations of the analytical reagent, $Na_3AsO_4 \cdot 12H_2O$ or $NaAsO_2$, was added (Table 2), which were then mixed with water until the mass ratio of water-to-binders (W/B) reached 0.35. The paste samples were stirred for 5 min and then cast into steel moulds (30 mm × 30 mm × 50 mm). All mortar and paste samples were cured in a moist cabinet at (40 ± 2) °C and (90 ± 1)% relative humidity (similar to the underground filling environment in Guangxi [27]) for 3 day, and then the samples were demoulded and placed under the same curing conditions until the appropriate degree of sample ageing was achieved.

2.3. S/S Performance and Spectroscopic/Microscopic Analysis

The S/S performance was evaluated in terms of the unconfined compressive strength (UCS) and As leachability of the MST samples. The UCS of the samples was assessed on the basis of the Chinese standard GB/T 17671-1999 method of testing cements. The determination of UCS on three replicates at each curing time, and the average value with standard deviations of less than 5% was reported. The leachability of MST was assessed using the Chinese standard method HJ 557-2010, wherein the ratio of leachant (deionised water) volume to solid specimen weight (crushed to size < 3mm) was maintained at 10:1 (L/kg). Leachate samples were oscillated horizontally at 110 ± 10 times/min for 8 h and then left to stand at room temperature for 16 h. The pH values were measured, and the leachate was filtered through a 0.45-μm polypropylene microfiltration membrane. Afterwards, the filtered leachate samples were acidified to below pH 2 using concentrated HNO_3 and then stored at 4 °C in the dark before As detection using a PerkinElmer Optima 8300 ICP-OES (Perkin Elmer, Cumberland, NJ, USA) or an Agilent 7500a ICP-MS (Agilent Technologies, Santa Clara, CA, USA), with all samples analysed in triplicate to ensure accuracy. Mean results for triplicate sample analysis (with standard deviations of less than 5%) and error bars, are presented in this study. The As curing rate was evaluated using Equation (1):

$$\text{Curing Rate} = (1 - C_{S/S}/C_T) \times 100\% \tag{1}$$

where $C_{s/s}$ is the As leaching concentration of MST, and μg/L; C_T is the As leaching concentration of untreated tailings.

Table 2. Mixture formulations of MST and MSB samples (wt%).

Notation	GGBFS	SSP	FGDG	Na_2SO_4	NaOH	$Ca(NO_3)_2$	$Ca(OH)_2$	$NaAsO_2$	$Na_3AsO_4 \cdot 12H_2O$	Tailing [1]
T	60	30	10	-	-	-	-	-	-	1:4
T-S1	57	28.5	9.5	5	-	-	-	-	-	1:4
T-S2	54	27	9	10	-	-	-	-	-	1:4
T-S3	51	25.5	8.5	15	-	-	-	-	-	1:4
T-H1	57	28.5	9.5	-	5	-	-	-	-	1:4
T-H2	54	27	9	-	10	-	-	-	-	1:4
T-H3	51	25.5	8.5	-	15	-	-	-	-	1:4
T-N1	57	28.5	9.5	-	-	5	-	-	-	1:4
T-N2	54	27	9	-	-	10	-	-	-	1:4
T-N3	51	25.5	8.5	-	-	15	-	-	-	1:4
T-C1	57	28.5	9.5	-	-	-	5	-	-	1:4
T-C2	54	27	9	-	-	-	10	-	-	1:4
T-C3	51	25.5	8.5	-	-	-	15	-	-	1:4
B	60	30	10	-	-	-	-	-	-	0
B(III)	57	28.5	9.5	-	-	-	-	5	-	0
B(V)	57	28.5	9.5	-	-	-	-	-	5	0
B-C3	51	25.5	8.5	-	-	-	15	-	-	0
B-C3(III)	48.45	24.225	8.075	-	-	-	14.25	5	-	0
B-C3(V)	48.45	24.225	8.075	-	-	-	14.25	-	5	0

[1] Binder-to-tailings mass ratio.

The hydration of pulverised and sieved paste samples was terminated using alcohol drenching at the specific testing age, with samples then dried at 50 °C under a vacuum for 24 h for further characterisation. Mineralogy was detected using a high-resolution powered Rigaku D/max-RB XRD with CuKα radiation, a voltage of 40 kV, current of 200 mA and 2θ scanning, ranging between 10° and 70°. The morphology and elemental mapping of the samples were analysed using SEM-EDS (JSM-6701F). Structural and chemical bond analyses were performed using GX Perkin-Elmer FTIR (Perkin Elmer, Cumberland, NJ, USA) between wavelengths of 400–4000 cm^{-1} and at a resolution of 0.125 cm^{-1}.

3. Results

3.1. The S/S Performance of MST Samples

The UCS results for MST samples (Figure 1a) demonstrate that MST combined with various additives performed differently. The 3, 7 and 28 day UCS results for T were 11, 13.8 and 16.3 MPa, respectively. An increase in concentrations of Na_2SO_4, NaOH and $Ca(NO_3)_2$ reduced the USC of MST at every curing time, especially for the addition of NaOH, which inhibits hydration reactions and reduced the USC of MST (the lowest 28 day USC (T-H3 3.7 MPa) was only 22.7% of T). The addition of $Ca(OH)_2$ enhanced mechanical properties (the highest 28 d UCS (T-C3 20.49 MPa) was 125.7% of T). The USC of all samples could meet the filling requirements (5 MPa, the black line in Figure 1a), except for T-H3.

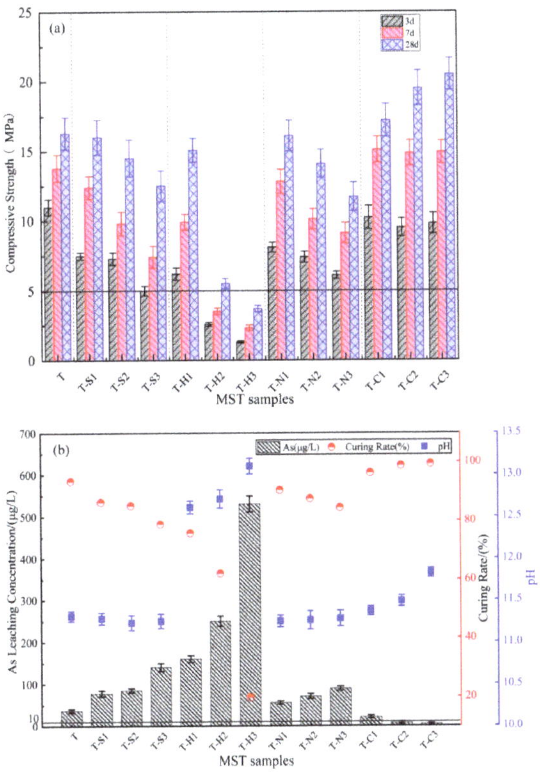

Figure 1. (**a**) UCS of MST samples at 3, 7 and 28 day; (**b**) As leachability of 28 day MST samples.

Figure 1b shows leachate As concentrations and the pH values of 28 day MST samples after the horizontal oscillation-leaching experiments. The leachate As concentration of T samples was 31 μg/L, which exceeded the prescribed As concentration limit value in the Chinese standard GB 5749-2006 for drinking water quality (10 μg/L, the black line in Figure 1b), indicating a relevant risk of environmental pollution. The addition of Na_2SO_4, NaOH or $Ca(NO_3)_2$ enhanced leachate As concentration, particularly that of T-H3, which reached 530 μg/L. The addition of $Ca(OH)_2$ reduced the leachate As concentration, particularly that of T-C2 and T-C3, which was less than 10 μg/L.

3.2. *Spectroscopic/Microscopic Analysis*

3.2.1. XRD Pattern Analysis

The XRD patterns of MSB paste hydrated for 3 and 28 day are presented in Figure 2a,b), respectively. The main mineral phases of the B sample, as shown in Figure 2a, are C-S-H gel, ettringite ($Ca_6Al_2(SO_4)_3(OH)_{12}\cdot 26H_2O$) and un-hydrated gypsum ($CaSO_4\cdot 2H_2O$), C_2F ($2CaO\cdot Fe_2O_3$), RO phase ($(MgO)_{0.239}(FeO)_{0.761}$), akermanite ($Ca_2MgSi_2O_7$) and C_2S ($2CaO\cdot SiO_2$) (Figure S2). With the addition of As(III) (sample B(III)), the peaks of ettringite decreased. However, when As(V) was added (B(V)), the peaks of ettringite disappeared due to the formation of a new phase, ($Ca_4Al_2SO_{10}\cdot 16H_2O$), and the peaks of Ca–As co-precipitation $Ca_4(OH)_2(AsO_4)_2\cdot 4H_2O$ appeared.

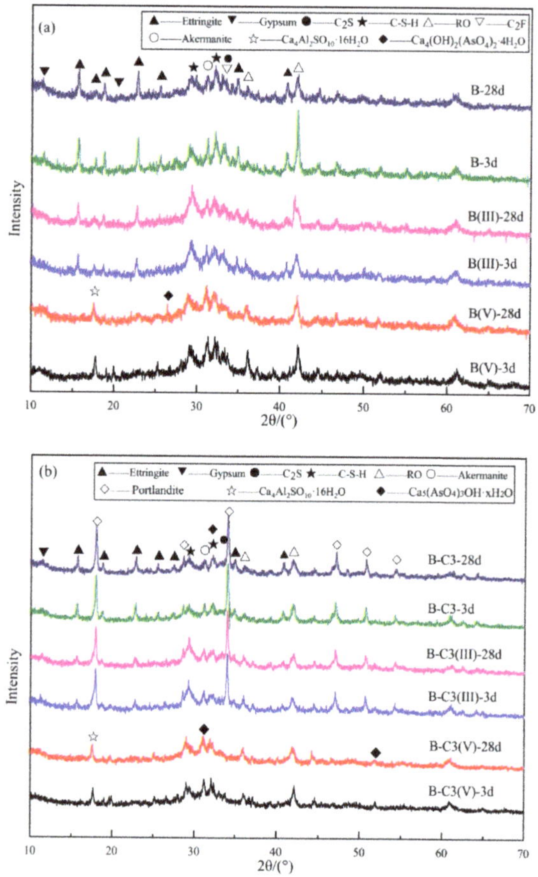

Figure 2. The XRD patterns of (**a**) B, B(III) and B(V) paste; (**b**) B-C3, B-C3(III) and B-C3(V) paste.

Compared with Figure 1a, the added $Ca(OH)_2$ characteristic peaks appeared in the XRD patterns of B-C3 and B-C3-(III); however, it did not appear in B-C3(V) (Figure 2b), and the peaks of Ca–As co-precipitation $Ca_5(AsO_4)_3OH{\cdot}xH_2O$ appeared.

3.2.2. FTIR Spectra Analysis

Figure 3a displays the FTIR spectra of the B, B(III) and B(V) samples. The spectra of the six samples are similar, presenting analogous adsorption bands. The pure B paste spectra (Figure 3a ① and ②) exhibit a small band at 3639 cm^{-1} associated with the O–H stretching vibration; bands at 3424 and 1646 cm^{-1} are related to the O–H stretching and bending modes of molecular interlayer water, respectively, and bands near 1450 (1478, 1428) cm^{-1} and at 876 cm^{-1} are related to the anti-symmetric stretching (v_3) and out-of-plane bending (v_2) modes of CO_3^{2-} ions, respectively. The bands at 973 cm^{-1} and near 450 (462, 422) cm^{-1} are associated with anti-symmetric Si–O(Al) stretching vibrations (v_3) and in-plane Si–O bending vibrations (v_2) in the SiO_4 tetrahedra of the C–(A)–S–H gel. The lack of a sharp band at 973 cm^{-1} indicates the wide distribution of SiQ^n(mAl) units. Moreover, the 3 day and 28 day infra-red spectra of B samples display large bands at 1114 cm^{-1} corresponding to the S–O_4 stretching mode (v_3), with a shoulder at 537 cm^{-1}, potentially due to out-of-plane Si–O bending vibrations (v_4).

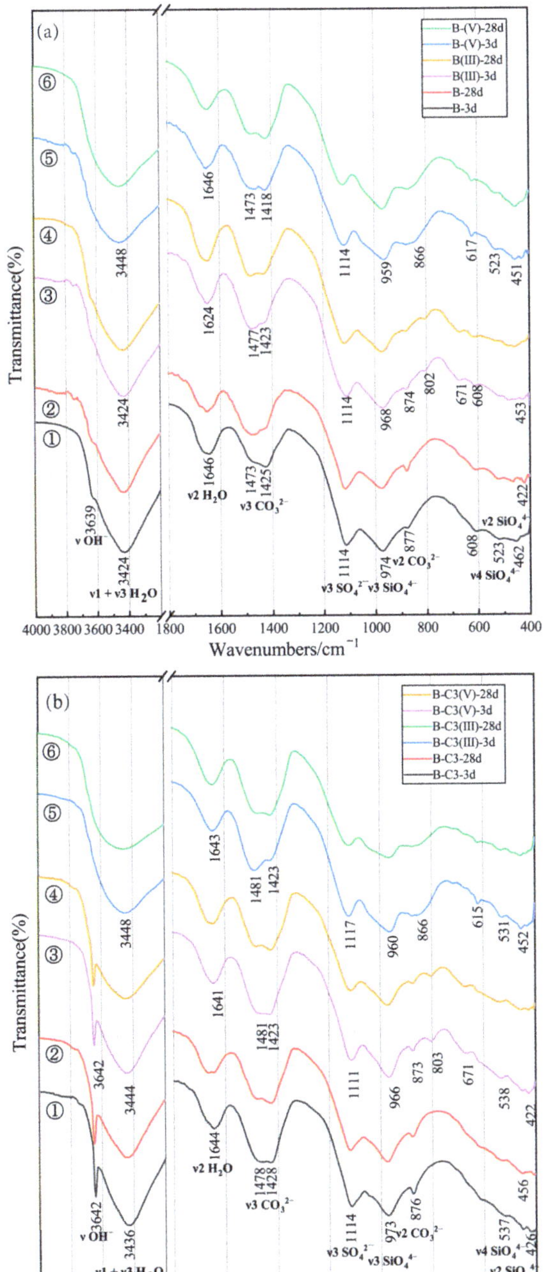

Figure 3. FTIR spectra of (**a**) ① and ② B sample, ③ and ④ B(III) sample and ⑤ and ⑥ B(V) sample; (**b**) ① and ② B-C3 sample, ③ and ④ B-C3(III) sample and ⑤ and ⑥ B-C3(V) sample.

Nevertheless, some differences were observed in the infrared spectra of B(III) and B(V) samples. The weak band at 3639 cm^{-1} in the B sample (Figure 3a ① and ②) disappeared in the B(V) (Figure 3a ⑤ and ⑥) samples, indicating that OH^- reacts with the added As(V), which is consistent with the result of XRD analysis (Figure 2a). The anti-symmetric Si–O(Al) stretching vibrations (v_3) of the C–(A)–S–H gel (973 cm^{-1} for B) decreased to 968 cm^{-1} for B(III) and 964 cm^{-1} for B(V), indicating that the added As(III) and As(V) were combined in the microstructure network and changed the degree of polymerisation of the C–(A)–S–H gel. In addition, the O–H stretching band (3424 cm^{-1} for B) increased to 3448 cm^{-1} and became broader for B(V) due to the absence of ettringite, leading to the change of interchannel molecular water, which combined with the result of XRD pattern (Figure 2a). Moreover, some new bands were observed, with the appearance of broadband at 802 and 671 cm^{-1} for B(III), and at 866 cm^{-1} for B(V) (Figure 3a ③–⑥), which was due to the As–O stretching vibration of the As(III) and As(V) species [28,29].

Figure 3b display the FTIR spectra of the B-C3, B-C3(III) and B-C3(V) samples, respectively. These spectra are all similar to those presented in Figure 3a, except that the spectra of the B-C3 and B-C3(III) samples exhibited sharp adsorption bands at 3642 cm^{-1} due to the addition of $Ca(OH)_2$. However, the sharp adsorption bands at 3642 cm^{-1} did not appear in B-C3(V), illustrating that $Ca(OH)_2$ reacted with As(V). Bands around 1110 cm^{-1}, corresponding to the S–O_4 stretching mode (v3) in Figure 3b ①–⑥, were offset at 1114 cm^{-1} in B-C3 sample, 1111 cm^{-1} in B-C3(III) and 1117 cm^{-1} in B-C3(V), indicating that, after As(III) and As(V) were added, the hydration products associated with SO_4^{2-} were affected, as shown by the XRD pattern (Figure 2b).

3.2.3. SEM-EDS Analysis

Figure 4 show SEM-EDS images for B, B(V) and B(III) at hydration ages of 3 and 28 day. Dense structures with amorphous gels and needle-like mineral phases are visible in the cement-free B paste in Figure 4a,b, which are calcium-sodium aluminosilicate hydrate (C–(N)–A–S–H) gels and ettringite, combined with the component analysis results in Table S2 (b-1, b-2 and b-3). After As(V) was added, the needle-like ettringite structure was undetected (Figure 4c,d), and a small amount of As(V) was found in the short columnar crystals of sodium sulphate, combined with the component analysis results (Table S2 (d-1)). Moreover, according to the SEM-EDS mapping images (Figure S3) of the red box in Figure 4d, the distribution of As(V) was almost similar to that of Ca, Si and Al, and the distribution of Na was similar to that of S. Therefore, the majority of the added As(V) entered the C–A–S–H gels through isomorphism substitution or was adsorbed and encapsulated by the gels, whereas the tiny minority of that entered the sodium sulphate to form sodium arsenic alunite. After As(III) was added, combined with the component analysis results (Table S2 (e-1,f-3)), different amounts of As(III) were detected in the C–A–S–H, ettringite, Na_2SO_4 and $Na_2Ca(SO_4)_2$ (Figure 4e,f).

Figure 5 show SEM-EDS images of B-C3, B-C3(V) and B-C3(III) at hydration ages of 3 and 28 day. In addition, amorphous C–A–S–H gels interwoven with needle-like ettringite structures were detected in the B-C3 paste. The images of B-C3(V) and B-C3(III) are similar to those for B(V) and B(III), except for the appearance of calcium hydroxide flakes in B-C3(V)-3d (Figure 5c) and calcium monosulfoaluminate hydrate (AFm) flakes in B-C3(III)-3d (Figure 5e), combined with the component analysis results in Table S3, which may be due to their content being lower than the detection limit of the XRD technique; both kinds of flakes disappeared as the hydration reaction proceeded.

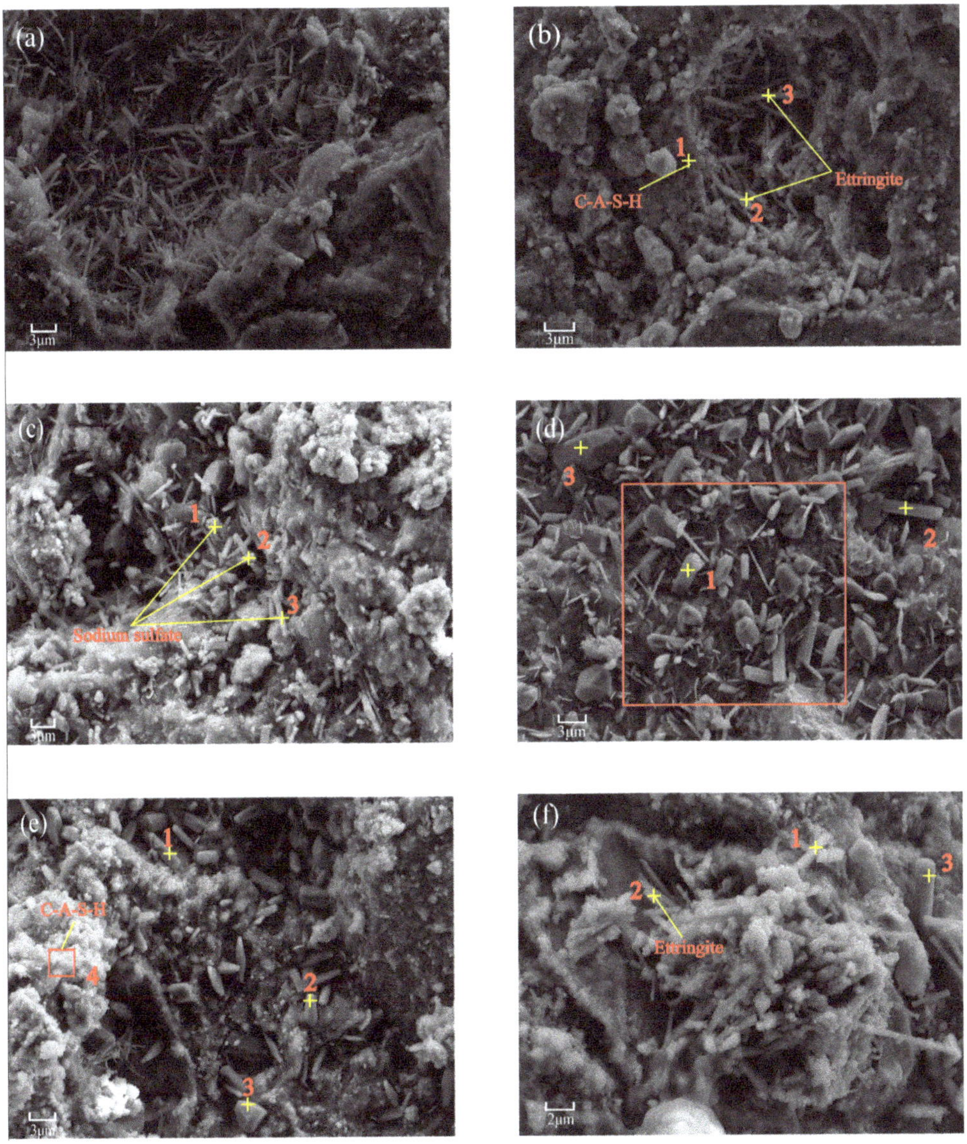

Figure 4. SEM-EDS result for (**a**) B-3d; (**b**) B-28 d; (**c**) B(V)-3 d; (**d**) B(V)-28 d; (**e**) B(III)-3 d; (**f**) B(III)-28 d.

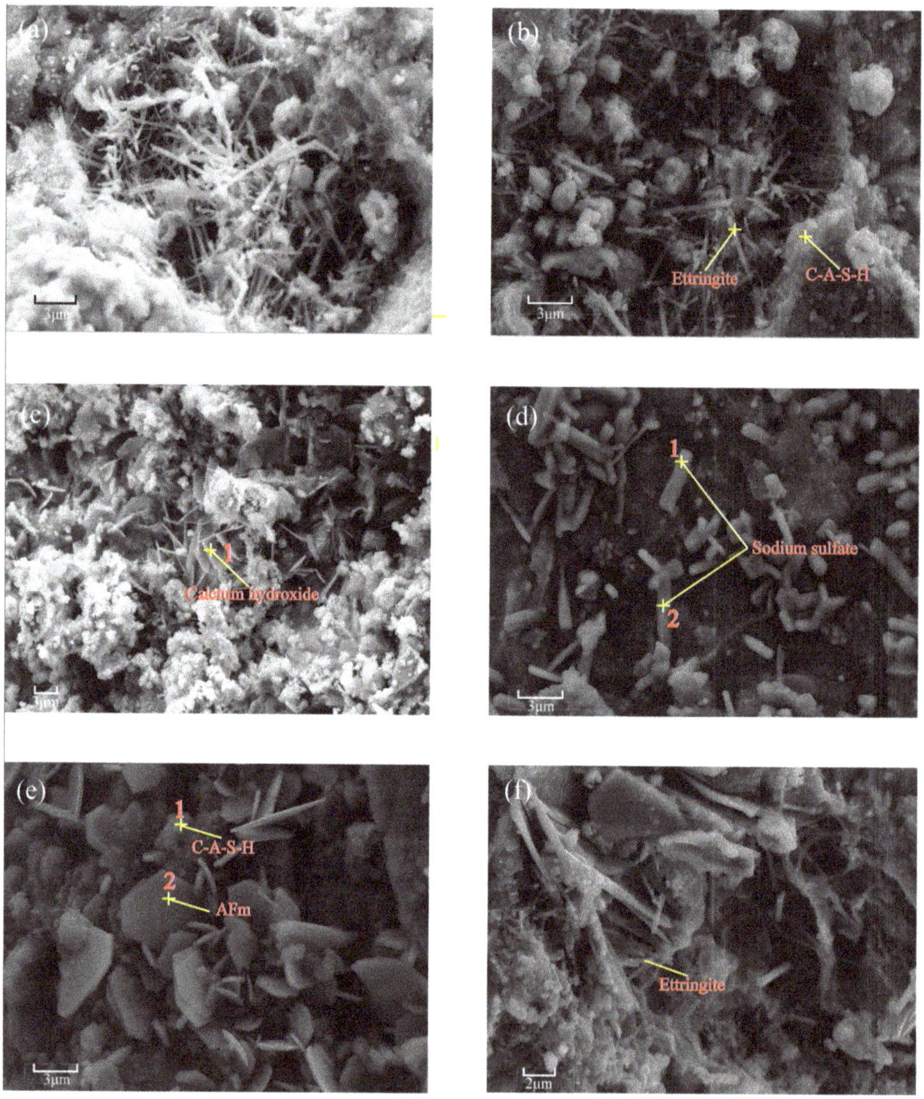

Figure 5. SEM-EDS result for (**a**) B-C3-3 d; (**b**) B-C3-28 d; (**c**) B-C3(V)-3 d; (**d**) B-C3(V)-28 d; (**e**) B-C3(III)-3 d; (**f**) B-C3(III)-28 d.

4. Discussion

This study demonstrates the different As S/S performance of MST samples after four additives (Na_2SO_4, NaOH, $Ca(NO_3)_2$ and $Ca(OH)_2$) were added, as shown in Table S4. The MSB is selective for additives and only $Ca(OH)_2$ exerted a positive effect on the As S/S performance. The addition of Na_2SO_4 had little effect on the pH values, UCS and As curing rate, probably because gypsum in the MSB system (10 wt% FGDG in B) was sufficient to provide sulphate for the hydration reaction. The addition of NaOH resulted in a pH value above 13, which inhibited the dissolution of Ca^{2+} and hindered the hydration reaction, leading to a sharp decrease in UCS [30]. The amphoteric nature of arsenic in an alkaline environment and repulsion by OH^- during the adsorption process of the hydration products led to the sharp decrease of the As curing rate [31,32]. Neither NaOH nor $Ca(NO_3)_2$ improved the As S/S performance, confirming that the individual addition of OH^- without Ca^{2+} would not improve As S/S efficiency and had a negative effect [33,34].

In addition, this study demonstrates that both the As(V) and As(III) S/S mechanism of MSB, and the effect of $Ca(OH)_2$ addition on As(V) and As(III) S/S are different. The addition of $Ca(OH)_2$ to MSB raised the pH of the reaction mixture (Figure 1), facilitated GGBFS dissolution and provided a source of Ca^{2+}, thus improving the As(V) and As(III) S/S related to the hydration products. Adsorption or solid solution formation of As(V) and As(III) occurred due to the high specific surface area and amorphous properties of C–A–S–H, based on the shift in anti-symmetric Si–O(Al) stretching vibrations (v3) in the discussion above (Figure 3) and SEM-EDS results (Figures 4 and 5).

Anion substitution in ettringite can occur via reaction with surface sites (ligand exchange), replacing Al/Ca coordinated surface OH^- or substituting for sulphate inside channels (isomorphic substitution) [35]. This occurs as the ettringite surfaces are negatively charged under alkaline pH range conditions and due to the column and channel-like structure of ettringite, which is composed of columns of $Ca_6[Al(OH)_6]_2{\cdot}24H_2O^{6+}$ and channels of $[(SO_4)_3{\cdot}2H_2O]^{6-}$ [36]. The extent of channel substitution of ettringite may be inversely proportional to the difference in size and electronegativity of the oxyanion compared with $SO_4{}^{2-}$ (0.29 Å), resulting in As(V) (0.47 Å) being likelier to be channel-substituted with sulphur than As(III) (0.69 Å) [37]. Thus, As(V) tended to form an inner-sphere complex in ettringite, whereas As(III) could form an outer-sphere complex [38], and the relatively larger size and charge of As(V) compared with $SO_4{}^{2-}$ restricts the substitution inside channels without affecting the ettringite structure under high loading of As(V), which is consistent with the XRD, IR and SEM-EDS analysis results.

Previous studies [39–41] have verified the formation of Ca–As precipitates under alkaline conditions. In the present study, the distinct Ca–As(III) mineral phase characteristic peak was not found in the XRD patterns, maybe because of the content being below the detection limit of XRD analysis or mainly existing as an amorphous or more complex structure. When $Ca(OH)_2$ was added, Ca–As(V) changed from $Ca_4(OH)_2(AsO_4)_2{\cdot}4H_2O$ ($K_{sp} = 10^{-27.49}$) to $Ca_5(AsO_4)_3(OH)$ ($K_{sp} = 10^{-40.12}$) (Figure 2), which restricted As leaching [42].

Based on the findings of this study, alkaline solid waste with high calcium ion equilibrium concentrations and high alkalinity can be used to replace $Ca(OH)_2$ added to MSB for the S/S of high As-containing solid wastes for low carbon and environmental protection.

5. Conclusions

In this study, the effects of four additives (Na_2SO_4, NaOH, $Ca(NO_3)_2$ or $Ca(OH)_2$) on the As S/S performance of MST to enhance the As S/S efficiency were investigated. As-containing analytical reagents were used to prepare pure MSB paste to study the S/S mechanisms of As(III) and As(V) and the mechanism of efficiency improvement using $Ca(OH)_2$, based on XRD, IR and SEM-EDS analysis. Based on the results obtained from this study, the following conclusions can be drawn:

- An MSB is selective for the additives, and only $Ca(OH)_2$ exerts a positive effect on the S/S performance of MST.
- As(V) can form an inner-sphere complex in ettringite, whereas As(III) can form an outer-sphere complex, and the relatively larger size and charge of As(V) compared with $SO_4{}^{2-}$ restrict its substitution inside channels without affecting the ettringite structure under high loading of As(V).
- The added $Ca(OH)_2$ facilitates the encapsulation and adsorption of As(V) and As(III) in MSB, and changes the Ca–As(V) from $Ca_4(OH)_2(AsO_4)_2{\cdot}4H_2O$ ($K_{sp} = 10^{-27.49}$) to $Ca_5(AsO_4)_3(OH)$ ($K_{sp} = 10^{-40.12}$).
- Cement-free MSB with $Ca(OH)_2$ added is feasible to dispose of high As-containing solid wastes with high efficiency and a low carbon footprint.

Supplementary Materials: The following are available online at https://www.mdpi.com/article/10.3390/met11091389/s1, Figure S1: (a) XRD pattern and (b) grain size results of tailings, Figure S2: The XRD pattern of raw materials: (a) ground granulated blast furnace slag (GGBFS), C_2S: dicalcium silicate; (b) steel slag powder (SSP), C_3A: tricalcium aluminate, RO phase: $(MgO)_{0.239}(FeO)_{0.761}$, C_2S: dicalcium silicate, C_2F: dicalcium ferrite; (c) flue gas desulfurisation gypsum (FGDG), Figure S3: The SEM-EDS electron image and mapping images of Figure 4 (d), Table S1: Leached concentration of raw materials and T in Table 2, Table S2: Component analysis result of point in the Figure 4, Table S3: Component analysis result of point in the Figure 5, Table S4: Assessment summary of the S/S performance effects of Na_2SO_4, NaOH, $Ca(NO_3)_2$ and $Ca(OH)_2$ at different doses (wt%).

Author Contributions: Formal analysis, W.G.; funding acquisition, P.F. and H.Y.; investigation, W.G. and Y.Z.; methodology, W.G.; project administration, S.Z.; resources, S.Z.; supervision, S.Z.; validation, W.G., Z.L., S.Z. and W.N.; writing—original draft, W.G.; writing—review and editing, Z.L., S.Z. and W.N. All authors have read and agreed to the published version of the manuscript.

Funding: This work was funded by the National Key Research and Development Program of China, grant number 2020YFC1807804 and the National Key Research and Development Program of China, grant number 2019YFC1803503.

Institutional Review Board Statement: Not applicable.

Informed Consent Statement: Not applicable.

Data Availability Statement: Not applicable.

Acknowledgments: Special thanks are addressed to Xuming Ma and Qihui Yan for their support with the interpretation of XRD results.

Conflicts of Interest: The authors declare no conflict of interest.

References

1. Wang, L.; Chen, L.; Tsang, D.C.; Zhou, Y.; Rinklebe, J.; Song, H.; Kwon, E.E.; Baek, K.; Ok, Y.S. Mechanistic insights into red mud, blast furnace slag, or metakaolin-assisted stabilization/solidification of arsenic-contaminated sediment. *Environ. Int.* **2019**, *133*, 105247. [CrossRef]
2. Ministry of Ecology and Environment of the People's Republic of China. Annual Report on Environmental Statistics. 2017. Available online: http://www.mee.gov.cn/hjzl/sthjzk/sthjtjnb/ (accessed on 8 August 2021).
3. Cheng, X.; Qi, W.; Huang, Q.; Zhao, X.; Fang, R.; Xu, J. Typical Geo-Hazards and Countermeasures of Mines in Yunnan Province, Southwest China. *IOP Conf. Ser. Earth Environ. Sci.* **2016**, *44*, 022008. [CrossRef]
4. Xiao, R.; Wang, S.; Li, R.; Wang, J.; Zhang, Z. Soil heavy metal contamination and health risks associated with artisanal gold mining in Tongguan, Shaanxi, China. *Ecotoxicol. Environ. Saf.* **2017**, *141*, 17–24. [CrossRef] [PubMed]
5. Kan, X.; Dong, Y.; Feng, L.; Zhou, M.; Hou, H. Contamination and health risk assessment of heavy metals in China's lead–zinc mine tailings: A meta–analysis. *Chemosphere* **2021**, *267*, 128909. [CrossRef] [PubMed]
6. Wang, Y. Status and countermeasures on soil heavy metals pollution control in nonferrous metals industry in China. *Nonferrous Met.* **2021**, *3*, 1–9. [CrossRef]
7. Shi, C.; Spence, R. Designing of Cement-Based Formula for Solidification/Stabilization of Hazardous, Radioactive, and Mixed Wastes. *Crit. Rev. Environ. Sci. Technol.* **2004**, *34*, 391–417. [CrossRef]
8. Habert, G.; Miller, S.A.; John, V.M.; Provis, J.L.; Favier, A.; Horvath, A.; Scrivener, K.L. Environmental impacts and decarbonization strategies in the cement and concrete industries. *Nat. Rev. Earth Environ.* **2020**, *1*, 559–573. [CrossRef]
9. Andrew, R.M. Global CO_2 emissions from cement production. *Earth Syst. Sci. Data* **2018**, *10*, 195–217. [CrossRef]
10. Li, J.-S.; Chen, L.; Zhan, B.; Wang, L.; Poon, C.S.; Tsang, D.C. Sustainable stabilization/solidification of arsenic-containing soil by blast slag and cement blends. *Chemosphere* **2021**, *271*, 129868. [CrossRef]
11. Wang, L.; Cho, D.-W.; Tsang, D.C.; Cao, X.; Hou, D.; Shen, Z.; Alessi, D.; Ok, Y.S.; Poon, C.S. Green remediation of As and Pb contaminated soil using cement-free clay-based stabilization/solidification. *Environ. Int.* **2019**, *126*, 336–345. [CrossRef] [PubMed]
12. Wang, L.; Chen, L.; Guo, B.; Tsang, D.C.; Huang, L.; Ok, Y.S.; Mechtcherine, V. Red mud-enhanced magnesium phosphate cement for remediation of Pb and as contaminated soil. *J. Hazard. Mater.* **2020**, *400*, 123317. [CrossRef]
13. Beiyuan, J.; Awad, Y.M.; Beckers, F.; Tsang, D.; Ok, Y.S.; Rinklebe, J. Mobility and phytoavailability of As and Pb in a contaminated soil using pine sawdust biochar under systematic change of redox conditions. *Chemosphere* **2017**, *178*, 110–118. [CrossRef] [PubMed]
14. Tsang, D.C.W.; Yip, A.; Olds, W.E.; Weber, P. Arsenic and copper stabilisation in a contaminated soil by coal fly ash and green waste compost. *Environ. Sci. Pollut. Res.* **2014**, *21*, 10194–10204. [CrossRef]

15. Li, J.; Zhang, Y.; Wang, F.; Wang, L.; Liu, J.; Hashimoto, Y.; Hosomi, M. Arsenic immobilization and removal in contaminated soil using zero-valent iron or magnetic biochar amendment followed by dry magnetic separation. *Sci. Total Environ.* **2021**, *768*, 144521. [CrossRef]
16. Criado, M.; Ke, X.; Provis, J.; Bernal, S.A. Alternative inorganic binders based on alkali-activated metallurgical slags. In *Sustainable and Nonconventional Construction Materials using Inorganic Bonded Fiber Composites*; Elsevier: Amsterdam, The Netherlands, 2017; pp. 185–220. [CrossRef]
17. Zhou, M.; Cheng, X.; Chen, X. Studies on the Volumetric Stability and Mechanical Properties of Cement-Fly-Ash-Stabilized Steel Slag. *Materials* **2021**, *14*, 495. [CrossRef]
18. Yildirim, I.Z.; Prezzi, M. Chemical, Mineralogical, and Morphological Properties of Steel Slag. *Adv. Civ. Eng.* **2011**, *2011*, 463638. [CrossRef]
19. Yang, S.; Zhang, B.; Yang, Y.; Liu, M.; Tang, G.; Liu, X. Research on the status of comprehensive utilization of steel slag. *Ind. Miner. Process.* **2021**, *4*, 31–35. [CrossRef]
20. Jiang, Y.; Ling, T.-C.; Shi, C.; Pan, S.-Y. Characteristics of steel slags and their use in cement and concrete—A review. *Resour. Conserv. Recycl.* **2018**, *136*, 187–197. [CrossRef]
21. Wang, S.; Wang, C.; Wang, Q.; Liu, Z.; Qian, W.; Jin, C.; Chen, L.; Li, L. Study on Cementitious Properties and Hydration Characteristics of Steel Slag. *Pol. J. Environ. Stud.* **2018**, *27*, 357–364. [CrossRef]
22. Li, Y.; Ni, W.; Gao, W.; Zhang, Y.; Yan, Q.; Zhang, S. Corrosion evaluation of steel slag based on a leaching solution test. *Energy Sources Part A Recover. Util. Environ. Eff.* **2018**, *41*, 790–801. [CrossRef]
23. Wang, L.; Geddes, D.; Walkley, B.; Provis, J.L.; Mechtcherine, V.; Tsang, D.C. The role of zinc in metakaolin-based geopolymers. *Cem. Concr. Res.* **2020**, *136*, 106194. [CrossRef]
24. Nguyen, H.; Carvelli, V.; Kunther, W.; Illikainen, M.; Kinnunen, P. Phase evolution and mechanical performance of an ettringite-based binder during hydrothermal aging. *Cem. Concr. Res.* **2021**, *143*, 106403. [CrossRef]
25. Gao, W.; Ni, W.; Zhang, Y.; Li, Y.; Shi, T.; Li, Z. Investigation into the semi-dynamic leaching characteristics of arsenic and antimony from solidified/stabilized tailings using metallurgical slag-based binders. *J. Hazard. Mater.* **2019**, *381*, 120992. [CrossRef] [PubMed]
26. Wang, K.; Wang, Y.; Wan, Y.; Mi, Z.; Wang, Q.; Wang, Q.; Li, H. The fate of arsenic in rice plants (*Oryza sativa* L.): Influence of different forms of selenium. *Chemosphere* **2020**, *264*, 128417. [CrossRef]
27. Zhang, Y.; Zhang, S.; Ni, W.; Yan, Q.; Gao, W.; Li, Y. Immobilisation of high-arsenic-containing tailings by using metallurgical slag-cementing materials. *Chemosphere* **2019**, *223*, 117–123. [CrossRef]
28. NanrRu, Y.; WenrHai, Y. *The Handbook of Inorganic Matalloid Material Atlas*; Wuhan University of Technology Press: Wuhan, China, 2000.
29. Zhang, G.-S.; Qu, J.-H.; Liu, H.-J.; Liu, R.-P.; Li, G.-T. Removal Mechanism of As(III) by a Novel Fe−Mn Binary Oxide Adsorbent: Oxidation and Sorption. *Environ. Sci. Technol.* **2007**, *41*, 4613–4619. [CrossRef]
30. Guan, X.; Chen, J.; Gao, Y.; Gao, J. Mechanical properties and microstructure of coal gangue mortar blocks excited by NaOH Alkali. *J. Xi An Univ. Sci. Technol.* **2020**, *40*, 658–664.
31. Yazdani, M.R.; Tuutijärvi, T.; Bhatnagar, A.; Vahala, R. Adsorptive removal of arsenic (V) from aqueous phase by feldspars: Kinetics, mechanism, and thermodynamic aspects of adsorption. *J. Mol. Liq.* **2016**, *214*, 149–156. [CrossRef]
32. Phearom, S.; Shahid, M.K.; Choi, Y.-G. Optimization of Arsenic Adsorption by Mill Scale–Derived Magnetite Particles Using Response Surface Methodology. *J. Hazard. Toxic Radioact. Waste* **2021**, *25*, 04021022. [CrossRef]
33. Yi, Y.; Shi, J.; Tian, Q.; Guo, X.-Y. Arsenic removal from high-arsenic dust by NaOH-Na 2 S alkaline leaching. *Chin. J. Nonferrous Met.* **2015**, *25*, 806–814.
34. Zhang, W.; Ma, B.; Wang, C. Thermodynamics analysis on formation of calcium arsenate slag. *Nonferrous Met.* **2019**, *9*, 61–66. [CrossRef]
35. Avalos, N.M.; Varga, T.; Mergelsberg, S.T.; Silverstein, J.A.; Saslow, S.A. Behavior of iodate substituted ettringite during aqueous leaching. *Appl. Geochem.* **2020**, *125*, 104863. [CrossRef]
36. Saslow, S.A.; Kerisit, S.N.; Varga, T.; Mergelsberg, S.T.; Corkhill, C.L.; Snyder, M.M.V.; Avalos, N.M.; Yorkshire, A.S.; Bailey, D.J.; Crum, J. Immobilizing Pertechnetate in Ettringite via Sulfate Substitution. *Environ. Sci. Technol.* **2020**, *54*, 13610–13618. [CrossRef]
37. Jiménez, A.; Prieto, M. Thermal Stability of Ettringite Exposed to Atmosphere: Implications for the Uptake of Harmful Ions by Cement. *Environ. Sci. Technol.* **2015**, *49*, 7957–7964. [CrossRef] [PubMed]
38. Guo, B.; Sasaki, K.; Hirajima, T. Selenite and selenate uptaken in ettringite: Immobilization mechanisms, coordination chemistry, and insights from structure. *Cem. Concr. Res.* **2017**, *100*, 166–175. [CrossRef]
39. Li, Y.-C.; Min, X.-B.; Chai, L.-Y.; Shi, M.-Q.; Tang, C.-J.; Wang, Q.-W.; Liang, Y.-J.; Lei, J.; Liyang, W.-J. Co-treatment of gypsum sludge and Pb/Zn smelting slag for the solidification of sludge containing arsenic and heavy metals. *J. Environ. Manag.* **2016**, *181*, 756–761. [CrossRef]
40. Lei, J.; Peng, B.; Min, X.; Liang, Y.; You, Y.; Chai, L. Modeling and optimization of lime-based stabilization in high alkaline arsenic-bearing sludges with a central composite design. *J. Environ. Sci. Health Part. A* **2017**, *52*, 1–10. [CrossRef]
41. Lei, J.; Peng, B.; Liang, Y.-J.; Min, X.-B.; Chai, L.-Y.; Ke, Y.; You, Y. Effects of anions on calcium arsenate crystalline structure and arsenic stability. *Hydrometallurgy* **2018**, *177*, 123–131. [CrossRef]
42. Cornelis, G.; Johnson, C.A.; Van Gerven, T.; Vandecasteele, C. Leaching mechanisms of oxyanionic metalloid and metal species in alkaline solid wastes: A review. *Appl. Geochem.* **2008**, *23*, 955–976. [CrossRef]

MDPI
St. Alban-Anlage 66
4052 Basel
Switzerland
Tel. +41 61 683 77 34
Fax +41 61 302 89 18
www.mdpi.com

Metals Editorial Office
E-mail: metals@mdpi.com
www.mdpi.com/journal/metals